Pulverized-Coal Combustion and Gasification

Theory and Applications for Continuous Flow Processes

Pulverized-Coal Combustion and Gasification

Theory and Applications for Continuous Flow Processes

Edited by
L. Douglas Smoot
Brigham Young University
Provo, Utah
and
David T. Pratt
University of Utah
Salt Lake City, Utah

SPRINGER SCIENCE+BUSINESS MEDIA, LLC

Library of Congress Cataloging in Publication Data

Main entry under title:

Pulverized-coal combustion and gasification.

Includes bibliographical references and index.
1. Coal, Pulverized. 2. Combustion engineering. 3. Coal gasification. I. Smoot,
Leon Douglas. II. Pratt, David T.

TP328.P84	662'.62	78-12564

ISBN 978-1-4757-1698-6   ISBN 978-1-4757-1696-2 (eBook)
DOI 10.1007/978-1-4757-1696-2

© 1979 Springer Science+Business Media New York
Originally published by Plenum Press, New York in 1979
Softcover reprint of the hardcover 1st edition 1979

Contributors

Clayton T. Crowe, Department of Mechanical Engineering, Washington State University, Pullman, Washington

M. Duane Horton, Department of Chemical Engineering, Brigham Young University, Provo, Utah

Philip C. Malte, Department of Mechanical Engineering, Washington State University, Pullman, Washington

David T. Pratt, Department of Mechanical and Industrial Engineering, University of Utah, Salt Lake City, Utah

Dee P. Rees, Department of Chemical Engineering, Brigham Young University, Provo, Utah

F. Douglas Skinner, Department of Chemical Engineering, Brigham Young University, Provo, Utah

Philip J. Smith, Department of Chemical Engineering, Brigham Young University, Provo, Utah

L. Douglas Smoot, Department of Chemical Engineering, Brigham Young University, Provo, Utah

J. Rand Thurgood, Department of Chemical Engineering, Brigham Young University, Provo, Utah

Sneh Anjali Varma, Department of Mechanical and Industrial Engineering, University of Utah, Salt Lake City, Utah

John J. Wormeck, Department of Mechanical and Industrial Engineering, University of Utah, Salt Lake City, Utah

Preface

The past technology for describing and analyzing pulverized-coal furnaces, combustors, MHD generators, and other reactors has relied largely on empirical inputs for the complex flow and chemical reactions that occur while more formally treating the heat transfer effects. Growing concern over control of combustion-generated air pollutants has revealed a lack of understanding of the relevant fundamental physical and chemical mechanisms. Fortunately, recent technical advances in computer speed and storage capacity, and in numerical prediction of recirculating turbulent flows, two-phase flows, and flows with chemical reaction have opened new opportunities for describing and modeling such complex combustion systems in greater detail. We believe that most of the requisite component models to permit a more fundamental description of pulverized-coal combustion processes are available or can presently be developed. At the same time, with the rapidly increasing use of coal, and with the greatly increased emphasis being placed on development of new coal processes in the United States, interest in modeling of coal reaction processes has increased dramatically. Such interest was demonstrated at the ERDA Coal Conversion Modeling Conference held in late 1976, which was attended by over 200 representatives of industry, government, and universities.

We have been working during the past six years on pulverized-coal combustion processes. This work has included both basic measurements and development of analytical models. In our modeling work, we have attempted to develop fundamental models based upon the general equations of conservation and to use the best literature information available as sources for model parameters. One of the major purposes of this book is to document our general modeling approach being used for pulverized-coal systems. This book emphasizes processes using finely pulverized coal which is entrained in a gaseous phase. Problems of particular interest include combustion in pulverized-coal furnaces and power generators, entrained coal gasifiers, coal-fired MHD power generators, and flame propagation in laminar and turbulent coal dust/gas mixtures. While many of the principles and approaches could be adapted to other coal conversion and combustion problems, we have not considered combustion or gasification in fluidized or fixed beds or *in situ* processes. In addition, we have not considered other fossil-fuel combustion problems associated with oil shale, tar sands, etc., even though many aspects of pulverized-coal combustion would relate to these problems.

For the case of pulverized-coal models, we have attempted to provide a detailed description of the model foundations. Parts I and II of this book emphasize general principles for describing reacting, turbulent or laminar, multiphase systems. General conservation equations are developed and summarized. The basis for computing thermochemical equilibrium in complex, heterogeneous mixtures is presented, together with techniques for rapid computation and reference to required input data. Rate processes are then discussed, including pertinent aspects of turbulence, chemical kinetics, radiative heat transfer, and gas–particle convective–diffusive interactions. Much of Part II deals with parameters and coefficients for describing these complex rate processes. This part of the book provides recommended values of coefficients and parameters for treating complex reacting flows. Parts I and II may well be suitable for use in an advanced course in reacting flows, and have been written partly with that in mind.

Part III deals with more specific aspects of pulverized-coal characteristics and rate processes. Following a general description of coal structure and constitution, coal pyrolysis and char oxidation processes are considered. Gas-phase combustion of volatile matter and pollutant aspects of pulverized-coal are next considered, and finally, a general modeling framework for pulverized-coal particle reactions completes this part of the book. Part III is based largely on a review and summary of published literature. Recommendations are included, where appropriate, for describing important coal reaction rates.

Part IV presents details of pulverized-coal models that have been or are being developed by the authors. A generalized model for predicting propagation in premixed gaseous or multiphase laminar flows is documented. Model predictions are compared with experimental data for gaseous and pulverized-coal flames, and mechanisms of propagation are discussed. Propagation in turbulent flames is also briefly treated.

A generalized, one-dimensional treatment of pulverized-coal combustion and gasification is then formulated. Model predictions for coal furnaces and entrained gasifiers are shown, and compared with experimental results. This model has potential use in systems analysis and data interpretation, but does not predict multidimensional variation of properties inside a coal reactor.

Finally, a generalized, multidimensional code for describing pulverized-coal combustion and gasification processes is outlined and a solution technique is described. Steady-state, two-dimensional solutions have been emphasized. The model considers turbulent, recirculating, multiphase, reacting flows.

The authors recognize the several uncertainties in constructing and applying models of this nature. Coal is a very complex heterogeneous substance whose structure and behavior are highly variable and are not well known. Pyrolysis and oxidation of coal are dependent upon coal type, size, size distribution, temperature history, etc., thus making generalization difficult. In addition, the complexities of turbulent recirculating flows, turbulent reacting flows, and turbulent two-phase flows are not fully resolved. Only recently has formal computation of recirculating flows been possible and the extent of comparison of measurements with predictions is limited. Suitable generalized parameters for the turbulence models are not well established. The interaction of turbulence with chemical reaction is even less well developed, and general computational techniques for reliably treating this complexity are simply not available. Significant uncertainties also pertain to the effect of gas turbulence on the motion of the particles. It is known that the random gas fluctuations can be a major force in dispersing particles, but treatments of this effect are in the formative stages. Finally, there has been little previous attempt to develop models which consider several of these aspects jointly. For this reason, models of pulverized-coal systems must be validated by comparison with experimental measurements. Measurements of outlet gas or char composition or temperature alone are not suitable for this model validation. Profiles of at least time-mean, local properties from within the reactor must be the basis for the model evaluation. As a part of the research program being conducted by the authors, such measurements are being obtained, some of which are presented in this book and compared with model predictions. These measurements include data for laminar, premixed coal–air and methane–air flames, as well as spatially resolved measurements of gas and char composition from inside entrained gasifiers and furnaces. Further, profile measurements have been and are being made for particle-laden, recirculating jet flows without chemical reaction. Comparison of these measurements with model predictions will permit evaluation of particle- and gas-flow effects in the absence of chemical reaction complications.

Because of the high level of interest in pulverized-coal combustion, and because research work is expanding in this area, it is anticipated that considerable additional results pertinent to the contents of this book will be published during the coming years. We can foresee the necessity of updating or altering the material included in this book. However, we hope that it has been a useful effort to provide this publication at a time when modeling of pulverized-coal processes, while perhaps not well developed, has a high level of technical interest.

Much of the work upon which this book has been based is being supported by federal and industrial agencies. We especially express our appreciation to the United States Bureau of Mines (Pittsburgh Mining and Safety Research Center), The United States Department of Energy (Fossil Energy Division), the Electric Power Research Institute (Fossil Fuels Division), and the National Science Foundation for contract or grant support related to measurements and model development for pulverized-coal systems. We further acknowledge the Chemical Engineering Department and Research Division of Brigham Young University for assistance in preparation of the document. We are also grateful to Elaine Alger, Michael King, and Scott Folster for typing of the manuscript and preparation of the illustrations. Finally, we thank the several participating authors, both professors and Ph.D. candidates, who conducted literature searches and analyses and who prepared book materials when there were many other demands upon their time. The participating authors are primarily those who are involved in this pulverized-coal research program at Brigham Young University and at the University of Utah. Professor M. Duane Horton is a senior investigator for the U.S. Bureau of Mines study on mine explosions. Dee P. Rees and J. Rand Thurgood are Ph.D. candidates in Chemical Engineering, working on the EPRI pulverized-coal combustion study, while F. Douglas Skinner is a Ph.D. candidate working on the DOE entrained coal gasification study. Professors Clayton T. Crowe and Philip C. Malte of Washington State University (Mechanical Engineering) are consultants to the project. Sneh Anjali Varma is a Ph.D. candidate in Mechanical Engineering at the University of Utah, working on the EPRI study, while Dr. John Wormeck is a Senior Research Engineer at the University of Utah conducting modeling studies on the DOE contract.

L. Douglas Smoot
David T. Pratt

Contents

PART II. RATE PROCESSES

Chapter 3

Turbulence

David T. Pratt

Chapter 4

Gas-Phase Chemical Kinetics

David T. Pratt

Chapter 5

Radiative Heat Transfer in a Pulverized-Coal Flame
Sneh Anjali Varma

Chapter 6

Gas–Particle Flow
Clayton T. Crowe

PART III. COAL CHARACTERISTICS AND RATE PROCESSES

Chapter 7

General Characteristics of Coal
L. Douglas Smoot

Chapter 8

Fast Pyrolysis
M. Duane Horton

Chapter 9

Heterogeneous Reactions of Char and Carbon
F. Douglas Skinner and L. Douglas Smoot

Chapter 10

Volatiles Combustion
J. Rand Thurgood and L. Douglas Smoot

Chapter 11

Mechanisms and Kinetics of Pollutant Formation during Reaction of Pulverized· Coal
Phillip C. Malte and Dee P. Rees

Chapter12

Modeling Pulverized-Coal Reaction Processes
L. Douglas Smoot and Philip J. Smith

PART IV. MATHEMATICAL MODELING OF COAL CONVERSION PROCESSES

Chapter 13

Modeling One-Dimensional Systems

L. Douglas Smoot and Philip J. Smith

Chapter 14

Modeling Multidimensional Systems

John J. Wormeck

PART V. APPENDIXES

Part I
Fundamentals of Heterogeneous Flow Systems

Multicomponent Equilibrium

David T. Pratt

1. Homogeneous Gas-Phase Equilibrium

Three schemes for determination of equilibrium states at prescribed temperature and pressure may be considered, namely, Gibbs function minimization, equilibrium constant formulation, and equating of forward and reverse rates.[1] The first two schemes are well known to be equivalent, "static equilibrium" formulations based on the Gibbs function minimization principle for a mixture at specified temperature and pressure. The third, a "dynamic equilibrium" scheme, is not of practical use, as it is essentially a limiting case of finite-rate kinetics based on infinite reaction time, and therefore has all of the difficulties associated with finite-rate chemistry solutions.

Of the two static equilibrium schemes, the equilibrium constant approach is the most familiar and is most easily formulated. In this approach, where NS equilibrium mole numbers are sought, NLM equations for conservation of atomic species are formulated, and the remaining (NS–NLM) required equations are formulated by postulating (NS–NLM) stoichiometrically independent elementary reactions, each of which has associated with it an "equation of reaction equilibrium" involving a temperature-dependent equilibrium "constant." Unfortunately, a scheme to solve any arbitrarily chosen set of these NS nonlinear algebraic equations does not exist. In addition, different sets of reactions are required as fuel–air equivalence ratios are varied from lean to rich.[2]

David T. Pratt • Professor of Mechanical Engineering, University of Utah, Salt Lake City, Utah

A vastly superior computational scheme is the Gibbs function minimization approach, which does not require a choice of individual reactions, and has been formulated by Gordon and McBride[3] in such a way that only NLM equations and unknowns are required, rather than NS. Further, a foolproof scheme for solution for any given reactants distribution has been devised. This formulation, adapted only slightly for present purposes from reference 3, will be derived in this chapter.

1.1. Gibbs Function Minimization

For a mixture of reacting gases at prescribed temperature and pressure, chemical equilibrium is obtained when the Gibbs function of the mixture is a minimum, subject to conservation of atomic species as specified in the set of initial reactant mole numbers. If the mixture enthalpy is specified instead of the temperature, an equation of conservation of thermal energy is required as an additional constraint.

For a mixture of ideal gases, the partial molar Gibbs function is given by

$$g_k \equiv h_k - Ts_k = h_k - Ts_k^0 + RT \log (n_k/n_m) + RT \log (P/P_0) \qquad (1)$$

where P_0 is standard atmospheric pressure, and where other symbols are as defined in Section 6, Notation. For an ideal gas, Eq. (1) is also equal to the chemical potential for species k in the mixture.

The mass-specific Gibbs function for the mixture is given by

$$G \equiv \sum_{k=1}^{NS} n_k g_k \qquad (2)$$

and the requirement for chemical equilibrium is

$$dG = \sum_{j=1}^{NS} (\partial G/\partial n_j) \, dn_j = 0, \qquad d^2 G > 0 \qquad (3)$$

subject to given P and T, and conservation of atomic species:

$$\sum_{k=1}^{NS} a_{ik}^L n_k - b_{i0} = 0, \qquad i = 1, NLM \qquad (4)$$

Subsequently, the symbol b_i will be employed for the first term on the left-hand side of Eq. (4).

If the initial mixture enthalpy H_0 is specified rather than T, the thermal energy equation is also required:

$$H_0 - \sum_{k=1}^{NS} h_k n_k - Q = 0 \qquad (5)$$

where H_0 is defined by

$$H_0 = \sum_{k=1}^{NS} h_k(T_0)n_{k0} \tag{6}$$

The Lagrange method of undetermined multipliers is now employed, as follows:

(1) Write out Eq. (3), with Eqs. (1) and (2) incorporated:

$$dG = \sum_{j=1}^{NS} (\partial/\partial n_j)\left\{\sum_{k=1}^{NS} n_k\left[g_k^0 + RT\log n_k - RT\log\left(\sum_{i=1}^{NS} n_j\right)\right.\right.$$

$$\left.\left. + RT\log(P/P_0)\right]\right\} dn_j$$

$$= \sum_{j=1}^{NS}\left\{\sum_{k=1}^{NS} [n_k(RTn_k^{-1}\delta_{jk} - RTn_m^{-1}) + g_k\delta_{jk}]\right\} dn_j = 0 \tag{7}$$

Simplifying Eq. (7),

$$dG = \sum_{k=1}^{NS} g_k\, dn_k = 0 \tag{8}$$

(2) Take the differential of each constraint equation [Eq. (4)] and multiply by an undetermined constant multiplier λ_i:

$$\lambda_i\left(\sum_{k=1}^{NS} a_{ik}^L\, dn_k\right) = 0, \qquad i = 1, \text{NLM} \tag{9}$$

(3) Add Eqs. (8) and (9), and combine coefficients of dn_k:

$$\sum_{k=1}^{NS}\left(g_k + \sum_{i=1}^{NLM} \lambda_i a_{ik}^L\right) dn_k = 0 \tag{10}$$

The coefficients of the dn_k's in Eq. (10) are required to vanish independently; that is, for arbitrary variations in $\{n_k\}$ about the equilibrium values $\{n_k^*\}$. Therefore, a set of NS equations must be satisfied,

$$g_k + \sum_{i=1}^{NLM} \lambda_i a_{ik}^L = 0, \qquad k = 1, \text{NS} \tag{11}$$

in addition to the NLM constraint Eq. (4).

At this point, there are (NS+NLM) equations for the (NS+NLM) unknowns (n_k, $k = 1$, NS) and (λ_i, $i = 1$, NLM), which is a greater number than the NS equations and unknowns required in the equilibrium constant formulation. The following section illustrates how the problem is reduced to an (NLM)-dimensional system of equations and unknowns.

1.2. Newton–Raphson Solution of the Extremum Equations

Defining nondimensional Lagrange multipliers $\pi_i = -\lambda_i/RT$, Eqs. (11) and (4) may be expressed as functionals:

$$f_k \equiv g_k/RT - \sum_{i=1}^{NLM} \pi_i a_{ik}^L, \qquad k = 1, NS \tag{12}$$

and

$$f_i \equiv b_i - b_{i0}, \qquad i = 1, NLM \tag{13}$$

These functionals must each vanish at the equilibrium solution $\{n_k = n_k^*, k = 1, NS\}$.

In addition, a functional

$$f_T \equiv -(H_0/RT) + \sum_{k=1}^{NS} (h_k/RT)n_k + (Q/RT) \tag{14}$$

must be defined from Eq. (5), if the mixture enthalpy H_0 and specific heat loss Q are specified instead of T.

A system of Newton–Raphson correction equations for correction variables Δx_j are defined by

$$\sum_{j=1}^{N} (\partial f_i/\partial x_j) \Delta x_j = -f_i, \qquad i = 1, N \tag{15}$$

for an N-dimensional system of equations f_i and unknowns x_j. Following Gordon and McBride,[3] a judicious choice of correction variables in the present case is ($\Delta \log n_k$, $k = 1$, NS), $\Delta \log n_m$, and $\Delta \log T$. The selection of $\Delta \log n_m$ as an independent correction variable requires yet another functional to vanish at equilibrium, namely,

$$f_M \equiv \sum_{k=1}^{NS} n_k - n_m \tag{16}$$

With this set of correction variables, and Eq. (12) substituted into Eq. (15), there results for conservation of the jth species,

$$\sum_{k=1}^{NS} (\partial f_j/\partial \log n_k) \Delta \log n_k + (\partial f_j/\partial \log n_m) \Delta \log n_m + (\partial f_j/\partial \log T) \Delta \log T$$

$$= -f_j, \qquad j = 1, NS \tag{17}$$

Writing out the species-j functional Eq. (12) in full,

$$f_j \equiv (g_j^0/RT) + \log n_j - \log n_m + \log (P/P_0) - \sum_{i=1}^{NLM} \pi_i a_{ij}^L, \qquad j = 1, NS \tag{18}$$

The motivation for the choice of $\Delta \log n_m$ as an independent correction variable is now apparent. Noting that

$$g_j^0 \equiv h_{f\,298j}^0 + \int_{298}^T C_{pj}\, dT' - T\left[s_{298j}^0 + \int_{298}^T C_{pj}(dT'/T') \right] \tag{19}$$

The partial derivative coefficients in Eq. (17) may be expressed fully as

$$(\partial f_j/\partial \log n_k) = \delta_{jk}, \qquad k = 1,\text{NS} \tag{20}$$

$$(\partial f_j/\partial \log n_m) = -1 \tag{21}$$

and

$$(\partial f_j/\partial \log T) = T(\partial f_j/\partial T) = -(h_j/RT) \tag{22}$$

Substitution of Eqs. (20)–(22) into Eq. (17) yields, for species k,

$$\Delta \log n_k - \Delta \log n_m - (h_k/RT)\,\Delta \log T = -(g_k/RT) + \sum_{i=1}^{\text{NLM}} \pi_i a_{ik}^L, \qquad k = 1,\text{NS} \tag{23}$$

The NLM atom-conservation correction equations are obtained from the functionals of Eq. (13):

$$f_i \equiv \sum_{j=1}^{\text{NS}} n_j a_{ij}^L - b_{i0}, \qquad i = 1,\text{NLM} \tag{13a}$$

The appropriate partial derivatives for the corresponding atom-i correction equations are

$$(\partial f_i/\partial \log n_k) = n_k a_{ik}^L \tag{24}$$

$$(\partial f_i/\partial \log n_m) = 0 \tag{25}$$

and

$$(\partial f_i/\partial \log T) = 0 \tag{26}$$

The correction equations for atom-i conservation are therefore

$$\sum_{k=1}^{\text{NS}} n_k a_{ik}^L \,\Delta \log n_k = -\sum_{k=1}^{\text{NS}} n_k a_{ik}^L + b_{i0}, \qquad i = 1,\text{NLM} \tag{27}$$

Because of the choice of $\Delta \log n_m$ as an independent variable, it is now possible to eliminate ($\Delta \log n_k, k = 1$, NS) as independent correction variables by substitution of Eq. (23) into the NLM atom-j correction equations, Eq. (27):

$$\sum_{k=1}^{\text{NS}} n_k a_{jk}^L \left[\Delta \log n_m + (h_k/RT)\,\Delta \log T - (g_k/RT) + \sum_{i=1}^{\text{NLM}} \pi_i a_{ik}^L \right]$$

$$= b_{j0} - \sum_{k=1}^{\text{NS}} n_k a_{ik}^L, \qquad j = 1,\text{NLM} \tag{28}$$

Rearranging terms, Eq. (28) may be rewritten as

$$\sum_{i=1}^{NLM} \left(\sum_{k=1}^{NS} a_{ik}^L a_{jk}^L n_k \right) \pi_i + \left(\sum_{k=1}^{NS} a_{jk}^L n_k \right) \Delta \log n_m$$

$$+ \left[\sum_{k=1}^{NS} a_{jk}^L n_k (h_k/RT) \right] \Delta \log T = (b_{j0} - b_j) + \sum_{k=1}^{NS} a_{jk}^L n_k (g_k/RT), \qquad j=1,NLM \tag{29}$$

The correction equation for reciprocal mixture molecular weight n_m is obtained by first taking partial derivatives of Eq. (16) with respect to the correction variables:

$$(\partial f_M/\partial \log n_j) = n_j (\partial f_M/\partial n_j) = n_j \sum_{k=1}^{NS} (\partial n_k/\partial n_j) = n_j, \qquad j=1,NS \tag{30}$$

$$(\partial f_M/\partial \log n_m) = n_m (\partial f_M/\partial n_m) = n_m(-1) = -n_m \tag{31}$$

and

$$(\partial f_M/\partial \log T) = 0 \tag{32}$$

Substitution of Eq. (16) and Eqs. (30)–(32) into Eq. (15) yields the correction equation for reciprocal mixture molecular weight, n_m:

$$\sum_{k=1}^{NS} n_k \Delta \log n_k - n_m \Delta \log n_m = n_m - \sum_{k=1}^{NS} n_k \tag{33}$$

As before, $\Delta \log n_k$ is eliminated by means of substitution from Eq. (23):

$$\sum_{k=1}^{NS} n_k \left[\Delta \log n_m + (h_k/RT) \Delta \log T - (g_k/RT) + \sum_{i=1}^{NLM} \pi_i a_{ik}^L \right]$$

$$- n_m \Delta \log n_m = n_m - \sum_{k=1}^{NS} n_k \tag{34}$$

Rearrangement of terms results in the n_m-correction equation:

$$\sum_{i=1}^{NLM} \left(\sum_{k=1}^{NS} a_{ik}^L n_k \right) \pi_i + \left(\sum_{k=1}^{NS} n_k - n_m \right) \Delta \log n_m$$

$$+ \left[\sum_{k=1}^{NS} n_k (h_k/RT) \right] \Delta \log T = n_m - \sum_{k=1}^{NS} n_k - \sum_{k=1}^{NS} n_k (g_k/RT) \tag{35}$$

The correction equation for temperature is obtained in a similar manner: First, take the partial derivatives of Eq. (14) with respect to ($\log n_k$, $k=1$, NS), $\log n_m$, and $\log T$. Second, substitute these derivatives and Eq. (14) into Eq. (15). Third, eliminate the terms ($\Delta \log n_k$, $k=1$, NS) by substitution of

Eq. (23) to yield the temperature correction equation:

$$\sum_{i=1}^{NLM}\left[\sum_{k=1}^{NS}a_{ik}^{L}n_k(h_k/RT)\right]\pi_i+\left[\sum_{k=1}^{NS}n_k(h_k/RT)\right]\Delta\log n_m$$

$$+\left\{\sum_{k=1}^{NS}n_k(C_{pk}/R)+\sum_{k=1}^{NS}n_k(h_k/RT)^2+(1/RT)\partial Q/\partial\log T\right\}\Delta\log T$$

$$=H_0-\sum_{k=1}^{NS}n_k(h_k/RT)-(Q/RT)+\sum_{k=1}^{NS}n_k(h_k/RT)(g_k/RT)\quad(36)$$

By judicious selection of correction variables, the number of independent variables was reduced from $(NS+NLM)$ to $(NLM+2)$, with a corresponding number of linear algebraic equations, Eqs. (29), (35), (36). Given an estimate set $(n_k^{(0)}, k=1, NS)$, $n_m^{(0)}$, and $T^{(0)}$, values for $(\pi_i, i=1, NLM)$, $\Delta\log n_m$, and $\Delta\log T$ are determined by solving the system of linear equations by pivotal Gaussian elimination. Substitution into Eq. (23) yields the corresponding corrections $(\Delta\log n_k, k=1, NS)$. New values of $(n_k, k=1, NS)$, n_m, and T are then obtained by the relation

$$n_k=n_k^{(0)}+\eta\exp(\Delta\log n_k),\quad\text{etc.}\quad(37)$$

where η is an underrelaxation or acceleration parameter. Iteration is continued until absolute values of all of the corrections are less than some suitably small number ε. It may be of interest to note that the values of the Lagrange multipliers π_i do not appear explicitly. They are of no interest physically, and their current values are merely utilized in Eq. (23) for the species-k correction, $\Delta\log n_k$.

2. Condensed Phases

Condensed phases can be considered within the formalism of Section 1 by increasing the number of species from NS to NC. For mole numbers of the condensed species, $(n_i, i=NS+1, NC)$, the partial molal Gibbs function has no dependence on mole fraction or pressure, and is simply

$$g_k=g_k^0,\qquad k=NS+1, NC\quad(38)$$

Additional correction variables $(\Delta n_k, k=NS+1, NC)$ are defined, and the Newton–Raphson correction equations for gas phase [Eqs. (29), (35), and (36)] must be altered as follows:

(a) Atom-j correction equations [Eq. (29)]: Add a term to the left-hand side of this equation:

$$\sum_{k=NS+1}^{NS}a_{jk}^{L}\Delta n_k\quad(39)$$

(b) n_m-correction equation [Eq. (35)]: No change.

(c) Add (NC–NS) new equations to the system:

$$\sum_{i=1}^{NLM} a_{ij}^L \pi_i + (h_j/RT) \Delta \log T = (g_j^0/RT), \qquad j = NS+1, NC \qquad (40)$$

(d) *T*-correction equation [Eq. (36)]: Add to the left-hand side of this equation a term

$$\sum_{k=NS+1}^{NC} (h_k/RT) \Delta n_k \qquad (41)$$

and to the left-hand side add a term within the brackets of the coefficient of $\Delta \log T$,

$$\sum_{k=NS+1}^{NC} n_k (C_{pk}/R) \qquad (42)$$

With the additional correction variables for condensed-phase mole numbers (Δn_k, $k = NS+1$, NC), together with the corresponding number of additional equations and additional terms in items (a)–(d) above, the Gibbs function minimization scheme for a homogeneous gas-phase system may be expanded to include condensed phases.[3]

3. Thermochemical Properties

Specific heat capacities for both gaseous and condensed phases at atmospheric pressure can be represented in the form of a polynomial curve fit over a wide range of temperatures:

$$C_p^0/R = Z_1 + Z_2 T + Z_3 T^2 + Z_4 T^3 + Z_5 T^4 \qquad (43)$$

Specific enthalpy and entropy can be similarly curve-fit, once datum states for these two extensive properties are established. Following conventional practice, the following datum states are assumed:

The *enthalpy of formation* (or "chemical enthalpy") is zero for all elements in their naturally occurring, dominant allotropic form at standard temperature and pressure, 298 K and 101 kPa (1 atm). For example, at stp, oxygen exists in overwhelming abundance as a diatomic gas O_2; therefore the enthalpy of formation of oxygen is taken to be zero for gaseous, *diatomic* oxygen at stp.

The representation of enthalpy is then completed by adding to the chemical enthalpy (enthalpy of formation) the *sensible* enthalpy, as follows:

$$h = h_{\text{chem}} + h_{\text{sens}} \qquad (44)$$

where

$$h_{chem} = \Delta h_{f\,298}^0 \quad \text{(0 denotes stp)} \quad \text{and} \quad h_{sens} = \int_{298}^{T} C_p^0 \, dT$$

The *zero of entropy* is taken to occur at 0 K and at "low pressure"; in practice, 101 kPa (1 atm). Therefore, ideal-gas entropy is represented by the sum of the absolute entropy at stp, plus the temperature-dependent and pressure-dependent contributions at other T and P, as follows:

$$s = s_{298}^0 + \Delta s_{temp} + \Delta s_{press} \tag{45}$$

where

$$\Delta s_{temp} = \int_{298}^{T} (C_p^0 \, dT'/T') \quad \text{and} \quad \Delta s_{press} = RT \log (P_0/P)$$

Adjustments to the enthalpy for phase change include corrections in Eq. (44) for h_{fg} or h_{sg}, the latent heat of vaporization or sublimation, as appropriate. Contributions from changes in pressure are always small for absolute pressures below 1 MPa for the compounds of interest here, and will be ignored.

For condensed substances, the pressure correction term in Eq. (45) is dropped, and phase-change correction made simply by (h_{fg}/T) or (h_{sg}/T), as required, in Eq. (45).

With the above datum states in mind, enthalpy and the temperature-dependent part of the entropy can be represented as follows:

$$h/RT = Z_1 + Z_2 T/2 + Z_3 T^2/3 + Z_4 T^3/4 + Z_5 T^4/5 + Z_6/T \tag{46}$$

and

$$s^0/R = Z_1 \log T + Z_2 T + Z_3 T^2/2 + Z_4 T^3/3 + Z_5 T^4/4 + Z_7 \tag{47}$$

where $(Z_i, i = 1, 5)$ are the coefficients in Eq. (43), and where the additional constants Z_6 and Z_7 provide corrections for the chemical enthalpy and absolute entropy, respectively.

4. Computational Techniques

In Section 1, an analytical technique was derived in detail which resulted in a coupled set of (NLM + 2) nonlinear algebraic equations, namely, the Newton–Raphson correction equations to determine the Lagrange multipliers, which in turn enable calculation of the mole number corrections, following reference 3. All that remains to be noted concerning computational techniques is some mention of estimates and of convergence criteria.

The principal element in the successful solution of the set of equations derived in Section 1 is the control of the underrelaxation parameter η in Eq. (37). Gordon and McBride[3] have devised an empirical, "self-adjusting" underrelaxation parameter, which is the key to a rapidly converging solution, even from very poor initial estimates. This scheme may be summarized as follows:

(i) For T and n_m, and for n_i for those species for which $n_i/n_m > 10^{-8}$ and $\Delta \log n_i > 0$, a number η_1 is defined as

$$\eta_1 = 2/\max(|\Delta \log T|, |\Delta \log n_m|, |\Delta \log n_i|), \qquad i = 1, NS \qquad (48)$$

This causes the corrections which appear in the denominator of Eq. (48) to be scaled so that the corresponding variable (T, n_m, or n_i for major species with positive corrections) will not be increased by more than exp (2) on any one iteration.

(ii) For species for which $(n_i/n_m) < 10^{-8}$ and $\Delta \log n_i > 0$, a number η_2 is defined as

$$\eta_2 = \min \left(\frac{\log (10^{-4}) - \log (n_i/n_m)}{\Delta \log n_i - \Delta \log n_m} \right) \qquad (49)$$

This constrains the minor mole fractions to increase to no greater than 10^{-4}.

(iii) The final choice of underrelaxation parameter n is defined by

$$\eta = \min (1, \eta_1, \eta_2) \qquad (50)$$

A new value of η is calculated at each iteration and applied to Eq. (37). Thus, whenever current values of estimates are far from the solution point, η will be less than unity, and the iterative move toward the solution will be damped accordingly. Near the solution point, η will equal 1 and the final iterations will be undamped. No attempt is made to accelerate the solution; η is never greater than 1.

Convergence of the iterative solution is assumed whenever the log mole number corrections ($\Delta \log n_i$, $i = 1$, NS) are all less than $\varepsilon = 0.01$. Since, for small increments,

$$\Delta \log n_i \cong (\Delta n_i/n_i) \qquad (51)$$

this criterion is equivalent to requiring convergence to within one percent for all mole numbers.

Because of the "self-adjusting" underrelaxation parameter described in the preceding paragraphs, a "garbage" estimate set may be employed, again following Gordon and McBride.[3] For prescribed pressure and mixture enthalpy, the temperature is initially estimated at 3800 K, and the mole numbers (n_i, $i = 1$, NS) are all simply set equal to (0.1/NS). Remarkably, Gordon and McBride[3] claim, and this author's experience confirms, that a converged solution is *always* obtained from the "garbage" estimates, and usually in 10 or fewer iterations.

5. Recommended Approach

The method of Gordon and McBride is so clearly superior to any other known to the author with respect to reliability and speed of convergence, that there can be no second choice worth mentioning. The rationale for this observation has been clearly spelled out in the text of this chapter, and will be only summarized here:

(1) Any other scheme requires a square system of nonlinear algebraic equations scaled on the number of chemical species to be considered, which can be over 100 in chemically complex systems. The recommended scheme requires solving a square system whose size is scaled on the number of chemical elements, almost always fewer than 10. The saving in execution time is enormous.

(2) The stability and rapidity of convergence of the recommended scheme enables use of either previous solutions or "garbage" estimates, with little penalty in execution time for use of coarse estimates, and with convergence practically guaranteed.

The only negative feature of the recommended approach is the apparent algebraic complexity of the Newton–Raphson difference equations. However, once the validity of these equations is established in the mind of the potential user, the actual coding of the equations is extremely efficient, and requires only modest storage for the execution algorithm.[4]

6. Notation

a_{ij}^L Number of kg-atoms of element i per kg-mole of species j [(kg-atom)$_i$ kmol$_j^{-1}$]

b_i kg-atom number of element i in mixture [(kg-atom)$_i$ kg^{-1}]

b_{i0} b_i of reactants [(kg-atom)$_i$ kg^{-1}]

C_{P_i} Constant-pressure specific heat capacity of ideal-gas species i (J kmol$_i^{-1}$ K^{-1})

g_k^0 $= h_k - Ts_k^0$: One-atmosphere ideal-gas partial molal specific Gibbs function of species k (J kmol$_k^{-1}$)

g_k $= h_k - Ts_k^0 + RT[\log(n_k/n_m) + \log(P/P_0)]$: ideal-gas partial molal specific Gibbs function of species k (J kmol$_k^{-1}$)

H_0 Reactant mixture specific enthalpy (J kg^{-1})

h_k Ideal-gas enthalpy of species k (J kmol^{-1})

$\dot{m}$ Mass flow rate (kg s^{-1})

n_i Mole numbers of species i in mixture (kmol$_i$ kg^{-1})

n_{i0} Mole numbers of species i in reactants (kmol$_i$ kg^{-1})

n_m $= \sum_{i=1}^{NS} n_i$: reciprocal of mixture mean molecular weight (kmol kg^{-1})

(NC − NS)   Number of condensed species

NLM   Number of distinct elements in mixture

NS   The number of distinct gaseous chemical species under consideration

P Pressure [Pa (N m^{-2})]

P_0 Pressure of one standard atmosphere (101.325 kPa)

Q Enthalpy heat loss term $(J\ kg^{-1})$

R Universal gas constant $(8314.3\ J\ kmol^{-1}\ K^{-1})$

s_i^0 One-atmosphere ideal-gas specific entropy of species i $(J\ kmol_i^{-1})$

s_i Ideal-gas specific entropy of species i $(J\ kmol_i^{-1})$

T Temperature (K)

V Volume (m^3)

ε Convergence interval criterion, relevant to Eq. (37)

η Underrelaxation factor, Eqs. (37) and (48)–(50)

λ_t Lagrange multipliers in Gibbs function minimization equations $(J\ kmol^{-1})$

π_i Dimensionless Lagrange multipliers in Gibbs function minimization equations

ρ Mixture mass density $(kg\ m^{-3})$

7. References

1. F. J. Zeleznik and S. Gordon, Calculation of complex chemical equilibria, *Ind. Eng. Chem.* **60**, 6 (1968).
2. R. F. Anasoulis and H. MacDonald, *A Study of Combustor Flow Computations and Comparison with Experiment*, Report No. EPA-650/2-73-045, U.S. Environmental Protection Agency (1973).
3. S. Gordon and B. McBride, *Computer Program for Calculation of Complex Chemical Equilibrium Compositions*, NASA SP-273 (1971).
4. D. T. Pratt and J. Wormeck, *CREK—A Computer Program for Calculation of Combustion Reaction Equilibrium and Kinetics in Laminar or Turbulent Flow*, Report WSU-TEL-76-1, Washington State University (1976).

Multicomponent Conservation Equations

Clayton T. Crowe and L. Douglas Smoot

1. Reynolds Transport Theorem

There are two fundamentally different ways to describe the motion of a fluid: Lagrangian or Eulerian. From the Lagrangian point of view, the flow field is regarded as the motion of numerous small, contiguous fluid elements which interact through pressure and viscous forces. The motion of each fluid element behaves according to Newton's second law of motion. Though appealing to the student of Newtonian mechanics, this approach is generally impractical to describe the flow of a continuum because of the large number of mass elements needed to achieve a reasonably accurate description of the flow field. On the other hand, the Lagrangian approach is worthy of consideration for dispersed two-phase flows (gas-particle, gas-droplet) in that the particles or droplets themselves naturally constitute individual mass elements.

By far the most common approach to fluid mechanic analyses utilizes the Eulerian description of the flow field. In contrast to following individual fluid elements, the Eulerian approach considers all fluid elements which pass a given point for all time. That is, the flow properties (density, velocity, temperature) are described at each point as a function of time. Flow-field solutions are obtained by integrating the governing equations over all points in the flow field. Generally, the reference frame is fixed in space and is chosen

Clayton T. Crowe • Professor of Mechanical Engineering, Washington State University, Pullman, Washington. *L. Douglas Smoot* • Professor of Chemical Engineering, Brigham Young University, Provo, Utah

to simplify the description of the boundary conditions. If the reference frame were chosen to move with a fluid particle, the Eulerian equations would reduce to the Lagrangian equations.

Fundamental to the utilization of the Eulerian approach is the concept of a continuum. A gas flow field consists of molecules moving through space with certain levels of ordered and random motion. In order to define the mass density, a sufficiently large volume is needed to obtain a stationary average of mass per unit volume. Strictly speaking, then, gas density cannot be defined at a mathematical point. However, a cube of nitrogen, each side 1 μm long, at standard conditions contains approximately 1,000,000 molecules, which is sufficient to establish a stationary average. As long as this averaging volume is much smaller than the flow field in question, which is generally the situation, then the gas can be considered a continuum and the Eulerian approach is viable.[1]

The question of a continuum, and the validity of the Eulerian approach, arises again for gas-particle flows. To define bulk density (mass of particles per unit volume of mixture), a volume sufficiently large to define a stationary average is needed. In lightly loaded, two-phase flows, the volume needed may not be compared to the flow-field dimensions, so the continuum assumption and Eulerian approach are not valid. Drew[2] has presented a very detailed mathematical derivation of the Eulerian form of the two-phase flow equations and indicates the magnitude of error to be expected by assuming a continuum. Besides density, average values must also be found for velocities and temperature. The fact that different-sized particles will have different temperatures and will be moving at different speeds complicates the averaging procedure; that is, the mass-averaged velocity will differ from the momentum-averaged velocity and so on. The complexity of the Eulerian form of the two-phase flow equations does not allow the direct application of solution schemes for single-phase flows. Because of the averaging problem, the Lagrangian approach for the particulate phase has been used to advantage in some numerical models.[3,4]

The relationship between the Lagrangian and Eulerian form of the governing equations is provided by the Reynolds transport theorem. Consider the mass element (Lagrangian approach) which is passing through a control volume (Eulerian approach) as shown in Figure 1. At time t, the mass element just occupies the control volume. Defining β as some intensive property of the fluid such as specific energy (energy per unit mass), the Reynolds transport theorem may be written as

$$d(m\beta)/dt = (d/dt) \int_{cv} \rho\beta \, dV + \int_{cs} \rho\beta \mathbf{v} \cdot d\mathbf{A} \qquad (1)$$

where m is the mass of the fluid element, ρ is the mass density, $\mathbf{v}$ is the velocity with respect to the control surface, and the differential area vector, $d\mathbf{A}$, is

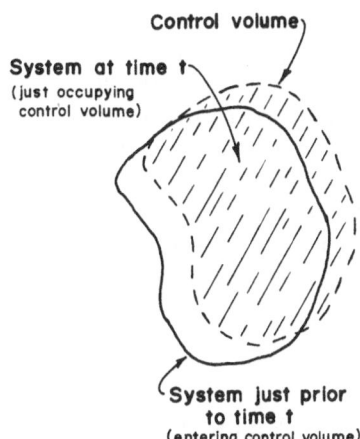

Control volume

System at time t
(just occupying
control volume)

System just prior
to time t
(entering control volume)

Figure 1. System passing through a control volume.

normal outward from the control surface. The Reynolds transport theorem states that the rate of change of some property of a system (a given amount of mass) is equal to the rate of change of the property within the control volume (cv) plus the net efflux of the property across the control surface (cs) at the very instant the mass element occupies the control volume. With the surface integral term converted to a volume integral, the right-hand side of Eq. (1) represents the Eulerian form of the governing equations.

The simplest application of the Reynolds transport theorem is the derivation of the continuity equation. The mass element (the system) in Eq. (1) consists always of the same matter so its mass is constant. The corresponding intensive property, mass per unit mass, is unity so that

$$\beta = 1 \tag{2}$$

and the Eulerian form becomes

$$0 = (d/dt) \int_{cv} \rho \, dV + \int_{cs} \rho \mathbf{v} \cdot d\mathbf{A} \tag{3}$$

Using Gauss' theorem to convert the surface integral to a volume integral yields

$$0 = \int_{cv} [\partial \rho / \partial t + \nabla \cdot (\rho \mathbf{v})] \, dV \tag{4}$$

and because the volume of integration is arbitrary,

$$\partial \rho / \partial t + \nabla \cdot (\rho \mathbf{v}) = 0 \tag{5}$$

where ∇ is the spatial gradient and is a vector. This is the standard form of the continuity equation. Following this same approach, utilizing Newton's second law and the first law of thermodynamics together with the Reynolds transport theorem, yields the momentum and energy equations, respectively, in Eulerian form.

The form of the Reynolds transport theorem represented by Eq. (1) applies to a continuous mass element which passes through a control volume. When the properties of a cloud of particles passing through a control volume are considered, there are many individual mass elements, each of which may be undergoing mass changes. The Reynolds transport theorem is still applicable, however, if the particles are identified by number, and β is defined as a particle property such as particle mass. The theorem can then be written as[5]

$$d(N\beta)/dt = \underbrace{d(N\beta)_{cv}/dt}_{\text{Lagrangian}} + \underbrace{(\dot{N}\beta)_{\text{out}} - (\dot{N}\beta)_{\text{in}}}_{\text{Eulerian}} \qquad (6)$$

where N is the number of particles and $\dot{N}$ is the number flow rate of particles. As with Eq. (1), this equation can also be written in integral form

$$d(N\beta)/dt = (d/dt) \int_{cv} \rho'_p b \, dV + \int_{cs} \rho'_p b \mathbf{v} \cdot d\mathbf{A} \qquad (7)$$

where b is the particle property per unit particle mass, ρ'_p is the mass of particle matter per unit volume of mixture, and $\mathbf{v}$ is the mass-averaged particle velocity with respect to the control surface. Equation (7) constitutes the fundamental relationship for deriving the Eulerian form of the particle cloud equations. As discussed previously, the concepts of a continuum and average properties must carefully be considered when expressing the particle cloud equations in Eulerian form.

It should be pointed out that the Lagrangian and Eulerian forms of the fluid-mechanic equations are simply two ways to express the basic laws of continuity, momentum and energy. The choice is solely a matter of convenience.

2. Continuity Equation

The continuity equation in fluid mechanics is an expression of the conservation of mass. In Section 2.1, the continuity equation will be derived utilizing the Reynolds transport theorem for a single-component (single-species) flow. For a flow consisting of a mixture of chemical species, it is useful to have a continuity equation for each species. Equivalently, for a two-phase flow, there is also a continuity equation for the gaseous and particulate phases. The addition of the component continuity equations yields the continuity equation for the mixture.

2.1. Gas-Phase Species Continuity

Fundamental to the derivation of the continuity equation for a chemical species is the definition of velocity. Because of diffusion, a given chemical species will have a velocity different from the mass-averaged or mole-average velocity of the mixture. Defining ρ_i as the mass density of species i [(mass of species i)/(volume of mixture)], the mass-averaged velocity $\mathbf{v}_g$ is given by

$$\mathbf{v}_g = \sum_i \rho_i \mathbf{v}_i / \rho_g \tag{8}$$

where $\mathbf{v}_i$ is the velocity of species i and ρ_g is the mass density of the mixture. An equivalent expression for the mass-averaged velocity is

$$\mathbf{v}_g = \sum_i \omega_i \mathbf{v}_i \tag{9}$$

where ω_i is the mass fraction ($\omega_i = \rho_i / \rho_g$) of species i.

In the same fashion, the mole-averaged gas velocity $\mathbf{v}_g^*$ is defined as

$$\mathbf{v}_g^* = \sum_i c_i \mathbf{v}_i / c_g \tag{10}$$

where c_i is the concentration or molar density of species i [(moles of i)/(volume of mixture)], and c_g is the molecular density of the gas mixture.

The mole-averaged velocity can also be written as

$$\mathbf{v}_g^* = \sum_i y_i \mathbf{v}_i \tag{11}$$

where y_i is the mole fraction of species i.

The diffusion velocity of species i is the difference between the actual velocity of the species and the mass- (or mole-) averaged velocity of the mixture:

$$\mathbf{v}_{Di} = \mathbf{v}_i - \mathbf{v}_g \tag{12}$$

or

$$\mathbf{v}_{Di}^* = \mathbf{v}_i - \mathbf{v}_g^* \tag{13}$$

The mass flux of species i, referenced to the mass-averaged velocity, is denoted as

$$\mathbf{j}_i = \rho_i \mathbf{v}_{Di} \tag{14}$$

and by definition

$$\sum_i \mathbf{j}_i = 0 \tag{15}$$

The same definition (with an asterisk) applies to molar-related quantities.

In a multicomponent system, the mass flux $\mathbf{j}_i$ is the sum of four components. Generally the largest component is that due to concentration gradient, but pressure gradients, external forces, and thermal gradients can also contribute to species mass flux.[6] A more complete discussion of diffusion in multicomponent systems and the pertinent diffusion coefficients are provided in Section 7.

Applying the Reynolds transport theorem for continuity of species i, the value of β in Eq. (1) is the mass fraction (or mole fraction) of species i. Thus the continuity equation for species i is

$$dm\omega_i/dt = (d/dt) \int_{cv} \rho_g \omega_i \, dV + \int_{cs} \rho_g \omega_i (\mathbf{v}_g + \mathbf{v}_{Di}) \cdot d\mathbf{A} \qquad (16)$$

The left-hand side is simply the rate of production of species i and can be written as

$$dm\omega_i/dt = \int_{cv} r_{ig} \, dV \qquad (17)$$

where r_{ig} is the reaction rate of species i per unit volume of mixture. Using Gauss' theorem to convert the surface to a volume integral gives

$$\int_{cv} r_{ig} \, dV = \int_{cv} [\partial \rho_i/\partial t + \nabla \cdot (\rho_i \mathbf{v}_g) + \nabla \cdot \mathbf{j}_i] \, dV \qquad (18)$$

Since the choice of control volume is arbitrary, there follows

$$r_{ig} - \nabla \cdot \mathbf{j}_i = \partial \rho_i/\partial t + \nabla \cdot (\rho_i \mathbf{v}_g) \qquad (19)$$

The overall continuity equation for the gas mixture is obtained by adding the species continuity equations, using Eq. (15) and noting that conservation of mass requires

$$\sum_i r_{ig} = 0 \qquad (20)$$

Thus, summed over all species, Eq. (19) becomes

$$\sum_i [\partial \rho_i/\partial t + \nabla \cdot (\rho_i \mathbf{v}_g)] = 0 \qquad (21)$$

or, in terms of mixture quantities,

$$\partial \rho_g/\partial t + \nabla \cdot (\rho_g \mathbf{v}_g) = 0 \qquad (22)$$

The equation is generally referred to as the gas continuity equation. In the present context, Eq. (22) will be referred to as the gas-mixture continuity equation.

2.2. Gas-Phase Continuity in a Gas–Particle Mixture

The presence of particles in the gas requires a new definition for gas-phase density. The volume occupied by the gas per unit volume of gas–particle mixture is identified as the void fraction, θ. Thus, the mass of the gas per unit volume of gas–particle mixture, the bulk gas density, is given by

$$\rho_g' = \theta \rho_g \tag{23}$$

where ρ_g is the material density of the gas. The mass flux of gas is now given by $\rho_g' \mathbf{v}_g$. The void fraction in flow systems typical of pulverized-coal combustion is usually greater than 0.99, so the bulk density can generally be represented by the material density, to good approximation.

The control volume for a gas–particle mixture enclosed by a surface S is shown in Figure 2. The portion of S which intersects the "boundary" particles is identified by δS. The surface adjacent to every interior particle is S_p and the surface surrounding the "boundary" particles interior to S is labeled δS_p. Thus the complete control surface for the gas is $S - \delta S + \delta S_p + S_p$. To abbreviate the notation, the portion of the control surface through the gas phase, $S - \delta S$, is referred to as S' and that adjacent to the particle surface as S_p'.

Applying the Reynolds transport theorem for conservation of gaseous mass yields

$$(d/dt) \int_{cv} \rho_g' \, dV + \int_{S'} \rho_g \mathbf{v}_g \cdot d\mathbf{A} + \int_{S_p'} \rho_s \mathbf{w} \cdot d\mathbf{A} = 0 \tag{24}$$

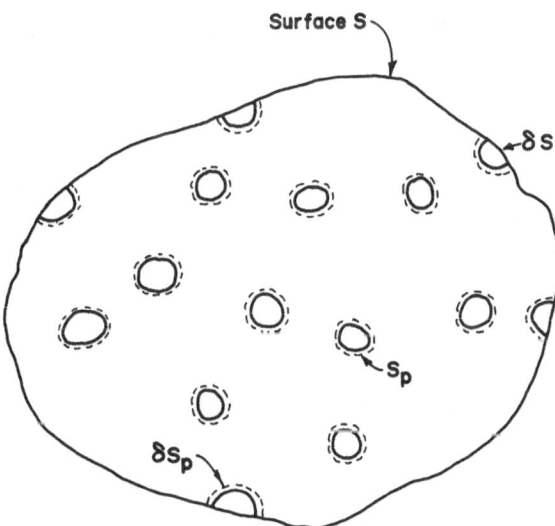

Figure 2. Gas–particle control volume.

where ρ_s is the gas density at the particle (control) surface and $\mathbf{w}$ is the gas velocity with respect to the control surface. The integral over S' can be written as

$$\int_{S'} \rho_g \mathbf{v}_g \cdot d\mathbf{A} = \int_S \rho'_g \mathbf{v}_g \cdot d\mathbf{A} \tag{25}$$

Also the integral over S'_p can be rewritten in the form

$$\int_{S'_p} \rho_s \mathbf{w} \cdot d\mathbf{A} = -\int_{cv} r_p \, dV \tag{26}$$

where r_p is the total mass of gas added per unit volume of mixture due to particle mass transfer. Using Gauss' theorem to convert the surface to a volume integral and setting the integrand equal to zero yields the gas-phase continuity equation for the gas–particle mixture:

$$(\partial \rho'_g / \partial t) + \nabla \cdot (\rho'_g \mathbf{v}_g) = r_p \tag{27}$$

2.3. Discrete-Phase Continuity

Applying the Reynolds transport theorem [Eq. (7)] for conservation of mass of a cloud of particles yields

$$d(Nm)/dt = (d/dt) \int_{cv} \rho'_p \, dV + \int \rho'_p \mathbf{v}_p \cdot d\mathbf{A} \tag{28}$$

where ρ'_p is the bulk density of the particulate phase and $\mathbf{v}_p$ is the mass-averaged particle velocity, or the velocity which describes the mass flux of particles across the control surface enclosing the mixture. The left-hand side can be written as

$$d(Nm)/dt = -\int_{cv} r_p \, dV \tag{29}$$

where r_p is the burning rate of particles per unit volume of mixture. A "positive" burning rate corresponds to mass removal from the particles. Using Gauss' theorem to convert the surface to a volume integral and setting the integrand equal to zero yields the differential form of the particle-cloud continuity equation:

$$(\partial \rho'_p / \partial t) + \nabla \cdot (\rho'_p \mathbf{v}_p) = -r_p \tag{30}$$

The mass-averaged particle velocity $\mathbf{v}_p$ is defined as

$$\rho'_p \mathbf{v}_p = \sum_k \rho'_{pk} \mathbf{v}_{pk} \tag{31}$$

where ρ'_{pk} and $\mathbf{v}_{pk}$ are the bulk densities and velocities of particle size class k, respectively.

2.4. Overall Continuity

Adding Eqs. (27) and (30) for gas-phase and discrete-phase continuity yields the overall continuity relationship for the gas–particle mixture:

$$(\partial/\partial t)(\rho'_g + \rho'_p) + \nabla \cdot (\rho'_g \mathbf{v}_g + \rho'_p \mathbf{v}_p) = 0 \tag{32}$$

where the terms due to particle mass addition have canceled.

Defining $(\rho'_g + \rho'_p)$ as the mixture density ρ_m, and $\mathbf{v}_m$ as the mixture velocity, then the expression

$$\rho_m \mathbf{v}_m = \rho'_g \mathbf{v}_g + \rho'_p \mathbf{v}_p \tag{33}$$

allows the overall continuity equation to be written as

$$\partial \rho_m / \partial t + \nabla \cdot (\rho_m \mathbf{v}_m) = 0 \tag{34}$$

3. Momentum Equation

The momentum equation (or equation of motion) in fluid mechanics is Newton's second law for a moving fluid, expressed in Eulerian form. In this section, the momentum equation for the flow of the gaseous species is derived, and the derivation of the momentum equations for the gaseous phase and particulate phase in a two-phase mixture is outlined. Finally, the two-phase mixture equation is obtained by addition of the momentum equations for the gas and particulate phases.

3.1. Gas-Phase Momentum

Application of the Reynolds transport theorem to the momentum change of a fluid element requires that the quantity β in Eq. (1) be the momentum per unit mass, or simply, the velocity. Thus the momentum equation can be written as

$$\mathbf{F} = (d/dt) \int_{cv} \rho \mathbf{v} \, dV + \int_{cs} \mathbf{v}(\rho \mathbf{v} \cdot d\mathbf{A}) \tag{35}$$

where $\mathbf{F}$ is the force on the fluid element at the instant it occupies the control volume. As noted in the preceding section, each gaseous species may have a different velocity; thus the momentum integrals on the right-hand side must be summed over all gaseous components of the mixture:

$$\mathbf{F} = (d/dt) \int_{cv} \sum_i \rho_i \mathbf{v}_i \, dV + \int_{cs} \sum_i \mathbf{v}_i (\rho_i \mathbf{v}_i \cdot d\mathbf{A}) \tag{36}$$

With the definition of mass-average velocity [Eq. (8)], the first integral in

Eq. (36) can be written simply as

$$(d/dt) \int_{cv} \sum_i \rho_i \mathbf{v}_i \, dV = (d/dt) \int_{cv} \rho_g \mathbf{v}_g \, dV \tag{37}$$

The second integral in Eq. (36) can be rewritten in terms of the mass-averaged and diffusional velocities:

$$\int_{cs} \sum_i \mathbf{v}_i (\rho_i \mathbf{v}_i \cdot d\mathbf{A}) = \int_{cs} \sum_i (\mathbf{v}_g + \mathbf{v}_{Di})[\rho_i (\mathbf{v}_g + \mathbf{v}_{Di}) \cdot d\mathbf{A}]$$

$$= \int_{cs} \mathbf{v}_g (\rho_g \mathbf{v}_g \cdot d\mathbf{A}) + \int_{cs} \sum_i \mathbf{v}_{Di} (\mathbf{j}_i \cdot d\mathbf{A}) \tag{38}$$

The integral for the momentum flux associated with the diffusional velocities is identified as the "apparent stress" due to diffusion.[1]

In the literature, this (second-order) tensor is either neglected or combined with the shear-stress tensor attributed to viscosity.[7,8]

The forces acting on the fluid element consist of surface and body forces:

$$\mathbf{F} = \mathbf{F}_s + \mathbf{F}_b \tag{39}$$

It is convenient to decompose the surface force per unit area into a pressure stress and a shear stress, given by $-p\mathbf{n} + \boldsymbol{\tau} \cdot \mathbf{n}$, where $\mathbf{n}$ is the unit outward normal vector from the control surface and $\boldsymbol{\tau}$ is the shear-stress tensor. Thus the total surface force on the control volume is

$$\mathbf{F}_s = \int_{cs} (-p\mathbf{n} + \boldsymbol{\tau} \cdot \mathbf{n}) \cdot d\mathbf{A} \tag{40}$$

The body forces (such as gravity) are proportional to the mass of each constituent. Taking $\mathbf{g}_i$ to be a body force per unit mass for species i and summing over the control volume yields

$$\mathbf{F}_b = \int_{cv} \sum_i \rho_i \mathbf{g}_i \, dV \tag{41}$$

if the body force per unit mass is constant ($\mathbf{g}_i = \mathbf{g}$), then

$$\mathbf{F}_b = \mathbf{g} \int_{cv} \rho_g \, dV \tag{42}$$

Finally, substituting the integrals for the forces into Eq. (36) and incorporating the "apparent stress" due to diffusion into the shear-stress term, $\boldsymbol{\tau}$, gives

$$(d/dt) \int_{cv} \rho_g \mathbf{v}_g \, dV + \int_{cs} \mathbf{v}_g (\rho_g \mathbf{v}_g \cdot d\mathbf{A}) = \int_{cs} (-p\mathbf{n} + \boldsymbol{\tau} \cdot \mathbf{n}) \cdot d\mathbf{A} + \mathbf{g} \int_{cv} \rho_g \, dV \tag{43}$$

Using Gauss' theorem to convert the surface integrals to volume integrals yields the following differential form of the momentum equation for the flowing mixture:

$$(\partial/\partial t)(\rho_g \mathbf{v}_g) + \nabla \cdot (\rho_g \mathbf{v}_g \mathbf{v}_g) = -\nabla p + \nabla \cdot \boldsymbol{\tau} + \rho_g \mathbf{g} \tag{44}$$

where it should be cautioned that $\nabla \cdot (\rho_g \mathbf{v}_g \mathbf{v}_g)$ and $\nabla \cdot \boldsymbol{\tau}$ are not simple divergences because of the tensorial nature of $\mathbf{v}_g \mathbf{v}_g$ and $\boldsymbol{\tau}$. Using the gas-mixture continuity equation [Eq. (22)], the momentum equation for a mixture of gaseous species can be rewritten as

$$\rho_g \partial \mathbf{v}_g / \partial t + (\rho_g \mathbf{v}_g \cdot \nabla)\mathbf{v}_g = -\nabla p + \nabla \cdot \boldsymbol{\tau} + \rho_g \mathbf{g} \tag{45}$$

which is simply the statement, in the classical form of Newton's second law, that the mass times acceleration of a fluid element is equal to the sum of the forces (pressure, shear stress, and body forces) acting on the fluid.

3.2. Gas-Phase Momentum in a Gas–Particle Mixture

As with the gas-phase continuity equation for a gas–particle mixture, the mass and momentum exchange between the particles and gas contribute to the gas-phase momentum. The gas-phase momentum equation for a gas–particle mixture is derived here treating the mixture of gaseous species as a single component. The details of the derivation are available in reference 5.

Consider once more the control surface for the gas shown in Figure 2. Applying the Reynolds transport theorem for the momentum of the gaseous phase yields

$$\mathbf{F} = (d/dt) \int_{cv} \rho'_g \mathbf{v}_g \, dV + \int_{S'} \rho_g \mathbf{v}_g \mathbf{v}_g \cdot d\mathbf{A} + \int_{S'_p} \rho_s(\mathbf{w}' + \mathbf{v}_p)\mathbf{w} \cdot d\mathbf{A} \tag{46}$$

where $\mathbf{w}$ is the gas velocity at the particle surface with respect to the particle's center of mass. If the particle is burning, $\mathbf{w}'$ is the vector sum of $\mathbf{w}$ and the surface regression velocity. Assuming a spherical particle and a uniform momentum flux over the particle surface allows the last integral in Eq. (46) to be written as

$$\int_{S'_p} \rho_g(\mathbf{v}_p + \mathbf{w}')\mathbf{w} \cdot d\mathbf{A} = -\int_{cv} \mathbf{v}_p r_p \, dV + \int_S \rho_s w' w (1 - \theta) \, dA \tag{47}$$

where w' and w are the magnitudes of $\mathbf{w}'$ and $\mathbf{w}$.

The surface forces acting on the gas are the pressure and shear forces on the surface S' and aerodynamic forces on the surface S'_p. The forces on the surface S' can be expressed as

$$\mathbf{F} = \int_S [-\theta p \, d\mathbf{A} + \theta \boldsymbol{\tau} \cdot d\mathbf{A}] \tag{48}$$

where $\boldsymbol{\tau}$ is the shear-stress tensor, including the "apparent stress" due to diffusion. The forces on the surface S'_p are given by

$$\mathbf{F} = \int_S \left[-(1-\theta)\bar{p}_s \, d\mathbf{A} + (1-\theta)\tau_a \cdot d\mathbf{A} \right] - \int_{cv} \mathbf{f}_p \, dV \qquad (49)$$

where $\tau_a \cdot \mathbf{n}$ is the shear force on the particle due to the local rate of strain in the fluid, p_s is the average pressure on the particle surface, and $\mathbf{f}_p$ is the aerodynamic force on the particles per unit volume of mixture.[7] The body force acting on the gas is simply

$$\mathbf{F}_b = \int_{cv} \rho'_g \mathbf{g} \, dV \qquad (50)$$

where $\mathbf{g}$ is the body force per unit mass of gas. The pressure change across the particle's boundary layer, assuming a thin boundary layer, is given by

$$p - \bar{p}_s = \rho_s w w' \qquad (51)$$

Incorporating Eqs. (47)–(51) into Eq. (46), using the divergence theorem to convert the surface to volume integrals, and equating the integrand to zero yields the momentum equation for the gas phase in a gas–particle mixture:

$$(\partial/\partial t)(\rho'_g \mathbf{v}_g) + \nabla \cdot (\rho'_g \mathbf{v}_g \mathbf{v}_g) = -\nabla p + \nabla \cdot (1-\theta)\tau_a + \nabla \cdot \theta \tau - \mathbf{f}_p + \rho'_g \mathbf{g} + \mathbf{v}_p r_p \qquad (52)$$

Setting θ equal to unity and $\mathbf{f}_p$ and r_p equal to zero reduces Eq. (52) to the gas-mixture continuity equation, Eq. (44). For application to most pulverized-coal combustion systems, θ is near unity, so that the term $\nabla \cdot (1-\theta)\tau_a$ may be neglected. Also, if there is a distribution of particle velocities associated with different particle sizes, the term relating to the momentum transfer due to mass addition [the last term in Eq. (52)] must be replaced by

$$\mathbf{v}_p r_p = \sum_k \mathbf{v}_{pk} r_{pk} \qquad (53)$$

where $\mathbf{v}_{pk}$ is the velocity of size class k and r_{pk} is the burning rate of particles in the same size class per unit volume of mixture.

3.3. Discrete-Phase Momentum

Consider a cloud of particles of differing sizes moving at differing velocities. The Reynolds transport theorem relating to the momentum of the cloud is given by

$$(d/dt) \sum_k m_k \mathbf{v}_{pk} = (d/dt) \int_{cv} \sum_k \rho'_{pk} \mathbf{v}_{pk} \, dV + \int_{cs} \sum_k \rho'_{pk} \mathbf{v}_{pk} \mathbf{v}_{pk} \cdot d\mathbf{A} \qquad (54)$$

where ρ'_{pk} is the bulk density of particle size class k, and the summation is performed over all size classes. The sum over m_k includes all particles completely inside the control volume and the portion of mass of the "boundary" particles inside the control volume. The left-hand side (Lagrangian) of the above equation can be expanded and rewritten as

$$(d/dt) \sum_k m_k \mathbf{v}_{pk} = \int_{cv} \left(-\sum_k r_{pk} \mathbf{v}_{pk} + \sum_k m_k \, d\mathbf{v}_{pk}/dt + \mathbf{T} \right) dV \qquad (55)$$

The first term represents the momentum change due to the particle mass transfer (burning), r_{pk} being the burning rate of the particle of size class k. The minus sign appears because a reduction in particle mass corresponds to a positive burning rate. Since the forces acting on a particle are equal to the product of the particle mass and acceleration (assuming a uniform momentum efflux from the particle surface), the second term is simply the sum of all the aerodynamic and body forces acting on the particles and can be conveniently rewritten as

$$\int_{cv} \left(\sum_k m_k \, d\mathbf{v}_{pk}/dt \right) dV = \int_{cv} (\mathbf{f}_p + \rho'_p \mathbf{g}) \, dV \qquad (56)$$

where $\mathbf{f}_p$ is the aerodynamic force on the particles per unit volume of mixture. The last term is the momentum exchange due to particle–particle collisions. This term is negligible for most pulverized-coal applications because the time between collisions is generally much larger than the aerodynamic response time for a particle.

The right-hand side of Eq. (54) can be rewritten using the mass-averaged velocity of the particles:

$$(d/dt) \int_{cv} \left(\sum_k \rho'_{pk} \mathbf{v}_{pk} \right) dV + \int_{cs} \sum_k \rho'_{pk} \mathbf{v}_{pk} \mathbf{v}_{pk} \cdot d\mathbf{A}$$

$$= \int_{cv} (\partial/\partial t)(\rho'_p \mathbf{v}_p) \, dV + \int_{cs} \rho'_p \mathbf{v}_p \mathbf{v}_p \cdot d\mathbf{A} + \int_{cs} \sum_k \rho'_{pk} \delta \mathbf{v}_{pk} \delta \mathbf{v}_{pk} \cdot d\mathbf{A} \qquad (57)$$

where

$$\rho'_p \mathbf{v}_p = \sum_k \rho'_{pk} \mathbf{v}_{pk} \quad \text{and} \quad \delta v_{pk} = \mathbf{v}_{pk} - \mathbf{v}_p$$

The last term is analogous to the "apparent stress" due to diffusion encountered in Eq. (38). It arises because the center of mass of the cloud moves at a different velocity than the center of momentum, if the particles move at different velocities. If the velocity of all particles is the same, the center of mass and center of momentum are coincident.

Finally, combining all the terms in the Reynolds transport theorem for particle momentum and converting the surface to volume integrals yields the following differential equation:

$$(\partial/\partial t)(\rho'_p \mathbf{v}_p) + \nabla \cdot (\rho'_p \mathbf{v}_p \mathbf{v}_p) + \sum_k \nabla \cdot (\rho'_{pk} \delta \mathbf{v}_{pk} \delta \mathbf{v}_{pk}) = -\sum r_{pk} \mathbf{v}_{pk} + \mathbf{f}_p + \rho'_p \mathbf{g} \qquad (58)$$

The last term on the right-hand side represents the momentum flux due to motion of the particles about the mass-averaged velocity. This term significantly complicates numerical modeling of dispersed-phase flows.[9]

If all particles are moving locally at the same velocity, the discrete-phase momentum equation becomes

$$(\partial/\partial t)(\rho_p' \mathbf{v}_p) + \nabla \cdot (\rho_p' \mathbf{v}_p \mathbf{v}_p) = -r_p \mathbf{v}_p + \mathbf{f}_p + \rho_p' \mathbf{g} \tag{59}$$

The same simplification concerning particle velocity was employed in the derivation of the equation for the gas-phase momentum in a gas–particle mixture [Eq. (52)].

3.4. Overall Momentum Equation

The overall momentum equation is obtained by adding the gas-phase equation [Eq. (52)] and the particle-cloud equation [Eq. (59)]. The result is

$$(\partial/\partial t)(\rho_g' \mathbf{v}_g + \rho_p' \mathbf{v}_p) + \nabla \cdot (\rho_g' \mathbf{v}_g \mathbf{v}_g + \rho_p' \mathbf{v}_p \mathbf{v}_p)$$
$$= -\nabla p + \nabla \cdot (1-\theta)\tau_a + \nabla \cdot \theta\tau + (\rho_p' + \rho_g')\mathbf{g} - \sum_k \nabla \cdot (\rho_{pk}' \delta \mathbf{v}_{pk} \delta \mathbf{v}_{pk}) \tag{60}$$

since the momentum terms due to force on the particles and mass addition cancel. The "apparent stress" due to diffusional velocities of the gaseous species has been incorporated into the shear-stress term, τ. Otherwise, the last term in Eq. (38) would be present with the diffusional velocities of the gaseous species.

4. Energy Equation

The energy equation in fluid mechanics is a statement of the first law of thermodynamics in Eulerian form. In this chapter the energy equation for the flow of gaseous species is derived and expressed in the total and thermal energy form. The energy equation for the gas phase and the particulate phase in a cloud of particles is also derived. Finally, the energy equation for the mixture, obtained by addition of the energy equations for the gas and particulate phases, is presented.

4.1. Gas-Phase Energy Equation

The energy of a fluid element includes two components: thermal and mechanical. The thermal energy is the internal (including chemical) energy of the fluid while the mechanical energy is simply the kinetic energy due to motion. The first law of thermodynamics for a fluid element is

$$dE/dt = \dot{Q} - \dot{W} \tag{61}$$

where E is the energy (sum of thermal and mechanical), $\dot{Q}$ is the heat transfer rate to the element, and $\dot{W}$ is the rate at which the element does work on the surroundings. Application of the Reynolds transport theorem to the energy equation yields

$$(d/dt) \int_{cv} \rho e \, dV + \int_{cs} e(\rho \mathbf{v} \cdot d\mathbf{A}) = \dot{Q} - \dot{W} \tag{62}$$

where e is the energy per unit mass. For a mixture of gaseous species, the energy per unit mass is

$$\rho e = \sum_i \rho_i(i_i + v_i^2/2) \tag{63}$$

where i_i is the internal energy and v_i is the speed of species i. Substituting the energy per unit mass into Eq. (62) yields

$$(d/dt) \int_{cv} \sum_i \rho_i(i_i + v_i^2/2) \, dV + \int_{cs} \sum_i \rho_i(i_i + v_i^2/2)\mathbf{v}_i \cdot d\mathbf{A} = \dot{Q} - \dot{W} \tag{64}$$

The work rate term consists of two components: that due to surface forces and that due to body forces. The work due to surface forces is the work against pressure force and shear force. The work rate due to surface forces on species i can be expressed as

$$\dot{W}_{si} = \int_{cs} (p_i \mathbf{v}_i - \tau_i \cdot \mathbf{v}_i) \cdot d\mathbf{A} \tag{65}$$

where p_i is the partial pressure of species i, and τ_i is the shear-stress tensor associated with species i. Summing over all species yields the overall work rate term due to surface forces:

$$\dot{W}_s = \int_{cs} \sum_i (p_i \mathbf{v}_i - \tau_i \cdot \mathbf{v}_i) \cdot d\mathbf{A} \tag{66}$$

The work rate due to body forces can be expressed as

$$\dot{W}_b = -\int_{cv} \sum_i \rho_i \mathbf{g}_i \cdot \mathbf{v}_i \, dV \tag{67}$$

The work rate terms substituted into the energy equation yield

$$(d/dt) \int_{cv} \sum_i \rho_i(i_i + v_i^2/2) \, dV + \int_{cs} \sum_i \rho_i(h_i + v_i^2/2)\mathbf{v}_i \cdot d\mathbf{A}$$
$$= \dot{Q} + \int_{cs} \sum_i (\tau_i \cdot \mathbf{v}_i) \cdot d\mathbf{A} + \int_{cv} \sum_i \rho_i \mathbf{g}_i \cdot \mathbf{v}_i \, dV \tag{68}$$

where the partial pressure of each species has been combined with the internal energy to form h_i, the enthalpy of species i.

Heat transfer occurs due to conduction and radiation. Employing Fourier's law for heat conduction, the heat transfer term can be written as

$$\dot{Q} = \int_{cs} (k\nabla T) \cdot d\mathbf{A} + \int_{cv} q_r \, dV \tag{69}$$

where k is the thermal conductivity of the mixture, T is the temperature, and q_r is the net absorption per unit volume due to radiation.

Incorporating the expression for heat transfer into the energy equation and converting the surface integrals to volume integrals yields the following differential form of the total energy equation:

$$(\partial/\partial t) \sum_i \rho_i(i_i + v_i^2/2) + \nabla \cdot \left[\sum_i \rho_i(h_i + v_i^2/2)\mathbf{v}_i \right]$$

$$= \nabla \cdot (k\nabla T) + q_r + \nabla \cdot \left(\sum_i \tau_i \cdot \mathbf{v}_i \right) + \sum_i \rho_i \mathbf{g}_i \cdot \mathbf{v}_i \tag{70}$$

It is convenient to introduce the following mixture quantities

$$\rho_g i_g = \sum_i \rho_i i_i \tag{71a}$$

$$\rho_g h_g = \sum_i \rho_i h_i \tag{71b}$$

$$\rho_g v_g^2 = \sum_i \rho_i v_i^2 \tag{71c}$$

and to write the velocity of species i as [see Eq. (12)]

$$\mathbf{v}_i = \mathbf{v}_g + \mathbf{v}_{Di}$$

which, when substituted into Eq. (6), yields

$$(\partial/\partial t)[\rho_g(i_g + v_g^2/2)] + \nabla \cdot [\rho_g \mathbf{v}_g(h_g + v_g^2/2)] + \nabla \cdot \left(\sum_i \mathbf{j}_i h_i \right) + \nabla \cdot \left(\sum_i \mathbf{j}_i v_i^2/2 \right)$$

$$= \nabla \cdot (k\nabla T) + q_r + \nabla \cdot \left(\sum_i \tau_i \cdot \mathbf{v}_i \right) + \sum_i \rho_i \mathbf{g}_i \cdot \mathbf{v}_i \tag{72}$$

The third and fourth terms on the left-hand side are the transfer of energy due to interdiffusion of the gaseous species. Frequently, the diffusional transport of enthalpy is regarded as a heat transfer effect and is combined with the heat conduction term. The diffusional transport of kinetic energy is small compared to the diffusional transport of enthalpy, and can be neglected. It is also consistent with the definition of shear stress as used in the momentum equation to write

$$\sum_i \tau_i \cdot v_i = \tau \cdot v_g \tag{73}$$

where τ is the shear stress on the gaseous mixture.[10]

If the body force is the same for all species (such as a gravitational force), the work rate due to body forces simplifies to

$$\sum_i \rho_i g_i \cdot v_i = \rho_g g \cdot v_g \tag{74}$$

Introducing these simplifications, the total energy equation becomes

$$(\partial/\partial t)[\rho_g(i_g + v_g^2/2)] + \nabla \cdot [\rho_g v_g(h_g + v_g^2/2)] = -\nabla \cdot q + q_r + \nabla \cdot (\tau \cdot v_g) + \rho_g v_g \cdot g \tag{75}$$

where

$$q = -k\nabla T + \sum_i j_i h_i$$

In many applications, the use of the total energy equation is cumbersome and the thermal energy equation is more appropriate. The thermal energy equation is obtained by subtracting the mechanical energy equation from the total energy equation. The mechanical energy relation is obtained by forming the scalar product of the velocity with the momentum equation [Eq. (45)]; the resulting equation is

$$(\partial/\partial t)(\rho_g v_g^2/2) + \nabla \cdot (\rho_g v_g^2 v_g/2) = -v_g \cdot \nabla p + v_g \cdot (\nabla \cdot \tau) + \rho_g g \cdot v_g \tag{76}$$

Subtracting the mechanical energy from the total energy equation yields

$$(\partial/\partial t)(\rho_g i_g) + \nabla \cdot (\rho_g h_g v_g) = -\nabla \cdot q + q_r + \tau : \nabla v_g + v_g \cdot \nabla p \tag{77}$$

which is the thermal energy equation for a gaseous mixture. The term $\tau : \nabla v_g$ is the viscous dissipation, which is always positive and arises from the irreversible work due to viscous forces. The last term represents the reversible work resulting from compression or expansion of the fluid element.

4.2. Gas-Phase Energy Equation in a Gas–Particle Mixture

In the following derivation, the gas-phase mixture is treated as a single-component gas. Thus the conductive heat transfer term includes both conduction and species interdiffusion.

Consider, once again, the control volume for the gas-phase illustrated in Figure 2. The energy per unit mass of the gas is $i_g + v_g^2/2$. Combining the Reynolds transport theorem for energy change of the system with the first law of thermodynamics gives

$$\dot{Q} - \dot{W} = (d/dt) \int_{cv} \rho'_g(i_g + v_g^2/2)\, dV + \int_{S'} \rho_g(i_g + v_g^2/2)\mathbf{v}_g \cdot d\mathbf{A}$$

$$+ \int_{S'_p} \rho_s(i_s + v^2/2)\mathbf{w} \cdot d\mathbf{A} \qquad (78)$$

where the subscript s refers to the control surface surrounding the particles and v is the speed of the gas at the particle control surface. Taking $\dot{\mathbf{r}}$ as the regression rate of the particle, v can be expressed as

$$v = |\mathbf{v}_p + \mathbf{w} + \dot{\mathbf{r}}|$$

Heat transfer to the gas takes place by conduction across the surfaces S' and S'_p as well as by radiative energy absorbed by the gas. Thus,

$$\dot{Q} = -\int_S \theta \mathbf{q} \cdot d\mathbf{A} - \int_{S'_p} \mathbf{q}_p \cdot d\mathbf{A} + \int_{cv} \theta q_{rg}\, dV \qquad (79)$$

where $\mathbf{q}_p$ is heat transferred by conduction from the gas to the particles. It is convenient to decompose $\mathbf{q}_p$ into two components: the average heat transfer rate around the particle and the deviation therefrom,

$$\mathbf{q}_p = \bar{q}\mathbf{n} + \delta\mathbf{q}_p \qquad (80)$$

where $\mathbf{n}$ is the unit normal vector inward toward the particle interior. The heat transfer term can now be expressed as

$$\dot{Q} = -\int_S \theta \mathbf{q} \cdot d\mathbf{A} - \int_{S'_p} \delta\mathbf{q}_p \cdot d\mathbf{A} + \int_{cv} (\theta q_{rg} - q_{cp})\, dV \qquad (81)$$

where q_{cp} is the heat transferred to the particles from the gas by conduction per unit volume of mixture.

The work rate due to surface forces is obtained by evaluating the integral of force times velocity over the surfaces S' and S'_p:

$$\dot{W}_s = \int_{S'} (p\mathbf{v}_g - \boldsymbol{\tau} \cdot \mathbf{v}_g) \cdot d\mathbf{A} + \int_{S'_p} [p(\mathbf{v}_p + \mathbf{w}') + \boldsymbol{\tau} \cdot (\mathbf{v}_p + \mathbf{w}')] \cdot d\mathbf{A} \qquad (82)$$

Regrouping terms of Eq. (82) and identifying the aerodynamic force on the particles gives

$$\dot{W}_s = \int_S \theta(p\mathbf{v}_g - \boldsymbol{\tau} \cdot \mathbf{v}_g) \cdot d\mathbf{A} + \int_{cv} \mathbf{v}_p \cdot \mathbf{f}_p\, dV + \int_{S'_p} (\bar{p}_s/\rho_s)\rho_s\mathbf{w} \cdot d\mathbf{A}$$

$$+ \int_S (1-\theta)(\bar{p}_s\mathbf{v}_p - \boldsymbol{\tau}_a \cdot \mathbf{v}_p) \cdot d\mathbf{A} + \int_{\delta S_p} (\delta p/\rho_s)\rho_s\mathbf{w} \cdot d\mathbf{A} - \int_{cv} \bar{p}_s s_v\, dV \qquad (83)$$

where s_v is the particle dilatation rate per unit volume of mixture.

The work rate due to body forces is simply

$$\dot{W}_b = -\int_{cv} \rho'_g \mathbf{v}_g \cdot \mathbf{g} \, dV \tag{84}$$

Finally, substituting the above expressions for heat transfer and work into Eq. (78) yields

$$(d/dt)\int_{cv} \rho'_g(i_g + v_g^2/2)\, dV + \int_S \rho'_g(i_g + v_g^2/2)\mathbf{v}_g \cdot d\mathbf{A}$$

$$= -\int_S [\theta(p\mathbf{v}_g - \tau \cdot \mathbf{v}_g) + (1-\theta)(\bar{p}_s\mathbf{v}_p - \tau_a \cdot \mathbf{v}_p)] \cdot d\mathbf{A}$$

$$+ \int_{cv} (\bar{p}_s s_v - \mathbf{v}_p \cdot \mathbf{f}_p + \rho'_g \mathbf{v}_g \cdot \mathbf{g})\, dV - \int_S \theta\mathbf{q} \cdot d\mathbf{A} + \int_{cv} (\theta q_{rg} - q_{cp})\, dV$$

$$+ \int_{cv} r_p\left(\bar{h}_s + \frac{v_p^2}{2} + \frac{w'^2}{2}\right) dV - \int_{\delta S_p} (\delta\mathbf{q}_p + \rho_s\mathbf{w}\,\delta h_s + \mathbf{r}\delta p) \cdot d\mathbf{A} \tag{85}$$

where $\bar{h}_s$ is the average enthalpy of the gas over the particle control surface and δh_s is the deviation therefrom. The integrand of the integral over δS_p can be identified as the heat transfer through the boundary particles[5] and is designated as $\mathbf{q}_s$. Applying Gauss' theorem to convert the surface to volume integrals and Eq. (51) to relate $\bar{p}_s$ and p yields the following total energy equation for the gas in a gas–particle mixture:

$$(\partial/\partial t)[\rho'_g(i_g + v_g^2/2)] + \nabla \cdot [\rho'_g(h_g + v_g^2/2)\mathbf{v}_g]$$

$$= -\nabla \cdot [\theta\mathbf{q} + (1-\theta)\mathbf{q}_s] - q_{cp} + \theta q_{rg} + r_p(\bar{h}_s + v_p^2/2 + w'^2/2)$$

$$-\nabla \cdot [(1-\theta)p\mathbf{v}_p] + \nabla \cdot [\theta\tau \cdot \mathbf{v}_g + (1-\theta)\tau_a \cdot \mathbf{v}_p] - \mathbf{v}_p \cdot \mathbf{f}_p + \rho'_g \mathbf{g} \cdot \mathbf{v}_g + \bar{p}_s s_v$$

$$\tag{86}$$

When this equation is applied to the combustion of pulverized coal, the terms with $(1-\theta)$ can often be neglected. Also, when θ is equal to unity and q_{cp}, r_p, $\mathbf{f}_p$ and s_v are set equal to zero, the energy equation for the gas phase devoid of particles [Eq. (75)] is recovered.

4.3. Discrete-Phase Energy Equation

The energy equation for a cloud of particles is obtained by using the equation for a single particle in the Reynolds transport theorem for a particle cloud. The derivation begins by obtaining the energy equation for a single particle.

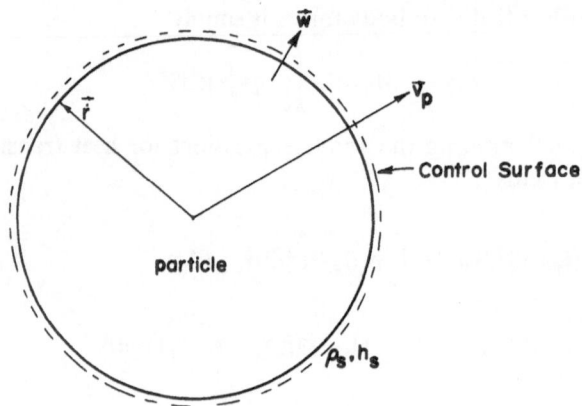

Figure 3. Control volume for a single particle.

The first law of thermodynamics applied to a particle is

$$\dot{Q}_p - \dot{W}_p = dE_p/dt \tag{87}$$

where E_p is the total energy of the particle, mechanical plus thermal. Taking the control surface adjacent to the surface of the particle as shown in Figure 3, the rate of energy change is given by

$$dE_p/dt = (d/dt) \int_{cv} \rho_p e_p \, dV + \int_{cs} e_s \rho_s \mathbf{w} \cdot d\mathbf{A} \tag{88}$$

where e_p is the energy per unit mass of the particle:

$$e_p = i_p + v_p^2/2 + e_\sigma \tag{89}$$

The last term is the surface energy due to surface tension, and e_s is the specific energy of the gas at the control surface.

The work done by the particle consists of flow work (the work done to overcome the pressure force in moving the vapor across the surface), the external or shaft work due to aerodynamic forces, and the work associated with particle volume change. Thus, the work rate term is

$$\dot{W}_p = \int_{cs} p_s \mathbf{w} \cdot d\mathbf{A} - \mathbf{F}_p \cdot \mathbf{v}_p + \dot{V}_p \bar{p}_s \tag{90}$$

where $\mathbf{F}_p$ is the force on the particle and $\dot{V}_p$ is the rate of change of particle volume.

Combining Eqs. (89) and (90) into the particle energy equation and assuming a uniform mass efflux over the particle surface yields

$$\dot{Q}_p = (d/dt)[m(i_p + e_\sigma)] - (\bar{h}_s + w'^2/2)(dm/dt) + \dot{V}_p \bar{p}_s \tag{91}$$

where, by using the particle momentum equation, the particle acceleration and force terms have been canceled. Assuming the surface tension is constant and taking $w'^2 \ll \bar{h}_s$, the energy equation can be simplified to

$$Q_p = m(di_p/dt) - (\bar{h}_s - h_p)(dm/dt) \tag{92}$$

which is commonly found in the literature.[6] The heat transfer to the particle is due to conduction and radiation.

The energy equation for a cloud of particles is obtained by using the Reynolds transport theorem, with the extensive property being the energy of the cloud of particles, namely,

$$E = \sum_k m_k(i_{pk} + v_{pk}^2/2 + e_{\sigma k}) \tag{93}$$

where i_{pk}, v_{pk}, and $e_{\sigma k}$ are the internal energy, speed, and surface energy of size class k, respectively. Application of the Reynolds transport theorem yields

$$dE/dt = (d/dt) \int_{cv} \sum_k \rho'_{pk}(i_{pk} + v_{pk}^2/2 + e_{\sigma k}) \, dV$$

$$+ \int_{cs} \sum_k \rho'_{pk}(i_{pk} + v_{pk}^2/2 + \sigma_k A_k/m_k)\mathbf{v}_{pk} \cdot d\mathbf{A} \tag{94}$$

The term on the left-hand side of the equation (Lagrangian) includes the rate of change of energy of the particles that are completely or partially inside the control volume.

Considerable care must be devoted to the "boundary" particles in order to formulate a precisely accurate energy equation. Consider the particle shown in Figure 4, which is penetrating the control surface. Let ε be a progress variable indicating the degree of penetration ($\varepsilon = -1$, penetration commencing; $\varepsilon = 1$, penetration complete). If, conceptually, the particle is separated along the control surface, as shown in Figure 4, the absence of the separated portion must be replaced by a heat transfer rate $q_s(\varepsilon)$. The energy equation for the portion of particle inside the control volume is

$$(d/dt)[m(\varepsilon)i_p(\varepsilon) + m(\varepsilon)e_\sigma(\varepsilon)] + \int_{-1}^{\varepsilon} (h_s + w'^2/2)(\rho_s \mathbf{w} \cdot d\mathbf{A})$$

$$= \bar{q}_p A(\varepsilon) - \int_{-1}^{\varepsilon} \delta\mathbf{q}_p \cdot d\mathbf{A} + q_s(\varepsilon) + \int_{-1}^{\varepsilon} p_s \mathbf{r} \cdot d\mathbf{A} \tag{95}$$

where the heat transfer term has been decomposed according to Eq. (90). The enthalpy of the gases at the particle surface is decomposed to the sum of the average value, h_s, and the deviation therefrom, δh_s. Taking $q_s(\varepsilon)$ equal to $q_s(-\varepsilon)$ along with δq_p, δh_s, and δp asymmetric functions of ε, and adding

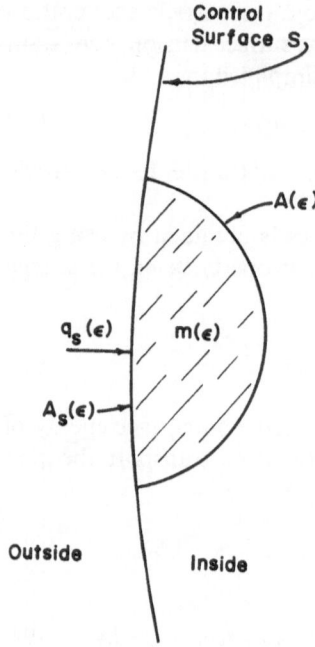

Figure 4. Particle penetrating a control surface.

the two energy equations for particles penetrated, ε and $-\varepsilon$, yields

$$(d/dt)[m(i_p+e_\sigma)]-(\bar{h}_s+w'^2/2)(dm/dt)=\dot{Q}_p-\bar{p}_sV_p$$

$$-2\int_{-1}^{\varepsilon}(\delta\mathbf{q}_p+\delta p\dot{\mathbf{r}}+\delta h_s\rho_s\mathbf{w})\cdot d\mathbf{A}+2q_s(\varepsilon) \qquad (96)$$

Subtracting the energy equation for a single particle [Eq. (91)] yields the equation for the heat transfer through the particle:

$$q_s(\varepsilon)=\int_{-1}^{\varepsilon}(\delta q_p+\delta p\dot{\mathbf{r}}+\delta h_s\rho_s\mathbf{w})\cdot d\mathbf{A} \qquad (97)$$

which was utilized in Eq. (86). Thus, the energy equation for the boundary particle is

$$(d/dt)[m(\varepsilon)i_p(\varepsilon)+m(\varepsilon)e_\sigma(\varepsilon)]+\rho_s w(\bar{h}_s+w'^2/2)A(\varepsilon)=\dot{Q}_p(\varepsilon)-\bar{p}_s\dot{r}A(\varepsilon) \qquad (98)$$

Summing the change of energy for all particles and portions thereof inside the control volume, substituting the result into Eq. (94), and converting the surface integrals to volume integrals yields the differential equation for the particulate phase

$$(\partial/\partial t) \sum_k \rho'_{pk}(i_{pk} + v^2_{pk}/2 + e_{\sigma k}) + \nabla \cdot \left[\sum_k v_{pk} \rho'_{pk}(i_{pk} + e_{\sigma k} + v^2_{pk}/2) \right]$$

$$= -\sum_k r_{pk}(\bar{h}_s + v^2_p/2 + w'^2/2) + \sum_k \mathbf{v}_{pk} \cdot (\mathbf{f}_{pk} + \rho'_p \mathbf{g}) + q_{cp} + q_{rp} - \bar{p}_s s_v \qquad (99)$$

where the heat transfer to the particle has been decomposed into conductive and radiative components.

Following conventional practice, it is convenient to define average energies for the composite mixture of particles:

$$\rho_p i_p = \sum_k \rho'_{pk} i_{pk} \qquad (100a)$$

$$\rho'_p v^2_p = \sum_k \rho'_{pk} v^2_{pk} \qquad (100b)$$

Using the relationship for the velocity introduced previously,

$$\delta \mathbf{v}_{pk} = \mathbf{v}_{pk} - \mathbf{v}_p$$

and combining the internal and surface energies, the left-hand side can be written as

$$(\partial/\partial t) \sum_k \rho'_{pk}(i_{pk} + v^2_{pk}/2) + \nabla \cdot \left[\sum_k v_{pk} \rho'_{pk}(i_{pk} + v^2_{pk}/2) \right]$$

$$= (\partial/\partial t)[\rho'_p(i_p + v^2_p/2)] + \nabla \cdot [\rho'_p \mathbf{v}_p(i_p + v^2_p/2)]$$

$$+ \nabla \cdot \left[\sum_k \rho'_{pk} \delta \mathbf{v}_{pk}(i_{pk} + v^2_{pk}/2) \right] \qquad (101)$$

The last term is analogous to the enthalpy transfer in the gas mixture due to species interdiffusion. If all particles moved with the same velocity, it would be zero. This term is seldom mentioned in the gas–particle flow literature. It is a very difficult term to treat in the numerical modeling of gas–particle flows.[9]

The final form of the total energy equation for a particle cloud is

$$(\partial/\partial t)[\rho'_p(i_p + v^2_p/2)] + \nabla \cdot [\rho'_p \mathbf{v}_p(i_p + v^2_p/2)] + \nabla \cdot \left[\sum_k \rho'_{pk} \delta \mathbf{v}_{pk}(i_{pk} + v^2_{pk}/2) \right]$$

$$= -\sum_k r_{pk}(\bar{h}_s + v^2_p/2 + w'^2/2)_k + \sum_k \mathbf{v}_{pk} \cdot \mathbf{f}_{pk} + \mathbf{g} \cdot \sum \rho'_{pk} \mathbf{v}_{pk} + q_{cp} + q_{pr} - \bar{p}_s s_v$$

$$(102)$$

where the origins of the various terms have already been discussed. One feature generally overlooked in the development of the particulate-phase energy equation is the absence of particle enthalpy in the convective term. The internal energy is correct in this situation, because the particles are incapable of doing flow work on each other unless particle–particle collisions are significant.

The thermal energy equation for the particulate phase can now be obtained by subtracting the mechanical energy equation (Section 4.3). The details are not included here, but the final result is

$$(\partial/\partial t)(\rho'_p i_p) + \nabla \cdot (\rho'_p \mathbf{v}_p i_p) + \nabla \cdot \left(\sum_k \rho'_{pk} \delta \mathbf{v}_{pk} i_{pk} \right)$$

$$= -\sum_k r_{pk}(\bar{h}_s + w'^2/2) + q_{cp} + q_{rp} - \bar{p}_s s_v \qquad (103)$$

In some applications, the thermal energy equation may be more useful.

The energy equation for the gas–phase in a gas–particle mixture [Eq. (86)] was derived assuming a uniform particle velocity for all particles. The corresponding energy equation for the particle cloud is

$$(\partial/\partial t)[\rho'_p(i_p + v_p^2/2) + \nabla \cdot [\rho'_p \mathbf{v}_p(i_p + v_p^2/2)]$$

$$= -r_p(\bar{h}_s + v_p^2/2 + w'^2/2) + \mathbf{v}_p \cdot \mathbf{f}_p + \rho'_p \mathbf{g} \cdot \mathbf{v}_p + q_{cp} + q_{rp} - \bar{p}_s s_v \qquad (104)$$

where the surface energy has been combined with the internal energy.

4.4. Overall Energy Equation

The overall energy equation is obtained by adding the energy equations for the gaseous and particulate phases [Eqs. (86) and (104)], yielding

$$(\partial/\partial t)[\rho'_p(i_p + v_p^2/2) + \rho'_g(i_g + v_g^2/2)]$$

$$+ \nabla \cdot [\rho'_p(i_p + p/\rho_p + v_p^2/2)\mathbf{v}_g + \rho'_g(h_g + v_g^2/2)\mathbf{v}_p] = -\nabla \cdot [\theta \mathbf{q} + (1-\theta)\mathbf{q}_s]$$

$$+ \theta q_{rg} + q_{rp} + \nabla \cdot [\theta \tau \cdot \mathbf{v}_g + (1-\theta)\tau_a \cdot \mathbf{v}_p] + \mathbf{g} \cdot (\rho'_p \mathbf{v}_p + \rho'_g \mathbf{v}_g) \qquad (105)$$

where the source terms due to the particles have canceled. The composite heat transfer consists of that due to thermal conduction through the gas and that through the particles. Typically, in combustion applications, the heat transfer through the particles can be neglected, which is achieved by taking $\theta = 1$. It must also be remembered that the heat transfer due to inter-diffusion of the species is included in $\mathbf{q}$. For consistency, it is conceivable that the internal energy flux due to particle velocity nonequilibrium may also be combined into the heat transfer term. In so doing, the heat transfer term *cannot* be modeled by Fourier's law.

The overall thermal energy equation is obtained by subtracting the mechanical energy equation corresponding to the mixture momentum equation. The resulting equation is very cumbersome and will not be discussed here. The reader is referred to reference 5 for details.

5. Summary of Governing Equations

Derivations of the governing equations for a gas–particle mixture have been presented in Sections 2–4 of this chapter. The final form of the continuity, momentum, and energy equations for the gas-phase, particle-phase, and mixture are summarized for convenience in Table 1. The particulate-phase equations correspond to locally uniform particle velocity. The terms accruing due to a distribution of particle velocities for differing particle sizes can be found by referring to the appropriate sections.

6. Macroscopic Form

In order to have the governing equations in a form amenable to analytical or numerical solution or to apply to a large control volume, it is often necessary to represent these equations in macroscopic form. This can be done by starting with the basic control volume approach or by integrating the differential equations over a finite-sized volume, assuming some distribution (usually uniform or linear) for the dependent variables. The latter approach will be utilized in this section assuming a uniform distribution of the dependent variables.

Table 1.   *Summary of Conservation Equations for Gas–Particle Mixtures*

Continuity equation

Gas phase: $\partial \rho'_g/\partial t + \nabla \cdot (\rho'_g \mathbf{v}_g) = r_p$

Particle phase: $\partial \rho'_p/\partial t + \nabla \cdot (\rho'_p \mathbf{v}_p) = -r_p$

Mixture: $(\partial/\partial t)(\rho'_p + \rho'_g) + \nabla \cdot (\rho'_g \mathbf{v}_g + \rho'_p \mathbf{v}_p) = 0$

Momentum equation

Gas phase: $(\partial/\partial t)(\rho'_g \mathbf{v}_g) + \nabla \cdot (\rho'_g \mathbf{v}_g \mathbf{v}_g) = -\nabla p + \nabla \cdot [\theta \boldsymbol{\tau} + (1-\theta)\boldsymbol{\tau}_a] - \mathbf{f}_p + \rho'_g \mathbf{g} + \mathbf{v}_p r_p$

Particle phase: $(\partial/\partial t)(\rho'_p \mathbf{v}_p) + \nabla \cdot (\rho'_p \mathbf{v}_p \mathbf{v}_p) = \mathbf{f}_p + \rho'_p \mathbf{g} - \mathbf{v}_p r_p$

Mixture: $(\partial/\partial t)(\rho'_g \mathbf{v}_g + \rho'_p \mathbf{v}_p) + \nabla \cdot (\rho'_g \mathbf{v}_g \mathbf{v}_g + \rho'_p \mathbf{v}_p \mathbf{v}_p)$
$$= -\nabla p + \nabla \cdot [\theta \boldsymbol{\tau} + (1-\theta)\boldsymbol{\tau}_a] + (\rho'_g + \rho'_p)\mathbf{g}$$

Energy equation (total)

Gas phase: $(\partial/\partial t)[\rho'_g(i_g + v_g^2/2)] + \nabla \cdot [\rho'_g \mathbf{v}_g(h_g + v_g^2/2)]$
$$= -\nabla \cdot [\theta \mathbf{q} + (1-\theta)\mathbf{q}_s] - q_{cp} + q_{rg} + r_p(\bar{h}_s + v_p^2/2 + w'^2/2) - \nabla \cdot [(1-\theta)p\mathbf{v}_p]$$
$$+ \nabla \cdot [\theta \boldsymbol{\tau} \cdot \mathbf{v}_g + (1-\theta)\boldsymbol{\tau}_a \cdot \mathbf{v}_p] - \mathbf{v}_p \cdot \mathbf{f}_p + \rho'_g \mathbf{g} \cdot \mathbf{v}_g + \bar{p}_s s_v$$

Particle phase: $(\partial/\partial t)[\rho'_p(i_p + v_p^2/2)] + \nabla \cdot [\rho'_p \mathbf{v}_p(i_p + v_p^2/2)]$
$$= -r_p(\bar{h}_s + v_p^2/2 + w'^2/2) + \mathbf{v}_p \cdot \mathbf{f}_p + \rho'_p \mathbf{g} \cdot \mathbf{v}_p + q_{cp} + q_{rp} - \bar{p}_s s_v$$

Mixture: $(\partial/\partial t)[\rho'_g(i_g + v_g^2/2) + \rho'_p(i_p + v_p^2/2)] + \nabla \cdot [\rho'_g \mathbf{v}_g(h_g + v_g^2/2) + \rho'_p \mathbf{v}_p(i_p + p/\rho_p + v_p^2/2)]$
$$= -\nabla \cdot [\theta \mathbf{q} + (1-\theta)\mathbf{q}_s] + \theta q_{rg} + q_{rp} + \nabla \cdot [\theta \boldsymbol{\tau} \cdot \mathbf{v}_g + (1-\theta)\boldsymbol{\tau}_a \cdot \mathbf{v}_p]$$
$$+ \mathbf{g} \cdot (\rho'_p \mathbf{v}_p + \rho'_g \mathbf{v}_g)$$

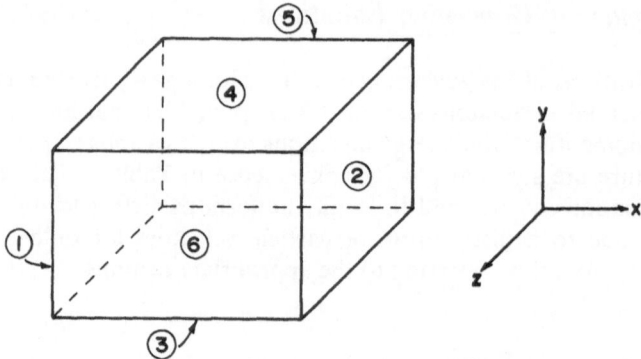

Figure 5. Control volume for macroscopic equations.

The volume to be used is the rectangular parallelepiped illustrated in Figure 5. The faces are designated 1–6 as shown on the figure. The three component directions, x, y, and z, are also indicated. The areas corresponding to the six faces are A_1 through A_6; the areas A_1 and A_2, which are normal to the x-direction, are designated as A_x and so on.

6.1. Continuity Equation

The integral form of the gas-continuity equation for a gas–particle mixture is given by Eq. (24). Substituting Eq. (26), which relates the mass efflux from the particles to a source of mass addition to the gas, into Eq. (24) yields the integral form of the gas-phase continuity equation:

$$(d/dt) \int_{cv} \rho'_g \, dV + \int_S \rho'_g \mathbf{v}_g \cdot d\mathbf{A} = \int_{cv} r_p \, dV \tag{106}$$

The integral of the bulk density over the volume is simply the mass of the gas in the volume, written in abbreviated form as

$$m_g = \int_{cv} \rho'_g \, dV \tag{107}$$

The mass addition over the volume due to particle burning is also abbreviated:

$$R_p = \int_{cv} r_p \, dV \tag{108}$$

The scalar product $\rho'_g \mathbf{v}_g \cdot d\mathbf{A}$ is the gas mass efflux from the volume through area $d\mathbf{A}$. The gas mass efflux across face i is represented by

$$G_{gi} = \int_{A_i} \rho'_g \mathbf{v}_g \, d\mathbf{A} \tag{109}$$

Substituting the above definitions into Eq. (106) allows the integral form of the gas-phase continuity equation to be expressed as

$$dm_g/dt + \sum_i G_{gi} = R_p \tag{110}$$

For a one-dimensional flow in the x-direction, the gas-phase continuity equation simplifies to

$$dm_g/dt + A_x(\rho_g' v_{gx}|_2 - \rho_g' v_{gx}|_1) = R_p \tag{111}$$

where $v_{g,x}$ is the mass-average gas velocity in the x-direction.

The integral form of the continuity equation for the particulate phase [Eq. (28)] is

$$(d/dt)\int_{cv} \rho_p' \, dV + \int_S \rho_p' \mathbf{v}_p \, d\mathbf{A} = -\int_{cv} r_p \, dV \tag{112}$$

Applying this expression to the control volume and using the same definitions for the integral of r_p over the control volume yields

$$dm_p/dt + \sum_i G_{pi} = -R_p \tag{113}$$

where m_p is the particulate mass in the control volume and G_{pi} is the particle mass efflux across the ith face.

The particulate-phase continuity equation for one-dimensional flow in the x-direction reduces to

$$dm_p/dt + A_x(\rho_p' v_{px}|_2 - \rho_p' v_{px}|_1) = -R_p \tag{114}$$

where v_{px} is the mass-average particle velocity in the x-direction.

The integral form of the overall continuity equation is obtained by adding Eqs. (110) and (113), yielding

$$d(m_p + m_g)/dt + \sum_i (G_{gi} + G_{pi}) = 0 \tag{115}$$

The one-dimensional form of this equation is obvious.

6.2. Momentum Equation

Integrating the momentum equation [Eq. (52)] for the gas phase in a gas–particle mixture over a control volume leads to

$$(d/dt)\int_{cv} \rho_g' \mathbf{v}_g \, dV + \int_S \mathbf{v}_g (\rho_g \mathbf{v}_g \cdot d\mathbf{A}) = -\int_S p \, d\mathbf{A} + \int_S [(1-\theta)\boldsymbol{\tau}_a + \theta\boldsymbol{\tau}] \cdot d\mathbf{A}$$
$$+ \int_{cv} (\rho_g' \mathbf{g} - \mathbf{f}_p + r_p \mathbf{v}_p) \, dV \tag{116}$$

For convenience, assume that particle velocity and body force per unit mass, g, are uniform over the control volume. Also, defining the cumulative aerodynamic force on the particles over the volume as $\mathbf{F}_p$ allows the last term in Eq. (116) to be expressed as

$$\int_{cv} (\rho'_g \mathbf{g} - \mathbf{f}_p + \mathbf{v}_p r_p) \, dV = m_g \mathbf{g} - \mathbf{F}_p + \mathbf{v}_p R_p \tag{117}$$

The momentum of the gas in the control volume is given by

$$m_g \mathbf{v}_g = \int_{cv} \rho'_g \mathbf{v}_g \, dV \tag{118}$$

Applying the integral equation for the gas-phase momentum in the x-direction to the parallelepiped control volume of Figure 5 results in

$$d(m_g v_{gx})/dt + \sum_i G_{gi} v_{gxi} = (p_1 - p_2) A_x$$
$$+ \sum_i [(1-\theta)(\tau_{ax})_i + \theta(\tau_x)_i] + m_g g_x - F_{px} + v_{px} R_p \tag{119}$$

where $(\tau_{ax})_i$ and $(\tau_x)_i$ are the shear stresses on the gas in the x-direction on the ith face. The integral equation for the gas-phase momentum in the remaining two directions is obtained by replacing x with y or z. Of course, the subscripts on the pressure force term must correspond to the chosen direction.

The gas-phase momentum equation for one-dimensional flow in the x-direction becomes

$$d(m_g v_{gx})/dt + A_x(\rho'_g v_{gx}^2|_2 - \rho'_g v_{gx}^2|_1) = (p_1 - p_2) A_x + T_x + m_g g_x - F_{px} + v_{px} R_p \tag{120}$$

where T_x is the composite force in the x-direction due to shear stress on all the six faces of the control volume. With a one-dimensional flow, the shear stress on faces 1 and 2 is usually neglected. If faces 3–6, inclusive, constitute a wall, T_x becomes

$$T_x = \tau \sum_{i=3}^{6} A_i \tag{121}$$

where τ is the shear stress on the fluid at the wall.

The integral form of the momentum equation for the particulate phase, assuming a locally uniform particle velocity as in Eq. (59), is

$$(d/dt) \int_{cv} \rho'_p \mathbf{v}_p \, dV + \int_S \mathbf{v}_p (\rho'_p \mathbf{v}_p \cdot d\mathbf{A}) = -\int_{cv} r_p \mathbf{v}_p \, dV + \int_{cv} \mathbf{f}_p \, dV + \int_{cv} \rho'_g \mathbf{g} \, dV \tag{122}$$

Applying the equation for particle momentum in the x-direction to the parallelepiped control volume gives

$$d(m_p v_{px})/dt + \sum_i G_{pi} v_{pxi} = -R_p v_{px} + F_{px} + m_p g_x \qquad (123)$$

The same equation applies to particle momentum in other directions as well by replacing the subscript x with y or z.

For one-dimensional flow in the x-direction, Eq. (123) reduces to

$$d(m_p v_{px})/dt + A_x[\rho'_p v_{px}^2|_2 - \rho'_{px} v_{px}^2|_1] = -R_p v_{px} + F_{px} + m_p g_x \qquad (124)$$

The momentum equation in the x-direction for the gas-particle mixture is obtained by adding Eqs. (119) and (123), yielding

$$(d/dt)(m_g v_{gx} + m_p v_{px}) + \sum_i (G_{gi} v_{gxi} + G_{pi} v_{pxi})$$

$$= (p_1 - p_2)A_x + \sum_i [(1-\theta)(\tau_{ax})_i + \theta(\tau_x)_i]A_i \qquad (125)$$

For pulverized-coal combustion, the asymmetric shear stress τ_{ax} can often be neglected, and θ can be taken equal to unity.

6.3. Energy Equation

The differential energy equation for the gas-phase of a gas–particle mixture, treating the mixture of gaseous species as a single component, is given by Eq. (86). Expressing this equation in integral form gives

$$(d/dt) \int_{cv} \rho'_g(i_g + v_g^2/2) \, dV + \int_S (h_g + v_g^2/2)\rho'_g \mathbf{v}_g \cdot d\mathbf{A}$$

$$= -\int_S [\theta \mathbf{q} + (1-\theta)\mathbf{q}_s] \cdot d\mathbf{A} - \int_{cv} (q_{cp} - \theta q_{rg}) \, dV$$

$$+ \int_{cv} r_p(\bar{h}_s + v_p^2/2 + w'^2/2) \, dV - \int_S (1-\theta)p\mathbf{v}_p \cdot d\mathbf{A} - \int_{cv} \mathbf{v}_p \cdot \mathbf{f}_p \, dV$$

$$+ \int_{cv} \rho'_g \mathbf{g} \cdot \mathbf{v}_g \, dV + \int_{cv} \bar{p}_s s_v \, dV + \int_S [\theta \tau \cdot \mathbf{v}_g + (1-\theta)\tau_a \cdot \mathbf{v}_p] \cdot d\mathbf{A} \qquad (126)$$

Using the parallelepiped control volume shown in Figure 5 as the volume of integration, the integral form of the gas-phase energy equation becomes

$$(d/dt)[m_g(i_g + v_g^2/2)] + \sum_i G_{gi}(h_g + v_g^2/2)_i$$

$$= -\sum_i [\theta q_i + (1-\theta)q_{si}]A_i - Q_{cp} + Q_{rg} + (\bar{h}_s + v_p^2/2 + w'^2/2)R_p - \sum_i (p/\rho_p)_i G_{pi}$$

$$+ \sum_i [\theta(\tau v_g)_i + (1-\theta)(\tau_a v_p)_i]A_i - (F_p v_p) + m_g(g v_g) + \bar{p}_s S_v \qquad (127)$$

where Q_{cp} is the heat transferred by conduction to the particles, Q_{rg} is the radiative energy absorbed by the gas, and S_v is the volume expansion rate of the particles in the control volume. The factor $(\tau v_g)_i$ is

$$(\tau v_g)_i = \sum_{j \neq i} (\tau_j)_i v_{gi} \tag{128}$$

where the summation is performed over all directions except i; that is, if i is equal to 1, then j is summed over 2 and 3. The term $(F_p v_p)$ is the work rate associated with the aerodynamic force on the particles and is equal to

$$(F_p v_p) = \sum_j F_{pj} v_{pj} \tag{129}$$

Similarly, the rate at which work is done against body forces is given by

$$(g v_g) = \sum_j g_i v_{gi} \tag{130}$$

Of course, considerable simplification results in Eq. (127) when θ is set equal to unity and terms with $(1 - \theta)$ are neglected.

For one-dimensional flow in the x-direction, in which the sides A_3–A_6 are walls, the gas-phase energy equation becomes

$$\begin{aligned}
(d/dt)[m_g(i_g + v_g^2/2)] &+ A_x[(\rho_g' v_{gx})_2(h_g + v_g^2/2)_2 - (\rho_g' v_{gx})_1(h_g + v_g^2/2)_1] \\
&= -[\theta q_x|_2 + (1-\theta)q_{sx}|_2 - \theta q_x|_1 - (1-\theta)q_{sx}|_1]A_x \\
&\quad + Q_w - Q_{cp} + Q_{rg} + (\bar{h}_s + v_p^2/2 + w'^2/2)R_p \\
&\quad - \sum_i (p/\rho_p)_i G_{pi} - F_{px}v_{px} + m_g g_x v_{gx} + \bar{p}_s S_v
\end{aligned} \tag{131}$$

where Q_w is the heat transferred to the gas from the confining walls.

The energy equation for a particle cloud assuming uniform discrete-phase particle velocity is given by Eq. (104). The integral form of this equation is

$$\begin{aligned}
(d/dt) \int_{cv} \rho_p'(i_p + v_p^2/2)\, dV &+ \int_S (i_p + v_p^2/2)\mathbf{v}_p \cdot d\mathbf{A} \\
&= -\int_{cv} r_p(\bar{h}_s + v_p^2/2 + w'^2/2)\, dV + \int_{cv} \mathbf{v}_p \cdot \mathbf{f}_p\, dV + \int_{cv} \rho_p' \mathbf{g} \cdot \mathbf{v}_p\, dV \\
&\quad + \int_{cv} q_{cp}\, dV + \int_{cv} q_{rp}\, dV - \int_{cv} \bar{p}_s s_v\, dV
\end{aligned} \tag{132}$$

where it must be remembered that the surface energy due to surface tension has been incorporated into the particle internal energy. Using the parallelepiped as the control volume gives the integral from

$$(d/dt)[m_p(i_p+v_p^2/2)] + \sum_i G_{pi}(i_p+v_p^2/2)_i$$

$$= -(\overline{h}_s + v_p^2/2 + w'^2/2)R_p + (F_p v_p) + m_p(g v_p) + Q_{cp} + Q_{rp} - \overline{p}_s S_v \qquad (133)$$

where Q_{rp} is the rate of radiative energy absorbed by the particles and, as with Eq. (127), $(g v_p)$ is the work rate associated with the body forces on the particles and is given by

$$(g v_p) = \sum_j g_j v_{pj} \qquad (134)$$

The corresponding energy equation for one-dimensional flow in the x-direction becomes

$$(d/dt)[m_p(i_p+v_p^2/2)] + A_x[(\rho_p' v_{px})_2(i_p+v_p^2/2)_2 - (\rho_p' v_{px})_1(i_p+v_p^2/2)_1]$$

$$= -R_p(\overline{h}_s+v_p^2/2+w'^2/2) + F_{px}v_{px} + m_p g_x v_{px} + Q_{cp} + Q_{rp} - \overline{p}_s S_v \qquad (135)$$

The thermal energy equation can be obtained by subtracting the mechanical energy equations from the above expressions (total energy equations) but will not be presented here. The most straightforward approach is to start with the differential forms of the mechanical energy equation, express them in integral form, and proceed according to the approach used in this section.

The overall energy equation in integral form for the gas-particle mixture is obtained by adding Eqs. (127) and (133), yielding

$$(d/dt)[m_g(i_g+v_g^2/2) + m_p(i_p+v_p^2/2)] + \sum_i [G_{gi}(h_g+v_g^2/2)_i + G_{pi}(i_p+p/\rho_p+v_p^2/2)_i]$$

$$= -\sum_i [\theta q_i + (1-\theta)q_{si}]A_i + Q_{rg} + Q_{rp} + \sum_i [\theta(\tau v_g)_i + (1-\theta)(\tau v_p)_i]A_i$$

$$+ m_g(g v_g) + m_p(g v_p) \qquad (136)$$

For application to pulverized-coal combustion, θ is approximately equal to unity, and terms with the factor $(1-\theta)$ may often be neglected, resulting in equations which are somewhat simpler. These equations represent the conservation equations in macroscopic form and are amenable to analytical or numerical solution.

7. Transport Coefficients

The multicomponent conservation equations outlined above contain the transport coefficients: viscosity (μ), thermal conductivity (k), and diffusivity (D). These molecular transport coefficients must be provided either from experimental data or from predictive techniques as input in order to solve the conservation equations. Since these coefficients are dependent upon local values of mixture composition and temperature (and in the case of diffusivity, upon pressure), predictive equations are required for incorporation into general computer codes. These coefficients are required for a number of additional purposes in models of reacting flow systems. For example, the gas-phase mixture conductivity is required to compute conductive and convective heat transfer between particles and the gas. The gas-phase mixture viscosity, which is a parameter in the Reynolds number, influences convective heat transfer effects to particles and walls, and can also affect the velocity distribution. The multicomponent diffusion coefficient is required for analysis of mass transfer between particles and gases, and for molecular diffusion in the gas phase, such as in diffusion flames and premixed propagating laminar flames.

In this section, techniques recommended for predicting values of k, μ, and D for the gas phase are summarized. The material that follows is not intended to be exhaustive. Liquid- or solid-phase transport coefficients have not been considered. Techniques shown are applicable to dilute (low pressure–high temperature) gas mixtures found in complex reacting systems. This section also includes a brief outline of the theoretical basis for gas-phase transport coefficients, but not attempt has been made to provide rigorous detail.

Reid and Sherwood[11] give an extensive treatment of techniques for estimating transport coefficients for gases and liquids. Bartlett et al.[12] have outlined an approximate and efficient scheme for predicting these coefficients in complex, multicomponent gas mixtures. Finally, these transport coefficients are basic properties of the molecular systems. Turbulent properties of complex gas and particle systems are treated in Chapter 3.

7.1. Technical Basis

For transport of property (i.e., mass, heat, or momentum) in the dilute gas phase, Foust et al.[13] show for a model gas that the flux of property is related to the gradient of concentration of property:

$$\psi_x = (-\bar{l}\bar{c}/6)(d\Gamma/dx) \tag{137}$$

where $\bar{l}$ is the mean free path of the molecules, $\bar{c}$ is the mean speed, and Γ is the concentration of property. This treatment assumes rigid spherical molecules of constant diameter (σ) in random motion, with no attractive or repulsive forces among molecules, which therefore undergo elastic collisions. The generalized transport coefficient, δ, is then defined as

$$\delta = \bar{l}\bar{c}/6 \tag{138}$$

For this same simple model gas, Jeans[14] shows that the mean free path between collisions is given by

$$1/\pi n_g \sigma^2 \tag{139}$$

where n_g is the number density of gas molecules. Further, the mean speed of the molecules, which have only three (translational) degrees of freedom, is[14]

$$\tfrac{1}{2}\bar{c}^2 = \tfrac{3}{2}RT \tag{140}$$

Substituting Eqs. (138) and (139) into Eq. (140) and employing the ideal gas law for n_g gives an equation for the generalized diffusivity for the model gas:

$$\delta = \bar{c}(T^3/M_{wg})^{1/2}/p\sigma^2 \tag{141}$$

According to Eq. (141) for the model gas at a given temperature and pressure, the transport coefficient is independent of the property being transferred. Thus

$$\delta = v = \alpha = D \tag{142}$$

Since $v = \mu/\rho$ and $\alpha = k/\rho C_p$, Eq. (142) can be rewritten for each of the transport coefficients of interest:

$$\mu = c_1 (M_w T)^{1/2}/\sigma^2 \tag{143}$$

$$k = c_2 (M_w/T)^{1/2}/\sigma^2 \tag{144}$$

$$D = c_3 (T^3/M_w)^{1/2}/p\sigma^2 \tag{145}$$

These equations contain the essential features of the predictive techniques used for pure gas transport coefficients. Consideration of the random velocity distribution of molecules (Maxwell–Boltzmann distribution) and forces of interaction among molecules[11] has led to the applicable equations that are summarized below.

Hirschfelder et al.[15] have reported the rigorous techniques for computation of transport properties of gas mixtures based on kinetic theory of dilute gases. From this result, Wilke[16] developed a simpler semiempirical technique which is suitable for most applications requiring viscosity and thermal conductivity. A summary of techniques recommended for application to complex, dilute gas mixtures is also given in the following section.

7.2. Transport Coefficient Equations

Viscosity. Table 2 summarizes equations recommended for predicting viscosities of pure components and gas mixtures. Results are shown in SI units. Values of the collision parameters (σ, ε/k, δ, Ω_v) required to predict pure component viscosity are tabulated in Appendix B. Lennard-Jones

Table 2. Summary of Recommended Equations for Predicting Viscosity of Dilute Bases

Classification	Reference	Equation (SI units)	Estimated accuracy[1]	Comments
A. Nonpolar, pure component	15, 19	$\mu = \dfrac{2.67(10^{-6})(M_w T)^{1/2}}{\sigma^2 \Omega_v}$	18 systems, 76 comparisons; avg. error: 1.2%, max. error: 7.6%	Chapman–Enskog equation; Lennard-Jones potential for molecular interactions; $\Omega_v = f(kT/\varepsilon)$; see Table B.2, Appendix B
B. Polar, pure component	11	$\mu = \dfrac{2.67(10^{-6})(M_w T)^{1/2}}{\sigma^2 \Omega_v}$	13 systems, 51 comparisons; avg. error: 1.7%, max. error: 10.0%	Chapman–Enskog equation; Stockmayer potential for molecular interactions; $\Omega = f(T/\varepsilon, \delta)$; see Table B.3, Appendix B
C. Gas mixtures, polar and nonpolar	16	$\mu_m = \sum_{i=1}^n \mu_i \left[1 + \sum_{\substack{j=1 \\ i\neq j}}^n \phi_{ij}\left(\dfrac{y_j}{y_i}\right)\right]^{-1}$ $\phi_{ij} = \dfrac{[1+(\mu_i/\mu_j)^{1/2}(M_{wi}/M_{wj})^{1/4}]^2}{\{8[1+(M_i/M_j)]\}^{1/2}}$	3 systems, 24 comparisons; avg. error: 1.0%, max. error: 4.6%	Wilke approximation to Chapman–Enskog equation

potential parameters are used for nonpolar gases while Stockmayer potential parameters are used for polar gases. Polar gases are those with a separation of ionic charge, or those with a dipole moment, such as H_2O, HCl, H_2S, NH_3, etc., while nonpolar gases are those without a dipole moment of charge, such as O_2, N_2, CH_4, Ar, CO_2, etc.

For gas mixtures, the Wilke approximation is a semiempirical approximation of the more rigorous result from kinetic theory,[15] and has been widely used. Bartlett *et al.*[12] have further simplified the Wilke result using the bifurcation technique of Bird[17]:

$$\mu_m = \sum_{i=1}^{n} y_i \mu_i [y_i + 1.385(RT/p\bar{D})(F_i \mu_i / M_{wi})(\mu_1 - y_i F_i)]^{-1} \tag{146}$$

where $\mu_1 = \sum_{i=1}^{n} y_i F_i$ and other symbols are as defined in the Notation section (Section 8), and are discussed by Bartlett *et al.*[12] This equation reduces the computation of mixture viscosity to the evaluation of a single summation and is thus efficient for computation of complex multicomponent properties.

Thermal Conductivity. Table 3 summarizes predictive techniques recommended for thermal conductivities of pure components and gas mixtures. Values for the collision parameters for conductivity (σ, ε/k, δ, Ω_v) are identical to those for viscosity and are tabulated in Appendix B. Lennard-Jones parameters are used for polar as well as nonpolar compounds with some increase in predictive error for the polar systems. The more recent techniques of Bromley,[18] who modified the Eucken form with temperature-dependent parameters, provide a more accurate but more complex method of predicting pure component thermal conductivity. According to Reid and Sherwood,[11] using the Bromley techniques, the average errors were 3.4% for 10 nonpolar, nonlinear systems (78 experiments), 3.7% for 10 linear systems (40 experiments), and 7.5% for 10 polar, nonlinear systems (31 experiments). While this result is significantly better than the Eucken correction (see Table 3), it has not been used as extensively because of the additional data input requirements.

For conductivity for gaseous mixtures, the technique shown in Table 3 is essentially identical to that of Table 2 for mixture viscosity but with a correction factor from Mason and Saxena.[20] Further simplification for mixture conductivities has been suggested by Bartlett *et al.*,[12] who developed a bifurcation method for ease of computation. Their result has the advantage of requiring only a single summation in evaluating conductivities for complex gas mixtures.

Molecular Diffusivity. Table 4 summarizes techniques recommended for evaluating binary and effective binary diffusion coefficients. Lennard-Jones parameters $[(\sigma_i + \sigma_j)/2, (\varepsilon_i \varepsilon_j)^{1/2}/k,$ and $\Omega_D]$ are summarized in Appendix B and are used for both polar and nonpolar compounds. According to Bird

Table 3. *Summary of Recommended Equations for Predicting Thermal Conductivity of Dilute Gases*

Classification	Reference	Equation (SI units)	Estimated accuracy[11]	Comments
A. Monatomic, pure component	11, 19	$k = \dfrac{0.0832(T/M_w)^{1/2}}{\sigma^2 \Omega_k}$		Chapman–Enskog theory, Lennard-Jones potential; $\Omega_k = \Omega_v = f(kT/\varepsilon)$; see Table B.2, Appendix **B**
B. Nonpolar, nonlinear polyatomic	11	$k = [C_p + 1.25(R/M_w)]\mu$ (Eucken) $\quad$ $k = [1.22\,C_p + 0.55(R/M_w)]\mu$ (modified Eucken)	Average error for Eucken correction: 8.2% nonpolar, nonlinear; 4.8% linear, 13.0% polar, nonlinear	Eucken correction predicts low values; modified Eucken correction predicts high results; see reference 11 for more accurate but more complex methods for k
C. Gas mixtures	20	$k_m = \sum_{i=1}^{n} k_i \left[1 + \sum_{\substack{j=1 \\ j \neq i}}^{n} A_{ij} \left(\dfrac{y_j}{y_i}\right) \right]$ $A_{ij} = 1.065\,\phi_{ij}$; see Table 2 for ϕ_{ij}		Wassiljewa equation with Mason–Saxena formulation for $A_{ij}^{[11]}$

Table 4. Summary of Recommended Equations for Predicting Diffusivity of Dilute Gases

Classification	Reference	Equation (SI units)	Estimated accuracy[1]	Comments
A. Binary, polar, and nonpolar components	11, 19	$$\mathscr{D}_{ij} = \frac{0.0188\{T^{3/2}[(M_{wi}+M_{wj})/(M_{wi}M_{wj})]^{1/2}\}}{p\sigma_{ij}^2\Omega_D}$$	Average error of 7.5% for 114 binary systems; maximum error up to 37% includes polar-nonpolar data	Chapman–Enskog theory for ideal polar and nonpolar gases with Lennard-Jones potential for Ω_D; see Reid and Sherwood[11] for modifications for polar–nonpolar and polar and polar combinations
B. Effective binary multicomponent system	6	$$\mathscr{D}_{im} = (1-y_i)\left[\sum_{\substack{j=1\\i\neq j}}^{n}\left(\frac{y_j}{\mathscr{D}_{ij}}\right)\right]^{-1}$$	Not established for general systems	Assumes i dilute in mixture or all but diffusing species stationary
C. Generalized multicomponent gas	15	$$D_{IJ} = \frac{1}{M_{wj}}\left(\sum_k y_k M_{wk}\right)\frac{K^{ji}-K^{ii}}{|K|}$$ $$k_{ji} = \frac{y_i}{[\mathscr{D}_{ij}]_1} + \frac{M_{wj}}{M_{wi}}\sum_{k\neq i}\frac{y_k}{[\mathscr{D}_{ik}]_1}$$ $K_{ii}=0$: $|K|$ is determinate of K^{ji}, and K^{ii} are minors[15]		See Hirschfelder et al.[15] for details of this generalized predictive technique
D. Binary, polar, and nonpolar compounds	12	$$D_{IJ} = \bar{D}/F_I F_J$$	Accuracy evaluated for selected multicomponent systems, Bartlett et al.[12] 17	Based upon bifurcation approximation of reference 17

et al.[6] for dilute, multicomponent gases,

$$\mathbf{j}_i = (\bar{c}^2/\rho) \sum_{j=1}^{n} M_{wi} M_{wj} D_{ij} \nabla X_j \qquad (147)$$

where D_{ij} are formally evaluated according to the equations of Table 4. However, because of the complexity of evaluating the D_{ij} values, multicomponent systems have often been treated as effective binary systems, such that

$$\bar{j}_i = c \mathcal{D}_{im} \nabla X_i \qquad (148)$$

where $\mathcal{D}_{im}$ can be evaluated as shown in Table 4. Even this approximate technique involves considerable computation for complex multicomponent mixtures, since $\mathcal{D}_{im}$ is a function of all of the $\mathcal{D}_{ij}$ values locally in the reacting gas, as well as of temperature and pressure.

Reduction in computation time for binary diffusivities can be achieved by using the result

$$\mathcal{D}_{im} = \bar{D}/F_i F_j \qquad (149)$$

as proposed by Bird[17] and outlined by Bartlett et al.[12] Here, the $\bar{D}$ values are self-diffusion coefficients which are principally functions of temperature and pressure, while the F_i diffusion factors are independent of concentration and pressure and nearly independent of temperature. Treatment of multicomponent systems by Bartlett et al. has then been based upon the Stefan–Maxwell equations, which require only the binary diffusion coefficients to compute the diffusion flux in a multicomponent mixture. This technique is computationally very efficient and appears to have significant advantages over the use of Eq. (147). However, the authors have had no specific experience with this more recent development. Using this approach, an equally efficient method has been outlined[12] for including thermal diffusion with little increase in computational time for complex mixtures.

Since combustion gases of interest in this book are generally at very high temperatures, and pressures of interest range from just below atmospheric pressure to several atmospheres, no corrections have been made to account for pressure effects on viscosity, conductivity, or the pressure-diffusivity product. However, where such corrections are required, appropriate techniques have been outlined by Reid and Sherwood.[11]

8. Notation

A	Area vector (m²)	A_{ij}	Thermal conductivity mixture
A_{11}^*	Multicomponent conductivity		parameter (1.065 ϕ_{ij})
	collision integral parameter	C_p	Heat capacity (J kg⁻¹ K⁻¹)

c Concentration or molar density ($k\ mol\ m^{-3}$)

$\bar{c}$ Mean molecular speed ($m\ s^{-1}$)

cv   Control volume

D_{ij} Multicomponent diffusion coefficient ($m^2\ s^{-1}$)

$\mathscr{D}_{ij}$ Binary diffusion coefficient ($m^2\ s^{-1}$)

$\mathscr{D}_{im}$ Effective binary diffusivity ($m^2\ s^{-1}$)

$\bar{D}$ Reference (self-) diffusivity ($m^2\ s^{-1}$)

E Energy [$J\ (N\ m)$]

e Specific energy ($J\ kg^{-1}$)

e_σ (Surface energy)/(particle mass) ($J\ kg^{-1}$)

$\mathbf{F}$ Force vector (N)

F_i Species diffusion factor

$\mathbf{f}_p$ (Force on particles)/volume ($N\ m^{-3}$)

G Mass flux ($kg\ s^{-1}$)

$\mathbf{g}$ (Body force)/mass ($N\ kg^{-1}$)

h Specific enthalpy ($J\ kg^{-1}$)

i Specific internal energy ($J\ kg^{-1}$)

$\mathbf{j}$ Mass flux ($kg\ m^{-2}\ s^{-1}$)

k Thermal conductivity ($W\ m^{-1}\ K^{-1}$)

$|K|$ Diffusivity determinant ($s\ m^{-2}$)

k_{ij} Diffusivity parameter ($s\ m^{-2}$)

K^{ji} Determinant minor ($s\ m^{-2}$)

$\mathcal{T}$ Mean free path (m)

M_w Molecular weight ($kg\ kmol^{-1}$)

m Mass (kg)

n Number of species in multicomponent mixture

$\mathbf{n}$ Unit outward normal vector

$\dot{N}$ Number flow rate of particles (s^{-1})

n_g Molecular gas density (m^{-3})

p Pressure [$Pa\ (N\ m^{-2})$]

$\dot{Q}$ Heat transfer rate [$W\ (J\ s^{-1})$]

q Heat transfer flux ($W\ m^{-2}$)

q_p (Heat transfer rate to particles)/volume ($W\ m^{-3}$)

q_s Heat transfer flux on "separated" particles ($W\ m^{-2}$)

R Gas constant ($J\ kmol^{-1}\ K^{-1}$)

$\dot{r}$ Particle regression rate ($m\ s^{-1}$)

S Control surface

S_v Dilatation rate of particles ($m^3\ s^{-1}$)

s_v (Dilatation rate of particle)/volume (s^{-1})

T Temperature (K)

t Time (s)

$\mathbf{T}$ Momentum transfer due to particle collision ($N\ m^{-3}$)

T_x Shear force on fluid in x-direction (N)

$\mathbf{v}$ Velocity ($m\ s^{-1}$)

$\mathbf{v}_D$ Diffusional velocity ($m\ s^{-1}$)

$\delta\mathbf{v}_p$ "Diffusional" particle velocity ($m\ s^{-1}$)

V Volume (m^3)

V_p Rate of change of particle volume ($m^3\ s^{-1}$)

$\mathbf{w}$ Gas velocity with respect to particle control surface ($m\ s^{-1}$)

$\mathbf{w}'$ Gas velocity at control surface w/r particle center ($m\ s^{-1}$)

X Space coordinate (m)

y_i Mole fraction

α Thermal diffusivity ($m^2\ s^{-1}$)

δ Dipole moment parameter

δ Generalized diffusivity ($m^2\ s^{-1}$)

ε Particle penetration variable

ε/k Energy interaction parameter (K)

ϕ_{ij} Mixture viscosity parameter

Γ Generalized property concentration

Ω Collision integral

μ Viscosity ($N\ s\ m^{-2}$)

μ_1, μ_2, μ_3 Multicomponent transport coefficient parameters

v Kinematic viscosity (viscous diffusivity) ($m^2\ s^{-1}$)

ρ Density ($kg\ m^{-3}$)

ρ' Bulk density ($kg\ m^{-3}$)

ρ_i Mass density, species i ($kg\ m^{-3}$)

θ Void fraction

σ Collision diameter (m, Å)

ψ Generalized property flux

τ Shear-stress tensor in gas ($N\ m^{-2}$)

τ_a Asymmetric shear-stress tensor on particle ($N\ m^{-2}$)

ω Mass fraction

Subscripts

b body

d diffusion

g	gas	p	particle
i	ith gas species, control surface	r	radiation
j	jth gas species	s	particle surface
k	particle size class	v	viscosity
m	gas-particle mixture	x	x-direction

9. References

1. C. Truesdell and R. Toupin, The classical field theories, in *Encyclopedia of Physics* (S. Flügge, ed.), Springer-Verlag, Berlin (1960).
2. D. A. Drew, Averaged field equations for two-phase media, *Stud. Appl. Math.* **50**(2), 133–136 (1971).
3. C. T. Crowe, M. P. Sharma, and D. E. Stock, The particle source in cell (PSI-cell) model for gas-droplet flows, *J. Fluids Eng.* **99**(2), 325–332 (1977).
4. W. J. Comfort, T. W. Alger, W. H. Giedt, and C. T. Crowe, Calculation of two-phase dispersed droplets in vapor flows including normal shock waves, *J. Fluids Eng.* (1978), in press.
5. C. T. Crowe, Conservation Equation for Vapor-Droplet Flows including Boundary Droplet effects, UCRL-52184, Lawrence Livermore Laboratories, California (1977).
6. R. B. Bird, W. E. Stewart, and E. N. Lightfoot, *Transport Phenomena*, John Wiley and Sons, New York (1960).
7. S. Penner, *Chemistry Problems in Jet Propulsion*, Pergamon Press, New York (1957).
8. W. Nachbar, F. Williams, and S. S. Penner, The conservation equations for independent coexisting continua and for multicomponent reacting gas mixtures, *Quart. Appl. Math* **17**(1), 43–54 (1959).
9. C. T. Crowe, M. P. Sharma, and D. E. Stock, Numerical modeling in design of multiphase flow systems, in *Proceedings of the First International Conference on Mathematical Modeling*, Vol. 3 (X. J. R. Avula, ed.), pp. 1249–1258, University of Missouri, Rolla, Missouri (1977).
10. P. D. Kelly, A reacting continuum, *Int. J. Eng. Sci.* **2**, 129–153 (1964).
11. R. C. Reid and T. K. Sherwood, *The Properties of Gases and Liquids—Their Estimation and Correlation*, 2nd ed., McGraw–Hill Book Co., New York (1966).
12. E. P. Bartlett, R. M. Kendall, and R. A. Rindal, An Analysis of the Coupled Chemically Reacting Boundary Layer and Charring Ablator, Part IV: A Unified Approximation for Mixture Transport Properties for Multicomponent Boundary Layer Applications, NASA CR-1063, Itec Corp., Vidya Division, Palo Alto, California (1966).
13. A. S. Foust, L. A. Wenzel, C. W. Clump, L. Maus, and L. B. Anderson, *Principles of Unit Operations*, John Wiley and Sons, New York (1960).
14. J. Jeans, *An Introduction to the Kinetic Theory of Gases*, p. 43, The University Press, Cambridge, England (1962).
15. J. O. Hirschfelder, C. F. Curtiss, and R. B. Bird, *Molecular Theory of Gases and Liquids*, p. 530, John Wiley and Sons, New York (1954).
16. C. R. Wilke, A viscosity equation for gas mixtures, *J. Chem. Phys.* **18**, 517 (1950).
17. R. B. Bird, Diffusion in multicomponent gas mixtures, *Kaguku Kogaku* **26**, 718 (1962).
18. L. A. Bromley, Thermal Conductivity of Gases at Moderate Pressures, University of California Radiation Lab, UCRL-1852, Berkeley, California (June 1952).
19. S. Chapman and T. G. Cowling, *Mathematical Theory of Nonuniform Gases*, 2nd ed., Cambridge University Press, England (1951).
20. E. A. Mason and S. C. Saxena, Approximate formula for the thermal conductivity of gas mixtures, *Phys. Fluids* **1**, 361 (1958).

Part II
Rate Processes

Turbulence

David T. Pratt

1. Characteristics of Turbulence

In laminar flows, velocity gradients are sufficiently mild so that the fluid can dynamically adjust to imposed shear stresses through molecular (viscous) forces. When imposed shear forces are too great for the fluid to adjust through molecular processes, the fluid is "torn" into largely coherent regions (turbulent eddies), which can rotate much like fluid "roller bearings" and thus relieve the shear forces caused by the imposed velocity differences. These turbulent eddies, in turn, undergo a chaotic sequence of events in which they are reduced in scale, until eventually a dimension is reached below which the laminar or molecular processes are sufficient to control the fluid response to the imposed shear forces.

The scale at which the eddies are first formed is called the *turbulent macroscale*, and is the same order of magnitude as the dimensions of the shearing system. There are a number of formally defined microscales,[1] but for present purposes it is sufficient to define a single microscale, namely, that of the smallest measured eddy size below which molecular forces are controlling.

An alternative way to describe the characteristics of turbulence is through energy considerations. A continuously imposed shear force on a moving fluid results in a continuous input of mechanical energy. In order to achieve a stationary state where energy input and output are balanced, the fluid must adjust its motion. In confined flows, the total mechanical energy

David T. Pratt • Professor of Mechanical Engineering, University of Utah, Salt Lake City, Utah

that the flowing stream can accept or absorb is often limited by other factors which are controlled by conservation laws of mass and momentum as well as energy. It is therefore necessary that the flow must dissipate much of the imposed mechanical energy, degrading it irreversibly to thermal energy. Again, if laminar shear stresses are insufficient to affect this dissipation, the fluid adjusts by means of the more vigorous secondary motions of turbulence to increase the rate of dissipation of mechanical energy to thermal energy. This dissipation of mechanical energy plays a strong role in the calculation or prediction of turbulence phenomena, as will be seen shortly.

One of the attributes of turbulence that leads to greatest difficulty in characterizing and predicting turbulent motion is the apparently chaotic nature of the motion. The fluid appears to seek a variety of states that will satisfy the governing conservation equations, subject to imposed boundary conditions. We are thus faced with an essentially unpredictable sequence of widely differing dynamic states, as opposed to the comparatively well ordered and predictable laminar flow condition.

2. Reynolds Equations and the Closure Problem

The conservation equations developed in Chapter 2 are correct at any instant of time. In a laminar field, in which the entire field of values are either time-steady or changing systematically, these same equations lead to correct predictions of time-mean values as well; thus they are often referred to as the "laminar equations of motion." However, turbulent flows consist of a randomly sequential series of solutions of these "laminar" equations. The interest is not in obtaining an enormous number of different solutions of a reacting flow field, but rather in obtaining the *mean* (in some sense) values of the independent variables of interest. In nonreacting incompressible flows, at least, some progress has been made toward this end.

In order to rewrite the instantaneous or "laminar" equations of motion in Chapter 2 in such a form that the independent variables of interest are the time-mean values, a process called *Reynolds decomposition* may be employed. This process will be illustrated in a conventional way, using the momentum equations for a nonreacting incompressible flow.

First, instantaneous values of the independent variables u_i (x_i component of velocity) are represented by the sum of a time-mean value $\bar{u}$, and an instantaneous fluctuating value u', the latter defined so that

$$u' \equiv u - \bar{u} \tag{1}$$

The equation of motion for the gas phase [Eq. (45) of Chapter 2] may then be rewritten for incompressible flow in Cartesian index notation as

$$(\partial/\partial t)(\bar{u}_j + u'_j) + (\partial/\partial x_i)[(\bar{u}_i + u'_i)(\bar{u}_j + u'_j)] = (\partial/\partial x_i)[-\delta_{ij}(\overline{P} + P') + (\bar{\tau}_{ij} + \tau'_{ij})]$$

(2)

where

$$\bar{\tau}_{ij} \equiv v[(\partial \bar{u}_i/\partial x_j) + (\partial \bar{u}_j/\partial x_i)] \quad \text{and} \quad \tau'_{ij} \equiv v[(\partial u'_i/\partial x_j) + (\partial u'_j/\partial x_i)]$$

and where body forces have been neglected.

Next, Eq. (2) is averaged over a sufficiently long time so that the time-mean values are meaningful:

$$\partial \bar{u}_j/\partial t + (\partial/\partial x_i)(\bar{u}_i\bar{u}_j + \overline{u'_iu'_j}) = (\partial/\partial x_i)(-\delta_{ij}\overline{P} + \bar{\tau}_{ij})$$

(3)

In reducing Eq. (2) to Eq. (3), recognition has been made from the definition of Eq. (1) that long-time averages of u' and P' are zero.

Equations (3) are called the *Reynolds equations* or the Reynolds-decomposed Navier–Stokes equations.[1] Comparison with the original momentum or Navier–Stokes equations shows that each term is similar, with overbars added to denote time-average values, except for the term $-\overline{u'_iu'_j}$, the so-called *Reynolds stress*, which has no counterpart in the laminar form of the equations. The appearance of this term, which represents an additional new variable, requires additional information to be able to solve Eq. (3). This difficulty is referred to as the *closure problem.* The field of turbulence modeling has been largely concerned for many years with closure assumptions, namely, tractable models that enable approximate representations of the Reynolds stresses in Eq. (3).

3. Turbulence Models

Very briefly, closure of the incompressible, nonreacting Reynolds equations can be considered at three levels of sophistication, referred to by Mellor and Herring[2] as MVF (*mean velocity field*), MTE (*mean turbulence energy*), and MRS (*mean Reynolds stress*) closures. Each of these closure assumptions will be reviewed briefly.

3.1. Mean Velocity Field Closure

The most conventional and simplest approach to the closure problem is due to Boussinesq (1877), and is expressed by analogy to molecular viscosity and the mean strain rates as

$$-\overline{u'_iu'_k} = \Gamma(\partial \bar{u}_i/\partial x_k + \partial \bar{u}_k/\partial x_i)$$

(4)

where Γ is the "eddy viscosity" or "effective turbulent viscosity." Equation (4) is used in both MVF and MTE closure.

This convenient assumption is incorrect theoretically for the following reasons[1]:

1. The "eddy viscosity," Γ, should be highly anisotropic and, therefore, should properly be represented as a fourth-order tensor, rather than as the isotropic scalar (zero-order tensor) as represented in Eq. (4). However, treatment of Γ as a fourth-order tensor would introduce 16 unknown scalar elements in a two-dimensional flow, which is hardly beneficial for closure.

2. Equation (4) implies that the length and time scales of the turbulent motion are of the same order of magnitude as the motion of the mean flow field. This is true only for flows dominated by shear forces; indeed, Eq. (4) works very well for pure shear flows. However, Eq. (4) must fail when forces other than shear play a co-equal or dominant role in determining the turbulent field motion; for example, buoyant body forces, which have their own characteristic length and time scales.

3. Equation (4) implies that the turbulence adjusts itself instantly to local changes in the mean flow field, since a universal relationship is assumed to exist between the Reynolds stresses and the mean strain rates. On the contrary, it is well established that both "history" and "action-at-a-distance" play essential roles in establishing local turbulence fields and, therefore, no such universal relationship can in fact exist.

Unfortunately, Eq. (4) has been used so widely and for so long a time that it is widely accepted in spite of the severe objections cited. It has been found satisfactory when applied to closely bounded flows (pipes, boundary layers, jets, and near wakes) with one-scale shear systems (pipe diameter, momentum thickness, nozzle or blockage diameter), in which the strong effect of the local shear system on the turbulent field dominates the "history" or transport mechanism. Equation (4) fails completely in unbounded flows such as the far wake or with free convection, because turbulence transport considerations dominate over shear systems in these flows, so that the governing characteristic length scales must be evolved by the turbulence itself, in the absence of the shear-generating solid boundaries.[3]

Mean velocity closure schemes all use Eq. (4), together with some algebraic equation relating the eddy viscosity Γ to the mean flow field variables. These equations may be characterized as "mixing length" hypotheses, typified by the *Prandtl hypothesis*

$$\Gamma = Cu' l_m \qquad (5)$$

where C is a constant of order 1, u' is the rms fluctuating velocity, $\overline{(u'u')}^{1/2}$, and l_m is a "mixing length," analogous to the mean free path (MFP) in elementary gas-phase kinetic theory. This analogy is subject to criticism because, in kinetic theory, the corresponding MFP equation relates two widely differing scales, one a microscopic (molecular) and one a macroscopic (mean flow field) scale. Turbulence, however, is entirely a macroscopic phenomena and the integral length scale or "mixing length" will be of the same scale as the mean flow field in shear flows.[1,2]

A typical application of Eq. (5) is to the spreading of a jet issuing from a slit or orifice into a semi-infinite fluid. From experiment, it has been found that values of $Cu' \simeq 0.01u_0$ and $l_m \simeq r_0$ (u_0 and r_0 being the inlet jet mean velocity and diameter, respectively) will model the near-jet turbulence field remarkably well.[1]

3.2. Mean Turbulent Energy Closure

In the MTE closure scheme, a more sophisticated approach is taken in relating the eddy viscosity to mean flow field properties. The Prandtl mixing-length hypothesis is replaced by the *Prandtl–Kolmogorov* relation

$$\Gamma = C_\mu k^2 / \varepsilon \tag{6}$$

where C_μ is a constant, $k = \overline{u'^2}/2$ is the specific turbulent kinetic energy, and ε is the rate of dissipation of k. A differential transport equation for k is employed, with a separate differential equation relating ε to flow variables. This closure scheme is often referred to as a "two-equation" model.[4,5]

An equation for the turbulent kinetic energy k may be derived from first principles as follows:

(1) By subtracting Eq. (2) from Eq. (3), an equation for the instantaneous velocity fluctuation u_i is obtained.
(2) By multiplying this equation by itself, dividing by 2 and time-averaging, an equation for k is obtained:

$$\underset{A}{\partial k/\partial t} + \underset{}{(\partial/\partial x_j)(\underset{B}{\overline{u_j k}} + \underset{C}{\overline{u'_j k}} + \underset{D}{\overline{u'_j P}} - \underset{}{\overline{u'_i \tau'_{ki}})} = \underset{E}{(-\overline{u'_j u'_i})(\partial \bar{u}_i/\partial x_j)} \underset{F}{- \overline{\tau'_{ji}(\partial u'_i/\partial x_j)}}$$
$$\tag{7}$$

The various terms in Eq. (7) are identified by Tennekes and Lumley[1] as follows: (A) advection of k; (B) turbulent transport of k by turbulent velocity fluctuations; (C) pressure gradient work; (D) transport of k by viscous stresses; (E) production of k from the mean flow, P_s; and (F) viscous dissipation of k, ε. In steady, homogeneous, pure shear flows, all terms but E and F are equal to zero, and the familiar equilibrium theory of turbulence results:

$$\text{production of } k = \text{dissipation of } k \quad \text{or} \quad P_s = \varepsilon \tag{8}$$

All of the terms in Eq. (7), except term A, must be related to mean flow field variables in order to achieve closure.

Tennekes and Lumley[1] give a lucid discussion and order-of-magnitude analysis of each term in Eq. (7). They point out that only term D, viscous transport of k, can be neglected, and then only when the turbulent Reynolds number $Re = (k^{1/2}l/v)$ is much greater than unity, where l is the *integral scale* or *macroscale* of turbulence.

Similarly, a transport equation for ε can be written, and the various terms in the equation can be modeled.[4,5] A summary of the modeled k and ε equations are as follows:

k equation:

$$(\partial/\partial x_j)(\bar{u}_j k) = (\partial/\partial x_j)[(\Gamma/\sigma_k)(\partial k/\partial x_j)] + P_s - \varepsilon \qquad (9)$$

ε equation:

$$(\partial/\partial x_j)(\bar{u}_j \varepsilon) = (\partial/\partial x_j)[(\Gamma/\sigma_\varepsilon)(\partial \varepsilon/\partial x_j)] + (c_1 P_s \varepsilon/k) - (c_2 \varepsilon^2/k) \qquad (10)$$

where c_1 and c_2 are "universal" constants, and σ_k and σ_ε are "turbulent Prandtl–Schmidt" numbers for k and ε, respectively.[5] With field values of k and ε determined from Eqs. (9) and (10), the Prandtl–Kolmogorov equation [Eq. (6)] is used to determine the eddy viscosity Γ, and finally Eq. (4) is used (as in MVF closure) to determine the Reynolds stress terms in Eq. (3).

3.3. Mean Reynolds Stress Closure

A more basic approach to the closure problem is to avoid the use of Eq. (4) entirely, and to model the Reynolds stress terms directly. A transport equation for $R_{ij} = \overline{(u_i' u_j')}$ is derived by taking the equation for fluctuating velocity u_i', multiplying by u_j', adding to this the same equation with subscripts i and j reversed, and time-averaging the result.

The result is the emergence of 15 new unknown correlation terms. Rather than "closing" Eq. (3), MRS closure, in fact, expands the problem further. Since MRS closure completely avoids the use of the eddy viscosity concept and rather solves a modeled transport equation for each of three (in two-dimensional "turbulence") Reynolds stresses, anisotropy is thus introduced into the turbulent field, which is necessarily isotropic with use of scalar eddy viscosity. In effect, the three $(\overline{u_i' u_j})$ equations are for component turbulent kinetic energies, so the k equation is redundant and is not used. However, the TKE (*turbulent kinetic energy*) dissipation equation [Eq. (10)] must be retained in order to compute length scales. Progress in MRS closure is not sufficiently advanced at the present time to merit its consideration.[4]

3.4. Turbulence Coefficients

Launder and Spalding[5] have determined a set of best-fit coefficients for the constants employed in the modeled MTE (*mean turbulent energy*) turbulence equations [Eqs. (9) and (10)]. They are summarized in Table 1.

Table 1. *Recommended Values of Constants for the $k - \varepsilon$ Model*[5]

Equation	Constant	Recommended value
(6)	C_μ	0.09
(9)	σ_k	1.1
(10)	σ_ε	1.3
(10)	c_1	1.44
(10)	c_2	1.3

4. Recommended Approach

For the present, it is necessary to assume the $k-\varepsilon$ model, expressed as the mean turbulent energy closure scheme in Section 3.2. In effect, since Eq. (4) is utilized, turbulent flow is treated as a *quasi-laminar* flow, with the molecular viscosity replaced by an effective turbulent viscosity in the transport equations.

This selection has been made largely on the extensive use of the $k-\varepsilon$ model by the Imperial College group.[5] While this model has been shown to be satisfactory for nonreacting incompressible flows, its application to reacting, necessarily compressible flows has been extremely limited. In addition, Bray[6] has shown that the properly Reynolds-decomposed conservation equations for a reacting flow should include additional terms in Eqs. (9) and (10) for contributions to the k-production and -dissipation terms, P_s and ε, due to heat release from chemical reaction and due to buoyant forces. Further, the modeling of turbulence has been predicated upon high turbulent Reynolds numbers in the flows, and the "lazy" flames encountered in certain types of pulverized-coal furnaces may not always satisfy this restriction. In short, this approach is adopted with all of its shortcomings simply because it is the best model available at the present time for calculating the gross features of the convective flow patterns.[7]

Turbulence effects on gas-phase reaction rates and particle dispersion are discussed in Chapters 4 and 6, respectively.

5. Notation

C Empirically determined constant in Eq. (5)

C_μ Empirically determined constant in Eq. (6)

c_1, c_2 Empirically determined constants in Eq. (10)

l Macroscopic turbulence scale (m)

l_m Mixing length [in Eq. (5)] (m)

k Mean turbulent kinetic energy $(\bar{u'^2}/2)\,(\mathrm{m^2\ s^{-2}})$

P Pressure (Pa)

P_s Production rate of k [term E in Eq. (7)] $(\mathrm{m^2\ s^{-3}})$

R_{ij} Reynolds stress, $\overline{u_i'u_j'}$ [in Eqs. (3) and (4)] $(\mathrm{m^2\ s^{-2}})$

t Time (s)

u_i Instantaneous ith component of velocity (m s^{-1})

$\bar{u}_i$ Time-mean ith component of velocity (m s^{-1})

u'_i $= (\bar{u}_i - u_i)$, instantaneous fluctuating value of ith component of velocity (m s^{-1})

x_i Spatial dimension, ith component (m)

δ_{ij} Kronecker delta ($=1$ when $i=j$; $=0$ when $i \neq j$)

ε Dissipation rate of k due to viscosity term F in Eq. (7)] (m^2 s^{-3})

Γ Eddy viscosity, or effective turbulent kinematic viscosity in Eq. (4) (m^2 s^{-1})

μ Molecular or dynamic viscosity (N s m^{-2})

ν Molecular kinematic viscosity, μ/ρ (m^2 s^{-1})

ρ Mass density (kg m^{-3})

$\sigma_k, \sigma_\varepsilon$ Turbulent Prandtl–Schmidt numbers in Eqs. (9) and (10), respectively

τ_{ij} Instantaneous value of shear-stress tensor (N m^{-2})

$\bar{\tau}_{ij}$ Time-mean value of shear-stress tensor (N m^{-2})

τ'_{ij} $= (\bar{\tau}_{ij} - \tau_{ij})$, instantaneous fluctuating value of shear-stress tensor (N m^{-2})

6. References

1. H. Tennekes and J. L. Lumley, *A First Course in Turbulence*, M.I.T. Press, Cambridge, Mass. (1972).
2. G. L. Mellor and H. J. Herring, A survey of the mean turbulent field closure models, *A.I.A.A. J.* **5**, 590 (1973).
3. J. L. Lumley and B. Khajeh-Nouri, Computational modeling of turbulent transport. *Adv. Geophys.* **18A**, 169 (1974).
4. W. C. Reynolds and T. Cebeci, *Calculation of Turbulent Flows, Topics in Applied Physics*, Vol. 12, 193–229, Springer-Verlag (1976).
5. B. E. Launder and D. B. Spalding, *Mathematical Models of Turbulence*, Academic Press, New York (1972).
6. K. N. C. Bray, Equations of Turbulent Combustion, I: Fundamental Equations of Reacting Flow, A.A.S.U. Report 330, University of Southampton, England (1973).
7. D. T. Pratt, Mixing and chemical reaction in continuous combustion, *Prog. Energy Comb. Sci.* **1**, 73–96 (1976).

4

Gas-Phase Chemical Kinetics

David T. Pratt

1. Homogeneous Rate Expressions

This chapter is concerned with basic aspects of chemical reactions in the gas phase. Rate data for homogeneous gaseous reactions are given in Chapter 10. Heterogeneous kinetics are discussed in Chapter 9, while applications are treated in Part IV.

Consider an elementary, physicochemical bimolecular exchange reaction represented by

$$A_1 + A_2 \underset{b}{\overset{f}{\rightleftharpoons}} A_3 + A_4 \tag{1}$$

The instantaneous, homogeneous forward rate may be expressed by the familiar expression

$$r_f = k_f [A_1][A_2] = k_f \rho^2 n_1 n_2 \tag{2}$$

where k_f is the forward reaction rate constant, expressed in a modified Arrhenius form as

$$k_f = 10^{B_f} T^{N_f} \exp(-T_f/T) \tag{3}$$

In Eq. (3), B_f, N_f, and T_f are parameters determined empirically or from theoretical approximation.[1] The symbols are defined in the Notation section (Section 6) at the end of this chapter.

The product $10^{B_f} T^{N_f}$ in Eq. (3) is often referred to as the "preexponential

David T. Pratt • Professor of Mechanical Engineering, University of Utah, Salt Lake City, Utah

factor," and, together with the concentration product $\rho^2 n_1 n_2$, may be thought of as the molecular collision frequency, of which a fraction, $\exp(-T_f/T)$, are "successful," or in other words, result in the forward reaction represented by Eq. (1). Typically, for most elementary reactions and in SI units, B_f will vary from 10 to 20, N_f from -2 to $+2$, and T_f from 0 to 100 K.

Expressions for the reverse (backward) rate r_b and rate constant k_b are identical to Eqs. (2) and (3), respectively, with subscripts b in place of f.

Termolecular recombination reactions of the form

$$A_1 + A_2 + M \underset{b}{\overset{f}{\rightleftharpoons}} A_3 + M \tag{4}$$

have homogeneous rate expressions of the form

$$r_f = k_f \rho^3 n_m n_1 n_2 \quad \text{and} \quad r_b = k_b \rho^2 n_m n_3 \tag{5}$$

The symbols M and n_m refer to any or all of the chemical species present; in other words, any molecule may serve as a gas-phase catalyst, as evidenced by the presence of the concentration of all species present, n_m in Eq. (5). (Strictly speaking, a multiplicative "third-body efficiency" factor should be used to allow for varying catalyst efficiencies of different molecules.)

In order to describe the detailed pyrolysis, ignition and oxidation of a complex molecule to the final, equilibrium products of combustion, a great many elementary reactions such as Eq. (1) act in series and in parallel. A complete specification of such a reaction scheme, together with the corresponding rate data for forward and backward rates of each reaction, is called a *mechanism*. Mechanisms for combustion of methane (CH_4) in air, to final products CO_2 and H_2O, are given in Chapter 10.

2. Inhomogeneous Rate Expressions: The Contact Index

Expressing the mass density ρ in Eq. (2) by the ideal gas law,

$$\rho = P/RTn_m \tag{6}$$

Eq. (2) may be rewritten as

$$r = 10^B T^{N-2} (P/Rn_m)^2 \exp(T_f/T) n_1 n_2 \tag{7}$$

where the subscript f has been omitted from r_f, B_f, and N_f for simplicity in operations to follow.

Equation (7) is valid for homogeneous, gas-phase reactions, as might be expected to occur in static reactors or in laminar flames and flow reactors. However, when the reacting flow becomes turbulent, local values of T, P, ρ, n_i, etc., fluctuate rapidly with time and position.[2,3] Thus, Eq. (7) describes only the local, instantaneous value of r, and further analysis is required to determine a correct expression for the time-mean value of r, $\bar{r}$.

In order to find an expression for $\bar{r}$, it is necessary to *Reynolds-decompose* Eq. (7); that is, to express the instantaneous values of independent variables by the sum of time-mean plus fluctuating values, as described in Chapter 3, Section 1; for example,

$$T = \bar{T} + T' \tag{8}$$

Ignoring fluctuating values of pressure and mean molecular weight, the Reynolds-decomposed Eq. (7) is

$$r = 10^B (T^{N-2} + T^{N-2'})(P/Rn_m)^2 \exp\left[-T_f/(\bar{T}+T')\right](\bar{n}_1+n'_1)(\bar{n}_2+n'_2) \tag{9}$$

Following Borghi,[3] the exponential term is represented as

$$A \equiv \exp\left[-T_f/(\bar{T}+T')\right]$$
$$= \exp\left(-T_f/\bar{T}\right)\exp\left[(T_f/\bar{T})(T'/\bar{T})/(1+T'/\bar{T})\right] \tag{10}$$

Two familiar series expansions are now introduced: an expansion for $\exp(x)$, which converges for all values of x, and an expansion for $(1+x)^\alpha$, which converges for x between (-1) and $(+1)$. With these substitutions and considerable manipulation, there results

$$\bar{A} = \exp\left(-T_f/\bar{T}\right)\left[1 + \sum_{n=2}^{\infty} (P_n \overline{T'^n}/\bar{T}^n)\right] \tag{11}$$

and

$$A' \equiv A - \bar{A}$$
$$= \exp\left(-T_f/\bar{T}\right)\left[P_1 + \sum_{n=2}^{\infty} (P_n/\bar{T}^n)(T'^n - \overline{T'^n})\right] \tag{12}$$

where the P_n are polynomials of degree n in (T_f/T) of the form

$$P_n \equiv \sum_{k=1}^{n} \frac{(-1)^{n-k}(T_f/\bar{T})^k (n-1)!}{(n-k)![(k-1)!]^2 k} \tag{13}$$

Utilizing again the series expansion for $(1+x)^\alpha$, the remaining temperature-dependent terms in Eq. (9) are now represented as

$$\overline{T^{N-2}} = \bar{T}^{N-2}\left[1 + \sum_{n=2}^{n} (Q_n \overline{T'^n}/\bar{T}^n)\right] \tag{14}$$

and

$$T^{N-2'} \equiv T^{N-2} - \overline{T^{N-2}}$$
$$= \bar{T}^{N-2}\left[Q_1(T'/\bar{T}) + \sum_{n=2}^{\infty} Q_n(T'^n + \overline{T'^n})/\bar{T}^n\right] \tag{15}$$

where

$$Q_n \equiv (N-2)(N-1)\cdots(N+1+n)/n! \tag{16}$$

Finally, substitution of Eqs. (10)–(16) into Eq. (9) and subsequent time-averaging results in

$$\bar{r} = 10^B \overline{T^{N-2}} \exp(-T_f/\overline{T}) \bar{n}_1 \bar{n}_2 X \tag{17}$$

where X, *the contact index*, is given by

$$X = 1 + \overline{n_1' n_2'}/\bar{n}_1 \bar{n}_2 + [(P_2 + Q_2 + P_1 Q_1)/\overline{T}^2]\overline{T'^2}$$

$$+ (P_1 + Q_1)(\overline{T' n_1'}/\overline{T n_1} + \overline{T' n_2'}/\overline{T n_2}) + P_1(\overline{T' n_1' n_2'}/\overline{T n_1 n_2})$$

$$+ P_2[\overline{T' T' n_1'}/\overline{T^2 n_1} + \overline{T' T' n_2'}/\overline{T^2 n_2}] + (P_3 + Q_3)(\overline{T'^3}/\overline{T^3}) + \cdots \tag{18}$$

In functional notation, Eq. (17) may be represented as

$$\bar{r}(T, n_1, n_2) = X r(\overline{T}, \bar{n}_1, \bar{n}_2)$$

Thus X is a correction term which relates the mean reaction rate to the rate expressed as a value of mean variables. Clearly, for uniform values of n_1, n_2, and T, X goes to unity.

Equation (18) for the correction term X is exact, within the assumption of constant pressure and mixture molecular weight, and for values of temperature fluctuation in the useful range, $|T'/\overline{T}| < 1$.

The terms in X are described as follows[3]:

(a) The coefficients Q_n decrease rapidly with increasing n, but the P_n do not, as the P_n are polynomials in $(T_f/\overline{T})$. If the activation temperature T_f is high, or the mean temperature $\overline{T}$ is low, P_n may increase very rapidly with increasing n. Therefore, a strong correction term X_f will result when slow reactions and/or low mean reaction temperatures exist, indicating a strong influence of turbulent mixing.

(b) The coefficient of the correlation terms in Eq. (18) are generally greater than zero, as $(T_f/\overline{T})$ is usually $\gg 1$. The signs of correlation terms themselves determine whether X is positive or negative, that is, whether or not $r(\bar{n}_1, \bar{n}_2, \overline{T})$ underpredicts or overpredicts the correct reaction rate $\bar{r}$.

The "correct" rate expression of Eq. (18) requires the modeling of eight third-order and lower correlation terms for each reaction considered. In order to close a system comprised of M conservation equations for N species, together with an energy equation for T, the number of third-order or lower correlation terms to be modeled will be:

(a) One each of the terms $\overline{T'^2}$ and $\overline{T'^3}$;
(b) N terms of the form $\overline{n_i' T'}$;
(c) N terms of the form $\overline{n_i' T' T'}$;
(d) At most $N(N-1)$ terms of the form $\overline{n_i' n_k'}$;
(e) At most $N(N-1)$ terms of the form $\overline{n_i' n_k' T'}$.

In any particular case, estimates (d) and (e) will be substantially reduced to the number of forward reactions having distinct pairs of reactants.

It is clear that closure of the system of equations for reacting turbulent flow is an even more formidable task than the closure problem for nonreacting turbulent flow, as discussed in Chapter 3. In fact, it is presently impossible to construct traditional model equations for the correlation terms because their values have not been measured.[4,5] However, some approximate means for predicting second- and third-order correlations, as well as direct estimates of the contact index, are in prospect. These and alternative descriptions of turbulent combustion are discussed in the following section.

3. Models of Turbulent Combustion

The predictive modeling of turbulent combustion is an extremely active and controversial research area at the present time, as evidenced by the number of recent reviews, workshops, and colloquia on the subject.[2,4-8] It is impossible and inappropriate to review the entire subject here. Rather, an abbreviated and necessarily simplistic presentation of currently and near-term potentially useful models are presented.

A predictive model for turbulent combustion must take into account the relative magnitudes of (at least) three finite-rate processes:

(a) *Macromixing,* or coarse-scale mixing. In the present context, convective recirculation and turbulent eddies account for the composition and *scale of segregation*[2,6,8] of adjacent fluid regions.

(b) *Micromixing,* or fine-scale mixing. After macromixing, scale reduction by turbulent breakup of large eddies and molecular diffusion is necessary so that molecular interactions may occur.

(c) *Chemical kinetics,* or finite-rate chemistry. When mixing processes (a) and (b) finally enable molecules A_1 and A_2 of Eq. (1) to collide, they may form product molecules A_3 and A_4 at a finite rate governed by Eq. (2).

In all three of the models to be discussed, it is assumed that finite-rate macromixing processes (a) are adequately predicted by the quasi-laminar model of turbulent flow described in Chapter 3. Basically, the three models to be described differ in that different combinations of the rates of processes (b) and (c) above are taken to be finite or infinite: Model 1 assumes (b) finite- and (c) infinite-rate; Model 2 assumes (b) infinite- and (c) finite-rate; while Model 3 treats both (b) and (c) as finite-rate.

3.1. Model 1. Locally Homogeneous, Finite-Rate Chemistry

For reacting flows in which the local rate of micromixing is greater than the rate of chemical reaction (for example, vigorous stirring by high-velocity inlet jets and slow chemistry), combustion occurs as a volume, rather than a flame sheet or surface, phenomenon. For this case, the chemistry is rate-limiting and the flow may be adequately modeled as an interconnected array of homogeneous perfectly stirred (or continuously stirred tank) reactors.[9−11] Since actual pulverized-coal-fired furnaces rarely, if ever, feature chemically rate-limited combustion in the gas phase, this model has utility for present purposes only if the contact index X of Eq. (18) can be predicted for forward and backward reactions.

3.2. Model 2. Finite-Rate Micromixing with Equilibrium Chemistry

Here the fuel and air are not premixed, and when fast chemistry, together with slow or "lazy" micromixing, occurs, chemical reactions may be taken to be infinite-rate or equilibrated. This is the well-known flame sheet or "if it mixes, it burns" analysis of turbulent diffusion flames first considered by Hawthorne *et al.*,[12] later by Vassilatos and Toor,[13] and more recently by many investigators.[4,7] The description given here is adapted largely from ideas of Spalding,[14] Elghobashi and Pun,[15] and Tamanini.[16]

Assumptions for this model may be summarized as follows:

(a) Fuel and air, both gaseous, are initially unmixed and enter through separate inlet streams.

(b) Fuel and air may not coexist, even instantaneously. Upon mixing, the reaction is instantaneously completed to the adiabatic, thermodynamic equilibrium state.

(c) Local variation in the fuel mixture fraction f (mass fraction of matter having entered via the fuel stream) is represented by a two-parameter *probability density function* $P(f)$, defined so that $P(f)\,df$ is the fractional time (over a sufficiently long time interval) during which the local mixture fraction is in the range $(f, f + df)$.[2]

The first two moments of the distribution $P(f)$ are the mean $\bar{f}$ and variance g, which are defined conventionally by

$$\bar{f} = \int_0^1 f P(f)\,df \tag{19}$$

$$g = \int_0^1 (f - \bar{f})^2 P(f)\,df \tag{20}$$

With these assumptions, local mean conversion and reaction rates are determined as follows:

(1) The two modeled equations for $\bar{f}$, the local time-mean value of mixture fraction, and g, the local variance of f, are due to Spalding.[14] In cylindrical polar coordinates, and for elliptic flows, these equations are[15]

$$\rho u(\partial \bar{f}/\partial x) + \rho v(\partial \bar{f}/\partial r) = (1/r)(\partial/\partial r)[(r\Gamma/\sigma_f)(\partial \bar{f}/\partial r)] + (\partial/\partial x)[(\Gamma/\sigma_f)(\partial \bar{f}/\partial x)]$$

$$(21)$$

and

$$\rho u(\partial g/\partial x) + \rho v(\partial g/\partial r) = (1/r)(\partial/\partial r)[(r\Gamma/\sigma_g)(\partial g/\partial r)] + (\partial/\partial x)[(\Gamma/\sigma_g)(\partial g/\partial x)]$$
$$+ C_{g_1}\Gamma[(\partial \bar{f}/\partial r)^2 + (\partial \bar{f}/\partial x)^2] - C_{g_2}\rho\varepsilon(g/k) \qquad (22)$$

In Eq. (22) the last two terms are modeled expressions representing production and dissipation, respectively, of fluctuations in f due to turbulence. Values of the "universal constants" in Eqs. (21) and (22) are given in Table 1.

(2) A "shape" for the $P(f)$ distribution must be assumed. For example, if it is assumed that local $f(t)$ varies randomly between two limits f_{min} and f_{max}, lying on the closed interval (0, 1), then the corresponding $P(f)$ is a "clipped Gaussian"—i.e., a Gaussian with the tail areas less than f_{min} and greater than f_{max} and compressed into Dirac deltas centered at f_{min} and f_{max}.[15] A somewhat simpler formulation results from assuming that local $f(t)$ varies *linearly* between f_{min} and f_{max} in a "sawtooth" manner. The resulting $P(f)$ is a "top hat" distribution. Details of this $P(f)$ have been developed by Tamanini[16] and are summarized in Table 2 and Figure 1. (For a detailed discussion of various shapes for $P(f)$, see reference 2 and D. B. Spalding's review article in reference 7.)

(3) Because of assumption (b), a complete solution for chemical equilibrium properties as a function of mixture fraction can be calculated once for all, namely, $T^*(f)$ and $[n^*(f), i = 1, NS]$. The local mean values of equilibrium-properties mole number are then given by

$$\bar{H} = \bar{f}H_{fu} + (1 - \bar{f})H_{air} \qquad (23)$$

$$\bar{T} = \int_0^1 T^*(f)P(f)\,df \qquad (24)$$

and

$$\bar{n}_i = \int_0^1 n_i^*(f)P(f)\,df, \qquad i = 1, NS \qquad (25)$$

Table 1. Recommended Values of Constants for the f-g Model[15]

Equation	Constant	Recommended value
(21)	σ_f	0.7
(22)	σ_g	0.7
(22)	C_{g_1}	2.8
(22)	C_{g_2}	2.0

Table 2. Dependence of the Limits of the Fluctuations, the Intermittencies, and the Probability-Density Function on the Values of $\bar{f}$ and g for Cases 1, 2, and 3 of Figure 1 $(0 < \bar{f} < \frac{1}{2})$ [16]

	Case number		
	1	2	3
Limits	$0 < g < \frac{1}{3}\bar{f}^2$	$\frac{1}{3}\bar{f}^2 < g < \frac{2}{3}(\bar{f} - \bar{f}^2)$	$\frac{2}{3}(\bar{f} - \bar{f}^2) < g < \bar{f}(1 - \bar{f})$
f_{max}	$\bar{f} + (3g)^{1/2}$	$\dfrac{3}{2}\dfrac{g + \bar{f}^2}{\bar{f}}$	1
f_{min}	$\bar{f} - (3g)^{1/2}$	0	0
α_1	0	0	$2\bar{f} - 1 + \alpha_0$
α_0	0	$\dfrac{g - \frac{1}{3}\bar{f}^2}{g + \bar{f}^2}$	$3g - (1 - \bar{f})(3\bar{f} - 1)$
P	$\dfrac{1}{2(3g)^{1/2}}$	$\dfrac{2}{3}\dfrac{(1 - \alpha_0)\bar{f}}{g + \bar{f}^2}$	$1 - \alpha_0 - \alpha_1$

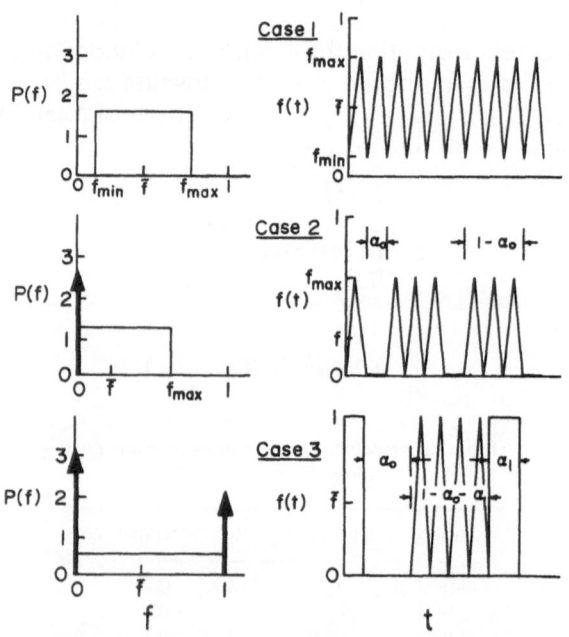

Figure 1. Probability-density function $P(f)$ for "sawtooth" $f(t)$, for cases of no intermittency (Case 1), intermittency of fuel only (Case 2), and intermittency of both fuel and air (Case 3). [16]

Then, from local solution of Eqs. (21) and (22) for $\bar{f}$ and g, corresponding values of f_{min} and f_{max} from Table 2, and appropriate substitutions for the "top hat" $P(f)$, Eq. (34) may be written, for Case 1 of Figure 1:

$$\bar{T} = P_{max} \int_{f_{min}}^{f_{max}} T^*(f)\, df = [2(3g)^{1/2}]^{-1} \int_{f-(3g)^{1/2}}^{f+(3g)^{1/2}} T^*(f)\, df \qquad (26)$$

For Case 2 of Figure 1 (intermittency of air),

$$\bar{T} = \alpha_0 T_{air} + P_{max} \int_0^{f_{max}} T^*(f)\, df \qquad (27)$$

where P_{max} and f_{max} are given in column 1 of Table 2. For case 3 of Figure 1 (intermittency of both fuel and air),

$$\bar{T} = \alpha_0 T_{air} + (1 - \alpha_0 - \alpha_1) \int_0^1 T^*(f)\, df + \alpha_1 T_{fu} \qquad (28)$$

Expressions identical to Eqs. (26)–(28) for $(\bar{n}_i, \; i = 1, NS)$ give the local mean mole numbers. (For values of $\bar{f}$ greater than $\frac{1}{2}$, the $P(f)$'s are simply mirror images of those given in Figure 1.)

3.2a. Estimates of the Contact Index

Estimates of the contact index X_f for the forward reaction

$$A_1 + A_2 \underset{b}{\overset{f}{\rightleftharpoons}} A_3 + A_4$$

may be constructed by consideration of each term in Eq. (18). For example, the second term in Eq. (18) may be calculated by

$$\overline{n_1' n_2'}/\bar{n}_1 \bar{n}_2 = (1/\overline{n_1 n_2}) \int_0^1 [n_1^*(f) - \bar{n}_1][n_2^*(f) - \bar{n}_2]P(f)\, df \qquad (29)$$

where $\bar{n}_1$ and $\bar{n}_2$ are determined by Eq. (25).

Further simplifying Eq. (29):

$$\int_0^1 [n_1^*(f) - \bar{n}_1][n_2^*(f) - \bar{n}_2]P(f)\, df$$

$$= \int_0^1 [n_1^*(f)n_2^*(f) - \bar{n}_1 n_2^*(f) - \bar{n}_2 n_1^*(f) + \bar{n}_1 \bar{n}_2]P(f)\, df$$

$$= \int_0^1 [n_1^*(f)n_2^*(f)]P(f)\, df - \bar{n}_1 \bar{n}_2 - \bar{n}_2 \bar{n}_1 + \bar{n}_1 \bar{n}_2$$

Substituting in Eq. (29),

$$\overline{n_1' n_2'}/\bar{n}_1 \bar{n}_2 = (1/\bar{n}_1 \bar{n}_2) \int_0^1 n_1^*(f)n_2^*(f)P(f)\, df - 1 \qquad (30)$$

Other terms in Eq. (18), such as

$$\overline{T'^2}/\overline{T}^2 = (1/\overline{T}^2) \int_0^1 [T^*(f)]^2 P(f)\, df - 1 \qquad (31)$$

are common to all X_f's and X_b's, and the binary products $n_i^*(f)n_j^*(f)$ can be precalculated, so that the computational work to evaluate the X's is really quite reasonable in spite of the apparent algebraic complexity.

3.2b. Modifications for Pulverized Fuel

The analysis above is valid only for gaseous fuel. When pulverized coal is considered as fuel in the inlet stream, the mixture fraction f alone does not correctly represent the amount and composition of pyrolyzate gases and gaseous products of char burnout that constitute the "fuel" for local gas-phase reactions. When these modifications are made to the model, much of the elegance and simplicity of this approach is lost; it is no longer possible to precalculate a single equilibrium solution set for temperature and mole numbers. Rather, equilibrium values of T and $(n_i, i = 1, NS)$ must be calculated locally for substitution in Eqs. (2)–(31). This is tractable only because of the use of the very efficient chemical equilibrium algorithm described in Chapter 1.

3.3. Model 3. Coalescence/Dispersion or the "Monte Carlo" Model

This approach appears to be the only potentially useful model of turbulent combustion which permits consideration of finite rates for *both* micromixing and chemical reaction. This approach originated from a stochastic analysis of liquid–liquid droplet coalescence and dispersion in a perfectly stirred reactor by Curl,[17] resulting in an integrodifferential equation for transient development of a mixture-fraction probability density function similar to the $P(f)$ of Section 3.2. Curl's equation was solved by Spielman and Levenspiel[18] by a Monte Carlo technique, which forms the basis of the present approach.

In this approach, the gas phase is assumed to be discretized into a manageable number of computational elements, which may be envisioned as somewhat gross fluid "particles" for purposes of this discussion. The calculation proceeds in a Lagrangian or "following-the-motion" framework, by tracking individual gas "particles" along time-mean stream tubes, predicted as described in Chapter 3. Particles are permitted to cross stream-tube boundaries at an exchange rate determined in the same manner as that for turbulent dispersion of solid particles (which is discussed in Section 1 of Chapter 6).

Molecular mixing among particles is determined by allowing pairs of fluid particles within a stream tube and at the same location to coalesce and redisperse,[2,17] thereby averaging their properties, at a rate

$$\beta = c_\beta k^2 / \varepsilon \tag{32}$$

where β is Corrsin's time constant for decay of a local concentration fluctuation,[2] and k and ε are locally predicted values of the turbulent kinetic energy and its dissipation rate, respectively, as described in Chapter 3.

The computational work and storage required for this model appear to be formidable, but not impossibly great. Extensions to allow mass transfer from the solid phase have not been fully examined, but should not present serious obstacles.

4. Recommended Approach

It can be seen from the previous section that the conceptual and computational problems associated with each of the three mixing/chemistry models are roughly inversely proportional to the desirability of potential results.

At the present time, a hybrid version of Models 1 and 2 appears to be practical, at least for flows which lack swirl and, therefore, have relatively weak backmixing. For this case, where stream-mixing is more important than backmixing, the diffusion flame approximation (Model 2) appears desirable.

By calculating the infinite-rate contact indices X_f and X_b for each reaction in a proposed reaction mechanism from Model 2, Model 1 can then be employed to account for the pollutant formation reactions, for which the flame-sheet approximation is known to be inadequate. Thus, by using Model 1 for finite chemistry, together with Model 2 to give estimates of the X_f's and X_b's, an approximate description of effects of both fast and slow chemistry may be achieved.[19]

5. Computational Techniques

Calculation of equilibrium states, required for Model 2, has been described in Chapter 1. Solution of the partial differential equations for field values $\bar{f}$ and g are described in Chapter 14.

As described in Chapter 14, the field equations for conservation of gaseous-species mole numbers and for the gas-phase temperature are solved simultaneously. The equations are expressed as Newton–Raphson functionals. The same correction variables, ($\Delta \log n_i$, $k = 1, NS$), $\Delta \log n_m$, and $\Delta \log T$, are employed as were used for the Gibbs function minimization equations described in Chapter 1.

The functionals for the species-i conservation equations are given by

$$f_i = A_p(n_i - n_{i0}) + \sum_{j=1}^{JJ} (\alpha'_{ij} - \alpha''_{ij})(r_j - r_{-j}) + {}_pS^{nk}, \qquad i = 1, NS \tag{33}$$

where the forward and reverse rates r_j and r_{-j} are given, as in Eq. (7), by

$$r_j = X_j 10^{B_j} T^{N_j} \exp(-T_j/T)(\rho n_m)^{\bar{\alpha}_j} \sum_{k=1}^{NS} (\rho n_k)^{\alpha'_{kj}} \tag{34}$$

and

$$r_{-j} = X_{-j} 10^{B-j} T^{N-j} \exp(-T_{-j}/T)(\rho n_m)^{\bar{\alpha}_j} \sum_{k=1}^{NS} (\rho n_k)^{\alpha_{kj}} \tag{35}$$

where the time-mean values are assumed.

The contact indices X_j and X_{-j} are taken to be constants in this formulation, having been estimated as described in Section 3.2.1.

The correction equations for species-i mole numbers are

$$\sum_{k=1}^{NS} (\partial f_i/\partial \log n_k) \Delta \log n_k + (\partial f_i/\partial \log n_m) \Delta \log n_m + (\partial f_i/\partial \log T) \Delta \log T = -f_i,$$

$$i = 1, NS \tag{36}$$

The partial derivative coefficients of the correction variables in Eq. (36) are

$$\partial f_i/\partial \log n_k = A_p n_i \delta_{ik} + \sum_{j=1}^{JJ} (\alpha'_{ij} - \alpha''_{ij})(r_j \alpha'_{kj} - r_{-j} \alpha''_{kj}) \tag{37}$$

$$\partial f_i/\partial \log n_m = \sum_{j=1}^{JJ} (\alpha'_{ij} - \alpha''_{ij})(r_{-j} n''_j - r_j n'_j) \tag{38}$$

and

$$\partial f_i/\partial \log T = \sum_{j=1}^{JJ} (\alpha'_{ij} - \alpha''_{ij})[r_j(N_j + T_j/T - \bar{\alpha}_j - n'_j) - r_{-j}(N_{-j} + T_{-j}/T - \bar{\alpha}_j - n''_j)] \tag{39}$$

Alternative treatments of the rate expressions, Eqs. (34) and (35), are possible. For example, if it is desired to construct the reverse rate data from the relation

$$k_{-j} = k_j [RT]^{(n''_j - n'_j)}/K_j^P \tag{40}$$

where K_j^P is the equilibrium constant for reaction j,

$$K_j^P \equiv \exp\left[\sum_{i=1}^{NS} (\alpha'_{ij} - \alpha''_{ij}) g_i^0/RT\right] \tag{41}$$

then the reverse rate r_{-j} is given by

$$r_{-j} = X_{-j} 10^{B_j} T^{N_j} \exp\left[-(T_j/T) - \sum_{i=1}^{NS} (\alpha'_{ij} - \alpha''_{ij}) g_i^0 / RT \right]$$

$$\times [RT]^{(n''_j - n'_j)} (\rho n_m)^{\bar{\alpha}_j} \sum_{k=1}^{NS} (\rho n_k)^{\alpha_{ij}} \tag{42}$$

and the partial derivative coefficients of $\Delta \log T$ in Eq. (39) become

$$\partial f_i / \partial \log T$$

$$= \sum_{j=1}^{JJ} (\alpha'_{ij} - \alpha''_{ij}) \left[(r_j - r_{-j})(N_j + T_j/T - \bar{\alpha}_j - n'_j) - r_{-j} \sum_{i=1}^{NS} (\alpha'_{ij} - \alpha''_{ij}) h_i / RT \right] \tag{43}$$

The functional for mixture reciprocal mole number, n_m, is the same as for the equilibrium formulation [Eq. (16) of Chapter 1]

$$f_m = \sum_{k=1}^{NS} n_k - n_m \tag{44}$$

and the corresponding correction equation is also the same as Eq. (33) of Chapter 1:

$$\sum_{k=1}^{NS} n_k \Delta \log n_k - n_m \Delta \log n_m = n_m - \sum_{k=1}^{NS} n_k \tag{45}$$

The functional for temperature (actually, for conservation of thermal energy) is the same as Eq. (14) of Chapter 1,

$$f_T \equiv (-H_0/RT) + \sum_{i=1}^{NS} (n_i h_i / RT) + Q/RT \tag{46}$$

and the corresponding correction equation is

$$\sum_{k=1}^{NS} (h_k n_k / RT) \Delta \log n_k + \left[\sum_{i=1}^{NS} (C_{p_i}/R) n_i + (1/RT) \partial Q / \partial \log T \right] \Delta \log T$$

$$= - \sum_{i=1}^{NS} (n_i h_i / RT) - Q/RT + \sum_{i=1}^{NS} (h_{i0} n_{i0} / RT) \tag{47}$$

5.1. Solution of the Kinetic Correction Equations

The basic approach is the same as described in Chapter 1 for the equilibrium correction equations. The details of the iteration procedure and control of the underrelaxation parameter described in Chapter 1 are identical for kinetic stationary states as for equilibrium states. However, convergence control and selection of estimates are somewhat more involved. The same convergence criterion η is used as described in Chapter 1. However, during the course of iterative solution of kinetic stationary states, oscillating nonconvergence is occasionally encountered, especially during ignition, when many species are undergoing relatively small changes, and the dominant chemical reactions are endothermic. One technique used by other investigators to avoid oscillation is simply to decouple the energy equation; that is, to delete the energy equation and solve the remaining system of equations for assigned temperature, and then iteratively correct the assigned temperature until the desired value of enthalpy is achieved. This practice, while it effectively avoids the oscillation problem, is successful only at the expense of increased computer execution time. Solution with the fully coupled energy equation, as in the present formulation, is considerably faster. The occasional case of oscillating nonconvergence can be effectively dealt with by a partial or "one-way" decoupling of the energy equation, in which the variations of temperature are not permitted to affect the distribution of species mole numbers, but the mole number changes in each iteration do affect the temperature. This is easily achieved within the present formulation by setting the species-temperature partial derivatives, Eqs. (39), equal to zero whenever oscillating convergence is encountered or suspected. The penalty is, at worst, a somewhat slower progress toward convergence than the fully coupled equations, although it is still considerably faster than full decoupling of the energy equation. By judicious use of partial decoupling, rapid, stable convergence can be achieved reliably over a wide range of fuel type, fuel–air ratio, and cell time constant A_p.[11]

5.2. Initial Estimates

The set of correction equations to be solved, Eqs. (36), (45), and (47), are $(NS + 2)$ in number. This is, unfortunately, a considerably larger system than the $(NLM + 2)$ number of equations required to be solved for chemical equilibrium. Further, the mole numbers $(n_i, i = 1, NS)$ are far more sensitive to variations in the cell time constant A_p and in T than are the smoothly varying Lagrange multipliers π_i. Efficiency of convergence, as well as stability or freedom from divergence, therefore, requires a choice of estimate set $(n_i^{(0)}, i = 1, NS)$ and $T^{(0)}$ which is quite close to the final solution. The most obvious choice of estimates is simply the stored values from a previous superiteration of the flow field. However, in the initial stage of the field solution, these estimates may not be available, or may be inadequate due to the rapidly changing field values of A_p and of the thermochemical variables. Therefore, a strategy is needed to automatically cope with these "problem cells" whenever they arise.

An example of a "problem cell" is a cell at which ignition occurs; that is, a cell which has largely "cold" or unreacted values of n_{i0} at upstream neighbor nodes, but which will ultimately have a "hot" or (partially) reacted set of values T and n_i at a stable solution condition. Possible solution states are illustrated in Figure 2. If adiabatic mixing (no reaction) is assumed, the curve "$A_p \to \infty$" (zero cell residence time), which represents the variation of H with T for the inlet mixture set (n_{i0}, $i = 1$, NS), is also the solution set. The intersection of this curve with the constant inlet enthalpy $H = H_0$ line determines the solution temperature, which is equal to the effective inlet temperature, T_0. If chemical equilibrium is assumed, then the curve "$A_p = 0$" (infinite cell residence time) corresponds to the set of mole numbers which minimize the mixture Gibbs function at each temperature. The intersection of the "$A_p = 0$" curve with $H = H_0$ determines the adiabatic equilibrium solution set (n_i^*, $n = 1$, NS) and the equilibrium temperature T^*.

For small values of A_p, such as in curve 1, which represent large but finite values of cell residence time, the solution point represented by the intersection of curve 1 with $H = H_0$ is unique, and corresponds to values of (n_i, $i = 1$, NS) and T, which differ only slightly from the equilibrium values; obviously, the equilibrium solution is an excellent estimate set for this kinetic stationary state. Increasing values of A_p (decreasing cell residence time) correspond to nonequilibrium solutions which are further away from equilibrium.

To complicate matters further, it is possible for multiple solutions to exist, as illustrated in curve 2. Of the three intersections of curve 2 with $H = H_0$, the lower and upper temperature values are physically stable

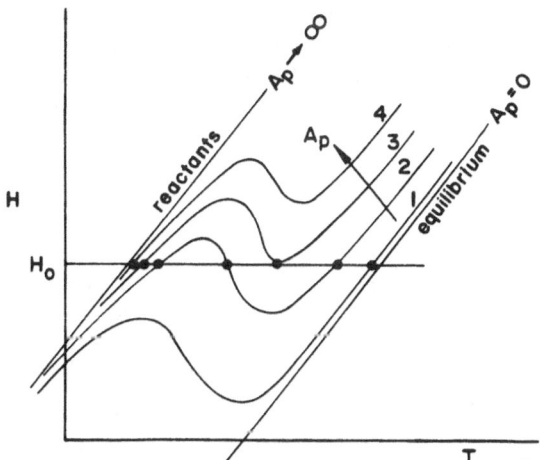

Figure 2. Schematic enthalpy–temperature diagram for chemical equilibrium and kinetic stationary states in a computational cell. $Q = 0$ is assumed for clarity. Parametric curves are for different values of cell convection/diffusion coefficient A_p.

stationary states, which correspond to slow oxidation and to combustion, respectively. The central branch is physically unstable, but satisfies all of the conservation equations. If an "upper-branch" or combustion solution is desired, estimates must obviously be chosen so that the desired state is approached from higher-temperature stationary states, that is, for values of A_p less than that of curve 2.

Curve 3 illustrates an upper-branch solution which represents incipient extinction or "blowout"; an increase in A_p, such as from curve 3 to curve 4 would cause a unique, lower-branch solution state to result. If the final solution for the cell should correspond to this condition, it is likely that intermediate solutions obtained for this cell during "superiteration" of the field solution may vary between upper-branch (combustion) and lower-branch (slow oxidation) solutions. A successful computational strategy, therefore, requires that, if convergence is not obtained from previous solution values as estimates within a very few iterations, an upper-branch solution should be approached from the right (high temperature); that is, from an equilibrium estimate, with systematically increasing values of the parameter A_p up to the data value. This strategy is necessary to avoid having a cold, nonreacting solution propagate throughout the field when a combustion solution is desired.

It is important to understand that the "cold" or lower-branch solution is as physically meaningful as is the combustion solution. In actual furnaces and combustors, for example, inflow of a given mixture of reactants does not automatically guarantee combustion: Some method of ignition such as a continuous igniter or pilot flame, or self-ignition by convective or diffusive recirculation of burned products, is essential. These computational difficulties, and the technique for their solution, simply reflect the actual physical processes which occur in a self-stabilized flame, in a finite volume as well as in a computational cell embedded in a combustion flow field.

6. Notation

A_i	Molecular symbol for species i	$\bar{f}$	Time-mean value of f, $\int_0^1 f P(f) df$
$[A_i]$	Molar concentration of species i (kmol, m^{-3})	g	variance of f, $\int_0^1 (f-\bar{f})^2 P(f) df$
A_p	Time constant for computational cell (kg s^{-1} m^{-3})	g_i	Partial molar Gibbs function of species i (see Chapter 1) (J kmol$_i^{-1}$)
B_f	Empirical constant in rate expression	h_i	Enthalpy of species i (see Chapter 1) (J kmol$_i^{-1}$)
f_i	Newton–Raphson functional		
f	Fuel mixture fraction (mass fraction of matter having entered via fuel stream)	h_{i0}	Mass flow-averaged h_i from adjacent computational cells (J kmol$_i^{-1}$)

H	Mixture enthalpy ($J\ kg^{-1}$)		from computational cell ($J\ kg^{-1}$)
k	Turbulent kinetic energy ($J\ kg^{-1}$)	r_f, r_b	Rates of forward and backward chemical reaction (general) ($kmol\ m^{-3}\ s^{-1}$)
k_f	Rate constant for forward reaction ($m^3\ kmol^{-1}\ s^{-1}$)	r_j, r_{-j}	Forward and backward rates of chemical reaction j (kmol $m^{-3}\ s^{-1}$)
K_j^P	Equilibrium constant for reaction j	$_pS^{n_k}$	Volumetric source term; rate of appearance of gaseous species k from volatilization or combustion of solid particles ($kmol_k\ m^{-3}\ s^{-1}$)
N_f	Empirical constant in rate expression		
n_i	Mole number of species i ($kmol_i\ kg^{-1}$)		
n_{i0}	Mass-flow averaged values of n_i from adjacent computational cells ($kmol_i\ kg^{-1}$)	T	Temperature (K)
$\bar{n}_i$	Time-mean value of n_i ($kmol_i$ kg^{-1})	T_f, T_j	Activation temperature for forward reaction; for reaction j (K)
n_m	$\Sigma_{i=1}^{NS}\ n_i$, equal to reciprocal of gas mixture molecular weight ($kmol\ kg^{-1}$)	X, X_f, X_j	Contact index (for forward reaction, for reaction j), defined by Eqs. (17) and (18)
n'_j, n''_j	$\Sigma_{i=1}^{NS}\ \alpha'_{ij}$ and $\Sigma_{i=1}^{NS}\ \alpha''_{ij}$, respectively	$\alpha'_{ij}, \alpha''_{ij}$	Stoichiometric coefficient of species i in reaction j (as a reactant, as a product)
NLM	Number of distinct elements in mixture	$\bar{\alpha}_j$	Third-body stoichiometric coefficient in reaction j (0 or 1)
NS	Number of distinct chemical species in mixture	β	Time constant for decay of a concentration fluctuation (s^{-1})
$P(f)$	Probability-density function for mixture fraction f: $P(f)\ df$ is the fractional time (at a point) during which f assumes values in the range $(f, f + df)$	δ_{ij}	Kronecker delta ($=1$ if $i=j$; otherwise 0)
P_n	Polynomial defined by Eq. (13)	ε	Dissipation rate of turbulent kinetic energy ($J\ m^{-3}\ s^{-1}$)
Q_n	Polynomial defined by Eq. (16)	π_i	Lagrange multiplier (see Chapter 1)
Q	Heat loss per unit mass		

7. References

1. G. L. Pratt, *Gas Kinetics*, John Wiley and Sons, New York (1969).
2. D. T. Pratt, Mixing and chemical reaction in continuous combustion, *Prog. Energy Comb. Sci.* **1**, 73–86, 213 (1976).
3. R. Borghi, Chemical reaction calculations in turbulent flows: Application to a CO-containing turbulent plume, *Adv. Geophys.* **18B**, 349 (1974).
4. S. N. B. Murthy (ed.), *Turbulent Mixing in Nonreactive and Reactive Flows*, Plenum Press, New York (1975).
5. R. Goulard (ed.), *Combustion Measurements*, Hemisphere Publishing Corp., Washington, D.C. (1976).
6. R. S. Brodkey (ed.), *Turbulence in Mixing Operations*, Academic Press, New York (1975).
7. F. V. Bracco (ed.), *Turbulent Reactive Flows*, Special Issue of *Comb. Sci. Tech.* **13**, 1–6 (1976).
8. D. T. Pratt, Theories of mixing in continuous combustion, in *Fifteenth Symposium (International) on Combustion*, p. 1339, The Combustion Institute, Pittsburgh, Pa. (1975).

9. J. J. Wormeck and D. T. Pratt, Computer modeling of combustion in a Longwell jet-stirred reactor, in *Sixteenth Symposium (International) on Combustion*, p. 1583, The Combustion Institute, Pittsburgh, Pa. (1977).

10. D. T. Pratt, Calculation of chemically reacting flows with complex chemistry, in *Studies in Convection, Volume II* (B. E. Launder, ed.), Academic Press, New York (1977).

11. D. T. Pratt and J. J. Wormeck, CREK: A Computer Program for Calculation of Combustion Reaction Equilibrium and Kinetics in Laminar or Turbulent Flows, Report WSU-ME-TL-76-1, Washington State University (1976).

12. W. R. Hawthorne, D. S. Weddell, and H. C. Hottel, Mixing and combustion in a turbulent gas jet, *Third Symposium on Combustion, Flame and Explosions*, p. 266, The Combustion Institute, Pittsburgh, Pa. (1949).

13. G. Vassilatos and H. L. Toor, Second-order chemical reactions in a nonhomogeneous turbulent fluid, *AIChE J.* **4**, 666 (1965).

14. D. B. Spalding, Concentration fluctuations in a round turbulent jet, *Chem. Eng. Sci.* **26**, 95 (1971).

15. S. E. Elghobashi and W. M. Pun, A theoretical and experimental study of turbulent diffusion flames in cylindrical furnaces, in *Fifteenth Symposium (International) on Combustion*, p. 1353, The Combustion Institute, Pittsburgh, Pa. (1975).

16. F. Tamanini, On the Numerical Prediction of Turbulent Diffusion Flames, paper given at the Eastern States Section/The Combustion Institute, Spring Meeting, April 1976.

17. R. Curl, Dispersed phase mixing 1: Theory and effects in simple reactors, *AIChE J.* **9**, 175 (1963).

18. L. A. Spielman and O. Levenspiel, A Monte Carlo treatment for reacting and coalescing dispersed phase systems, *Chem. Eng. Sci.* **20**, 247 (1965).

19. J. J. Wormeck, private communication (1977).

5

Radiative Heat Transfer in a Pulverized-Coal Flame

Sneh Anjali Varma

1. Historical Background

As early as the 1920s, it was recognized that thermal radiation is a significant mode of heat transfer in an industrial furnace. In 1925, Wohlenberg and Marrow[1] presented a simple, yet comprehensive analysis of radiation in a pulverized-coal-fired furnace. Their heat-transfer model consisted of energy transfer between gaseous and solid particle constituents of the pulverized-coal flame and refractory wall and cold tubes carrying water or steam. In addition, a correction term was included for convective heat transfer between furnace walls, cold surfaces, and the moving high-temperature gases along these surfaces. Their work was expanded in a paper by Wohlenberg and Lindseth,[2] in which comparison was made among various operating variables in stocker-fired and pulverized-coal-fired furnaces. In the above-mentioned work, the approach was somewhat empirical, and the results could only be applied to particular types of tested furnaces and coals.

In 1928, Haslam and Hottel[3] recognized the need for a more general and analytical approach to the problem of radiative heat transfer in furnace design. They devised an improved model for energy transfer by radiation from suspended particles in pulverized-coal flames, from non-luminous and luminous gas flames between finite solid surfaces, and from clouds of gases

Sneh Anjali Varma • Graduate Research Assistant, Department of Mechanical and Industrial Engineering, University of Utah, Salt Lake City, Utah

and suspended solids. Their general energy-transfer equations were

$$Q_F = P_c[K_1(T_F^4 - T_R^4) + K_2(T_F^4 - T_c^4) + Z_c] \qquad (1)$$

Heat trans- mitted from flame	Heat from flame in direction of refractory	Heat from flame in direction of cold surfaces	Heat to cold surface by convection

Heat transmitted by radiation

and

$$K_1(T_F^4 - T_R^4) = K_3(T_R^4 - T_c^4) \qquad (2)$$

Heat from flame   Heat from refrac-
to refractory   tory in direction
                        of cold surface

where Z_c can be expressed as

$$Z_c = S \cdot Z \cdot \Delta T \qquad (3)$$

The approach to solving the problem of radiative heat transfer in industrial furnaces remained essentially unchanged until the 1950s when the advent of application of the theory of electromagnetic radiation to problems of thermal radiation, such as that adopted by Hottel and Cohen,[4] gave a new start to the analysis of radiative heat transfer in industrial furnaces. As a result, some very accurate and extensive models for radiative heat transfer such as Hottel's zone model, the flux model, the Monte Carlo method, and the neutron-diffusion approximation were adopted.

2. Heat Transfer in an Industrial Furnace

Due to the highly emitting and absorbing nature of components of a pulverized-coal flame, radiation becomes the most significant or dominant mode of heat transfer in a pulverized-coal furnace. The finite velocity of the flame and the presence of boundary layers do contribute to heat transfer by convection and conduction, but still 95% of the heat transfer in a pulverized-coal flame is due to radiation.[5,6] For this reason, convection and conduction are neglected in the analysis of heat transfer in a pulverized-coal furnace which follows.

3. Basic Concepts of Radiation

In presenting this material, it is assumed that the reader has a basic understanding of thermal radiation phenomena. This material is therefore only a brief review. For a detailed understanding of radiation heat transfer, references 7, 8, and 9 are recommended.

3.1. Radiation

Radiation is electromagnetic energy in transport. An energy input to a particle results in a temporary excitation or increased energy level of the particle. Return of the particle to a lower energy level results in the emission of one or more photons of radiation. When the excitation comes by virtue of temperature, radiation is termed thermal radiation.

Radiation is transmitted by electromagnetic waves within a wavelength range of 1 nm to 1 km, corresponding to x-rays and radio waves, respectively. Thermal radiation covers a range of 0.1 to 100 μm in wavelength. The range of wavelength of thermal radiation in a combustion process is usually between 0.5 and 10 μm.[10]

3.2. Intensity

The measure of radiation is usually its intensity, which has a spectral character. The amount of energy, de_v, in frequency interval v and $v+dv$, which is transported across an arbitrarily oriented element of area dA (located at a point P and having outward unit normal vector $\mathbf{n}$) during a time interval dt and confined to an element of solid angle $d\Omega$ about the direction of unit vector S, is related to the intensity I_v at point P by

$$U_v = \lim_{dA, d\Omega, dt, dv \to 0} \left| \frac{de_v}{\cos\theta \, dA \, d\Omega \, dt \, dv} \right| \tag{4}$$

The term $\cos\theta$ appears because the intensity is not in the direction of the unit normal $\mathbf{n}$. Being a function of location of point P, the intensity in a radiation field is written as

$$I_v = I_v(\mathbf{r}, \mathbf{S}, t) \tag{5}$$

or

$$I_v = I_v(x, y, z; l, m, n; t) \tag{6}$$

where (x, y, z) and (l, m, n) define the point location and the direction in which I_v is directed (direction cosines), respectively.

It is a preferred practice to base intensity of radiation on frequency rather than wavelength, since the latter changes while passing from one medium to another (due to changing index of refraction), while the frequency remains the same.

A radiative field is called isotropic and homogeneous if the intensity is independent of direction at a point, and if the intensity is the same at all points and in all directions. A number of related definitions follow:

Integrated Intensity (I)

$$I = \int_0^\infty I_v \, dv \tag{7}$$

where the intensity I_v is integrated over all frequencies.

Due to the macroscopic nature of this formulation, the parameters describing the state of polarization of the radiation field will not be considered.

Average Intensity (I_v)

$$I_v = \int_{4\pi} I_v \, d\Omega \tag{8}$$

where I_v is averaged over all directions.

Radiation Flux (F_v)

The net flow of radiation energy in a given direction **S** per unit area, time, and frequency interval due to radiation from all directions is known as the radiation flux vector and is expressed as

$$\mathbf{F}_v = \int_{\Omega = 4\pi} I_v \overline{S} \, d\Omega \tag{9}$$

The spectral radiation flux F_v is defined as the net flow of radiant energy per unit of area, time, and frequency due to contributions from all directions:

$$F_v = \mathbf{n} \cdot \mathbf{F}_v = \int_{\Omega = 4\pi} I_v \mathbf{n} \cdot \mathbf{S} \, d\Omega = \int_{4\pi} I_v \cos\theta \, d\Omega \tag{10}$$

The spectral radiation flux depends on the choice of outward normal **n** to the surface across which the flow of radiant energy is being considered.

Radiant Energy Density (U_v)

Radiant energy is propagated at a finite, though very large, velocity, namely, the velocity of light. For this reason, a finite amount of energy is present in a finite space. The amount of energy per unit volume and frequency in the neighborhood of the point in the course of transit can be expressed as spectral energy density at a given point, U_v:

$$U_v = (1/c) \int_{4\pi} I_v \, d\Omega \tag{11}$$

3.3. Emission Coefficient

Emission is a source or mode of creation of radiation and takes place at the expense of other forms of energy. The nature of emission is characterized by its wavelength and depends on the material of the body and its temperature.

The amount of energy de_v radiated by the matter within an elementary volume is

$$de_v = Ke_v \, dv \, d\Omega \, dV \, dt \tag{12}$$

where Ke_v is the volumetric emission coefficient. The mass emission coefficient can be obtained by dividing the volumetric coefficient by ρ, the mass density of the emitting matter.

A surface is called isotropic if it emits radiation with equal intensity in all directions over the solid angle 2π. Such a surface, commonly known as a "diffuse" surface, is a good approximation for many solid surfaces, including the combustor walls.

3.4. Absorption and Scattering Coefficient

A pencil of radiation traversing a continuous medium will be weakened by its interaction with matter. The interaction between the radiation field and matter is usually expressed in terms of the absorption and scattering coefficients. The pencil of radiation, when incident on a body, may be partially reflected at the surface. The remaining flux entering the surface may then be transmitted, absorbed, or scattered. The flux emerging from the body in the same direction as the incident flux is the transmitted flux.

The flux not transmitted is due to attenuation, also known as extinction or interception. The attenuation can take place in two ways: (1) absorption, and/or (2) scattering. The absorbed flux is that which does not re-emerge from the body as either a transmitted or scattered flux. The absorbed flux is converted into stored thermal internal energy in the body, thus raising its temperature, The ability to absorb is related to the ability to emit, as is given by Kirchhoff's law: For a gray body at thermal equilibrium or even under the conditions of nonzero net interchange, the absorptivity is equal to the emissivity[20]:

$$\alpha_\lambda = \varepsilon_\lambda \tag{13}$$

Since the spectral distribution does not affect the absorptivity and emissivity, in a gray body $\alpha = \varepsilon$.[7] The scattered flux is that portion of the attenuated flux that re-emerges from the body in different directions than that of the incident flux. Thus, while absorption is transformation of radiant energy into thermal energy, scattering refers to redistribution of attenuated radiant flux into other directions.

The effect of attenuation is explained by the Bouger–Lambert law, which states that a beam of monochromatic radiation I_v, confined in a small angle

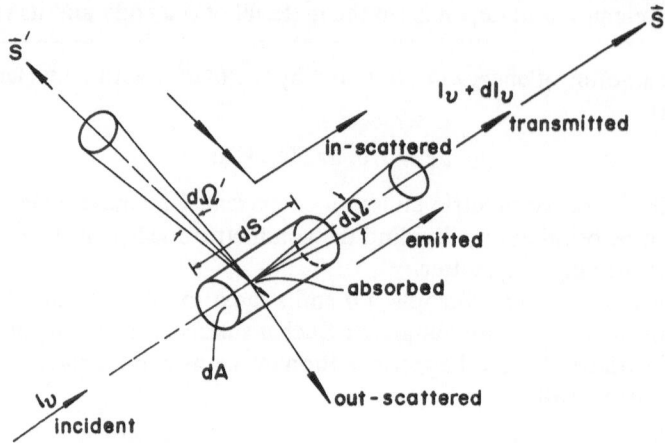

Figure 1. Radiation energy balance in a specified direction.

of divergence $d\Omega$ during its passage through a medium of thickness dS, experiences a decrease in intensity, also known as attenuation. This fractional decrease in intensity is proportional to the intensity itself and to the distance traversed:

$$-dI_v/I_v = K_{tv}\, dS \tag{14}$$

or

$$I_v = I_{0v} \exp\left(-K_{tv}S\right) \tag{15}$$

where I_{0v} is the intensity of incident radiation and K_{tv} is the extinction coefficient, which is the sum of the absorption and scattering coefficients:

$$K_{tv} = K_{av} + K_{sv} \tag{16}$$

For a gray body, in the absence of spectral dependence:

$$I = I_0 \exp\left(-K_t S\right) \tag{17}$$

The nonabsorptive decrease in intensity of radiation is due to "out-scattering"; that is, a part of the incident beam is scattered in other directions, thereby reducing its intensity. In a similar way, the radiant intensity of the beam can be increased by radiant energy scattered by surrounding interceptors into the direction of the beam considered. This is known as "in-scattering" as shown in Figure 1. This increase in the radiant energy of the beam in direction S (confined in solid angle $d\Omega$) due to in-scattering of radiation by the matter within the element of volume from all possible directions, is

$$K_{sv}\, dS\left[(1/4\pi)\int_{\Omega'=4\pi} P_v(S' \to S)I_v(S')\, d\Omega'\right] dA\, d\Omega\, dv\, dt \tag{18}$$

where the phase function $(1/4\pi)\, P_o(S' \to S)\, d\Omega'$ represents the probability that an incoming pencil of rays $(S', d\Omega')$ will be directed into direction $(S, d\Omega)$. Since the sum of the probabilities over all directions must equal unity, the phase function is defined such that[9]

$$(\tfrac{1}{4}\pi) \int_{\Omega'= 4\pi} P_o(S' \to S)\, d\Omega' = 1 \qquad (19)$$

Scattering is termed "isotropic" when the distribution of scattered intensity is uniform in all directions. In this case, the in-scatter term in Eq. (18) may be evaluated by a Gaussian quadrature formula.[8]

When the scattered energy distribution is not uniform in all directions, but tends to intensify in one certain direction, the scattering is considered to be anisotropic. This phenomenon exists when the intercepting particles are larger in size than the wavelength of radiation.

The intensity of scattered radiation by a single particle may be computed by the solution of Maxwell's wave equation. This method, developed by Gustav Mie, is known as "Mie theory."[6,8] At low particle densities, the scattered intensity is equal to the scattered intensity from a single particle, multiplied by the number of particles. Such an effect is called "single scattering." As the particle density increases, the intensity of radiation, which has been scattered two or three times, becomes significant and this effect is known as multiple scattering.[11] The dominant nature of radiative heat transfer as encountered in pulverized-coal flames is both multiple and anisotropic.

4. Optics of Particle Scattering

The subject of scattering is very complex and involved, so that a complete discussion is outside the scope of this book. Only a brief review of the subject is presented in this section.

Scattering by particles whose radius is greater than about 0.8 times the wavelength of light is called "Mie scattering." Complex in its nature, Mie theory has been developed into many forms and expressions to serve different fields of application. In this presentation, the basic Mie theory is presented without derivations, along with the concepts and functions that are used in heat transfer analysis. Because particles in a pulverized-coal flame cover a broad size range, and because further complexities are introduced by the polydispersions of char, ash, and soot, the simplest possible approach to particle scattering will be adopted.

Scattering is a result of interaction between electromagnetic waves and the electric charge that constitutes matter. In contrast to a gas molecule, where only a single dipole is involved, a particle consists of many closely packed, complex molecules that constitute an array of multipoles. These multipoles, when excited by an incident wave, create oscillating multipoles which give rise to secondary electric and magnetic waves called partial waves. Scattered intensity at a particular observation angle is the squared summation of slowly converging series of amplitudes of these partial waves. When these partial waves combine to form scattered waves, interferences are caused by phase differences between the partial waves. These interferences depend on the wavelength of the incident light, the size and refractive index of the particle, and the angle of observation. These interferences increase as the particle size increases. As the particle size increases, the main features of Mie scattering can be characterized as follows[26]:

(1) A complicated dependence of scattered light intensity on the angle of observation, the complexity increasing as the ratio of particle size to wavelength increases.

(2) An increasing ratio of forward scattering to backscattering as the particle size increases, with resulting growth of the forward lobe as shown in Figure 2.

(3) Little dependence of scattering on wavelength when particle size relative to wavelength is large.

A simple geometrical representation of Mie scattering is given in Figure 3. The scattering center at point O may be either a single particle or a unit volume containing N particles. The incident light is assumed to travel in the $+x$-direction. The observation direction OD lies in the xz-plane and at angle θ from the forward direction of incident light. When the incident light is plane-polarized, the orientation of the plane containing the electric vector is specified by angle ψ measured from the observation or xz-plane.

The parameters associated with scattering are summarized in the following sections.

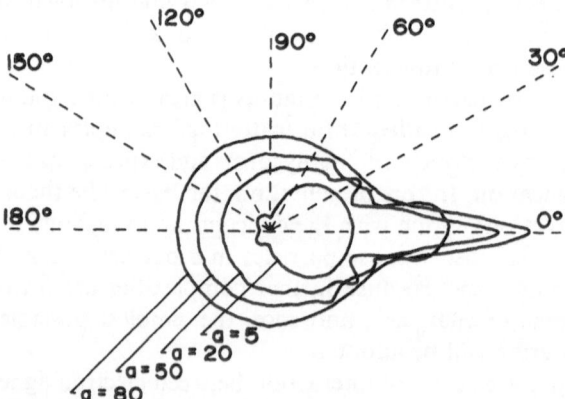

Figure 2. The angular distribution of scattered intensity for char particles at wavelength of $2\ \mu m$, a is the size parameter, $a = \pi D/\lambda$.

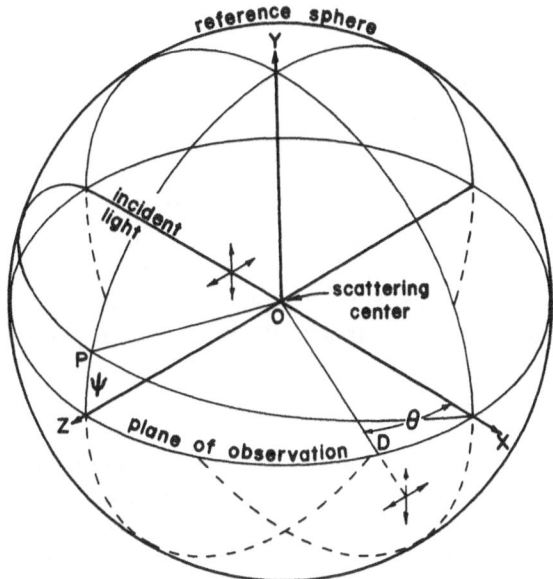

Figure 3. Geometry of Mie scattering, θ is the angle of observation.

4.1. Particle Size Parameter

The most important governing factor of Mie scattering is neither the absolute value of particle size nor the absolute value of wavelength. Rather it is the ratio of the two values or relative size that is defined as

$$a = \pi D / \lambda \tag{20}$$

where λ is the wavelength in the medium surrounding the particle. The qualitative nature of scattering depends largely on this size parameter. For a size parameter $a < 1$, scattering is defined as Rayleigh scattering, whereas, for large particle size parameter $a > 5$, scattering is defined as Mie scattering. For intermediate size parameter, scattering needs special treatment, which is neither totally Rayleigh nor Mie scattering. Detailed discussion of this matter is beyond the scope of this book, and readers are referred to references 27 and 28.

Since the particle size spectrum in a pulverized-coal flame is dominated by char and ash particles, which clearly fall into the large particle category, the dominant nature of scattering is Mie scattering. Scattering by soot particles is often neglected.

4.2. Shape

The shape of particles is another factor that decides the nature of the scattering. Mie theory applies only to isotropic and spherical particles. Char particles (after devolatilization) take the shape of spongy spherical balls and Mie theory can be applied to these particles with little error.

Scattering by nonspherical particles is more complex than that by spheres and is discussed by Van De Hulst,[27] Kerker,[28] etc. Ash particles are nonspherical in shape and can be given exact treatment, although this is beyond the scope of this work. For simplicity, it is assumed that ash particles are spherical and thus can be treated the same as char particles.

4.3. Refractive Index

Refractive index is the relative index of refraction of a particle with respect to the index of refraction of the surrounding medium. Thus, it is apparent that refractive index of a particle is subject to change with wavelength. Fortunately, the optical behavior of particles loses its wavelength dependence as particle size increases and this eliminates the need for finding the refractive index at each wavelength. For example, char particles have an index of refraction of 1.93 $(1-0.53i)$ at wavelength of $2\,\mu m$, which is the average wavelength of radiation in pulverized-coal combustion.[10] This value of refractive index for char particles can be used throughout the combustion process.

When a light wave passes through a particle whose refractive index is greater than the surrounding medium, an optical discontinuity occurs, which is known as attenuation. If the particle does not absorb any of the flux, then only scattering occurs and attenuation is expressed in terms of total scattering coefficient. The refractive index in such a case is expressed by a real number. If the particle also absorbs, the two processes are additive and produce an extinction coefficient. The effect of absorption on angular scattering is not a simple matter, and again is outside the scope of this work. When absorption is significant, the refractive index of the material is represented by a complex number:

$$m(\lambda) = n(\lambda) - in_i(\lambda) \tag{21}$$

The imaginary part is identical to a quantity called absorption index k of the material and is related to the Lambert absorption coefficient K_a by

$$k \equiv n_i = K_a \lambda / 4\pi \tag{22}$$

For ash particles, the absorption index is very small and thus almost negligible. This is due to the constituents of ash which usually are MgO, Al_2O_3, CaO, etc. All of these compounds have negligible absorption

indices.[37] When the absorption index is zero, particles are known as dielectric particles.

4.4. Angular Scattering Cross Section

The angular scattering cross section of a particle is defined as that cross
section of an incident wave, acted on by the particle, having an area such that
the power flowing across it is equal to the scattered power per steradian at an
observation angle θ. In symbols it may be represented as

$$X_S(\theta)E_0 = I_0(\theta) \tag{23}$$

4.5. Total Scattering Cross Section

The total scattering cross section is defined as that cross section of an
incident wave, acted on by the particle, having an area such that the power
flowing across it is equal to the total power scattered in all directions:

$$X_S = \int_0^{4\pi} X_S(\theta)\, d\Omega \tag{24}$$

4.6. Efficiency Factor for Scattering

The efficiency factor, Q_s, also known as the Mie coefficient, is the ratio
of scattering to the geometric cross section:

$$Q_s = (X_S)/(\pi r^2) = (2/r^2) \int_0^{\pi} X_S(\theta) \sin \theta\, d\theta \tag{25}$$

4.7. Efficiency Factor for Absorption

The difference between the flux removed from the incident beam and the
totally scattered flux is due to absorption by the particle. Absorption occurs
when the refractive index is complex, and the absorbed energy goes into
heating the particle.

Extinction or attenuation is the additive effect of absorption and
scattering:

$$Q_{ex} = Q_a + Q_s \tag{26}$$

or

$$Q_a = Q_{ex} - Q_s \tag{27}$$

The efficiency factors for char particles as determined by Mie theory are
shown in Figure 4.

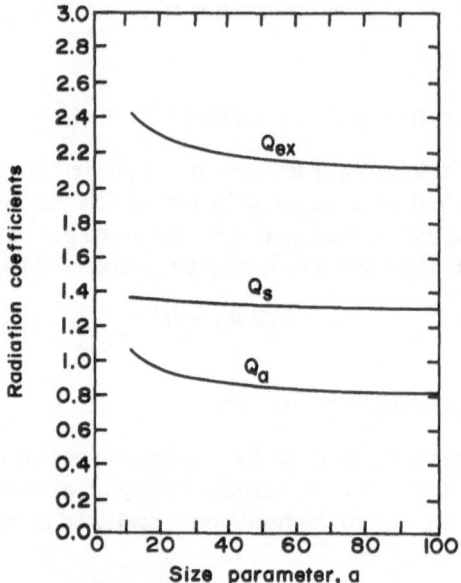

Figure 4. The absorption, scattering and extinction efficiencies of char particles at wavelength of 2 μm and refractive index of $1.93[1 - i(0.53)]$.

4.8. Scattering and Absorption Coefficients

The scattering or absorption coefficient expresses the total amount of flux removed from a beam by a unit volume of particle suspension per unit irradiance of the volume:

$$K_s = NX_s = N\pi r^2 Q_s \tag{28}$$

$$K_a = N\pi r^2 Q_a \tag{29}$$

and, at ultraviolet and infrared wavelength,

$$K_e = N\pi r^2 Q_{ex} \tag{30}$$

The above equations are valid only if the distribution of particles is monodisperse, i.e., particle size is constant throughout. However, in a pulverized-coal flame, not only is the size of a particle constantly changing, but also in a given volume there is a wide range of sizes and types of particles. For such a polydispersed medium, N has to be represented by a size distribution function. In case the size distribution function is not defined, the coefficient K can also be found by numerical summation over known sizes:

$$K = \sum_j N_j Q_j \pi r^2 \tag{31}$$

5. Equation of Radiative Transfer

Radiative transfer theory deals with the propagation of intensity and was initiated by Shuster in 1903. Basic radiative transfer is governed by an integrodifferential equation called the equation of transfer and is equivalent to Boltzmann's equation in the kinetic theory of gases and neutron transport theory. This equation has also been successfully employed in engineering heat transfer studies.

The basic radiative transfer equation includes the scattering and absorption of radiation by a participating medium and is written in the form[17]

$$(S \cdot \nabla)I_v = -(K_{av} + K_{sv})I_v + K_{av}I_{bv}(T) + (K_{sv}/4\pi) \int_{\Omega' = 4\pi} P_v(S' \to S)I_v(S') \, d\Omega'$$

(32)

The term $P_v(S' \to S)$ is the phase function, source function, or scattering function and is of considerable significance if the scattering is anisotropic. The phase function is defined such that for isotropic scattering, it can be normalized to unity:

$$(1/4\pi) \int_{\Omega' = 4\pi} P_v(S' \to S) \, d\Omega' = 1$$

(33)

A pulverized-coal flame is a multicomponent, nonuniform, emitting/ absorbing/scattering gas–particle cloud which generates heat and is surrounded by nonuniform emitting/absorbing/reflecting solid surfaces (walls). The nonuniformity of the system and the difficulty of finding equilibrium conditions at each point, together with the lack of knowledge about radiative properties of the components of the system, makes the problem of heat transfer in the combustor an extremely complex one unless a number of approximations are made to simplify the problem.

6. Simplification of the Problem

In a pulverized-coal flame, the pulverized-coal particles, which are the main component of radiation, undergo continuous change in mass, shape, size, optical properties, and number throughout the combustion process. A single coal particle, when introduced into a stream of hot gases, undergoes a process of devolatilization and fragmentation, in addition to combustion which changes its shape, size, and mass continuously, as well as breaking it into small particles. In addition, soot particles are formed continuously. To take into account the exact behavior of all the particles is virtually impossible at this stage. In order to develop a viable model for heat transfer, simplification is achieved by assuming that a particle retains a spherical shape, diameter, and mass during the time it travels through a small control volume.

There exists a temperature gradient within a coal particle which is a function of particle size, thermal diffusivity, and thermal energy exchanges due to radiation, convection, and conduction with surrounding gases. The complete treatment of such a complex situation is simply not possible and an analytical solution is possible only under the assumption that the temperature gradient inside the particle is sufficiently small to be neglected.

The different components within a small control volume may also differ in temperature; the temperature of a burning coal particle may be several hundred degrees higher than that of the carrier gas.[10] In addition, the emissivity and absorptivity of the control volume cannot strictly be defined unless the control volume is at local thermodynamic equilibrium.[12] It is therefore necessary to assume that local thermodynamic equilibrium exists between particles and gas of the medium for certain calculations.

Another area that requires simplification is the spectral dependence of emissivities of gas–particle clouds. It is shown in reference 10 that soot and combustion gases are not gray, and that the char particles are assumed to be gray only in the absence of evidence to the contrary. However, it may be assumed that multiplicity of components increases the chance of overall grayness. This makes it possible to simplify the problem further by assuming that the gas–particle medium is gray.

The temperature field that exists in the flame is nonuniform since the temperature of the gas–particle cloud is changing continuously. The non-uniformity of the temperature field requires a three-dimensional solution of the problem of heat transfer, but axial symmetry of the flow and temperature field may be assumed to simplify the problem.

The final simplification is obtained by assuming that quasi-stationary conditions exist due to the fact that velocities of the flame are small compared to the velocity of light.

All the above-mentioned approximations may be summarized as follows:

(1) The particles are spherical in shape.
(2) Local thermodynamic equilibrium exists between the components of the flame.
(3) The gas–particle medium is gray.
(4) The temperature field is axially symmetric.
(5) Quasi-stationary conditions exist.

7. Solution of the Radiative Transfer Equation

Due to the integrodifferential nature of the equation of radiative transfer, an exact analytical solution is extremely difficult to obtain. However, certain approximations related to the physical situation of the problem, as discussed in Section 6, can lead to a simpler form of Eq. (32). The main area of difficulty in Eq. (32) is the last term (in-scattering) on the right-hand side. This term, also known as the source term, is in integral form, and must be reduced to a differential form in order to simplify the problem. This is achieved by substituting the Gaussian quadrature formula for the integral term, as suggested by Chandrasekhar.[8] This method is equivalent to approximating the integrand by a polynomial of order $2n - 1$, where n is the order of Gaussian quadrature.

Due to the presence of "large" particles such as char, ash, etc., the scattering in a pulverized-coal flame is anisotropic. An effort is made in this work to consider anisotropic scattering in solving the radiative transfer equation, rather than making the assumption of isotropic scattering or no in-scattering at all, which has been adopted as a rough approximation in all the existing analyses of radiative heat transfer in industrial flames. A brief review of the existing approaches to the problem of radiative transfer in industrial flames follows.

7.1. Hottel's Zone Method

Hottel's zone method is a classic, somewhat complex, but commonly used method. It offers a complete solution for the gray-gas system, and also incorporates approximations to treat nongray effects. However, this method has been used for predictions of heat transfer where the heat release and flow patterns are either assumed or measured.[14,22] A very accurate analysis of radiative heat transfer can be achieved by combining Hottel's radiation solution, which is in integral form, with the complete analytical flow and mixing solutions developed by Gosman *et al.*[15] The combined radiative heat transfer and flow equations are then solved by a grid network using the finite-difference approach.[14] Since the zones can "see" each other and n simultaneous nonlinear equations have to be solved for n number of zones, the solution becomes uneconomical with respect to computer time,[5] even though the method offers good generality and high accuracy.

7.2. Monte Carlo Method

In the Monte Carlo method, a "bundle" of intensities are generated within a control volume and "fired" in all directions. Individual rays are followed for their "lifetime" as they undergo simulated absorption, emission, and scattering. This is a powerful and flexible technique for solving the radiative-transport equations[16] and, like the zone method, offers a complete solution to gray-gas problems, and with some approximations, for nongray gases.[5] This method has an advantage of being applicable to combustors with complex geometry. It also has poor economy in computer time, but high generality, flexibility, and accuracy.

7.3. Diffusion Approximation Method

Representation of specific intensities by a two-term series of spherical harmonics permits reduction of the transport equation to the classical diffusion equation. The results of this method are quite encouraging, making it an increasingly popular method.[16] Whitacre and McCann[14] have demonstrated that the accuracy offered by the diffusion approximation method is quite comparable to Hottel's zone method. In a comparison test for different methods for solving the radiative transport equation, they found that the diffusion approximation method predicted radiation flux to walls quite accurately, whereas temperature predictions were somewhat low at the centerline.[14] In its simplicity, this method compares well with the flux method, to be discussed next.

7.4. Flux Method

The angular distribution of intensities is replaced by a number of discrete intensity vectors in different directions, thus reducing the complexity of the integrodifferential equation that arises due to angular dependence of intensity. The energy transfer in each direction is represented by a closed, first-order, ordinary differential equation known as a "flux equation," which is obtained by integrating the transfer equation with respect to solid angle. This method was originated by and is commonly used by astrophysicists for the one-dimensional case. Different two-flux models developed for one-dimensional cases are reviewed by Siddal.[17] The one-dimensional or two-flux models are not of much use in situations where the fluid flow and transport equations are generally solved for two or three dimensions. Gosman and Lockwood[18] have developed a four-flux model to be used in a two-dimensional furnace computation, but in this model, scattering is assumed to be isotropic. This model can be used with sufficient accuracy in a furnace where the solid particles present in the flame are too small in size (e.g., soot) to cause anisotropic scattering. However, in a pulverized-coal flame where ash and char particles definitely fall into the category of "large" particles $(\pi d/\lambda \geqslant 5)$, scattering becomes significant in the heat transfer equations and can neither be ignored nor considered isotropic without causing some inaccuracy. For this reason, Gosman and Lockwood's four-flux model loses its value in radiative-transfer analyses in a pulverized-coal flame. In an experimental demonstration, Whitacre and McCann[14] have shown that the four-flux model developed by Gosman and Lockwood, when compared with Hottel's zone method and the diffusion approximation method, predicted the temperature quite well, whereas radiation fluxes at many places were underestimated. It was then suggested that through some improvements and modifications, the flux method can yield improved efficiency.[14]

Lockwood and Shah[19] have developed an improved "six-flux" model, but again this model is not applicable to a case of anisotropic scattering, since in this model they have ignored the in-scattering term in the transport equation.

Theising[29] formulated a two-flux model for anisotropic scattering that represented the scattered intensity in forward and backward directions only. Fritze[30] also developed a flux model with preference to forward scattering and an isotropically distributed component. In an anisotropic situation, models of Theising and Fritze should yield more accurate results than the classical diffusion equation.[31]

Chu and Churchill[13] have developed a six-flux model for anisotropic and multiple scattering that is believed to be sufficiently accurate for most practical applications. In comparison with the increasingly popular diffusion equations, the flux model developed by Chu and Churchill is of comparable accuracy and easier to use.

It is proposed to utilize a four-flux model based on Chu and Churchill's six-flux model, reduced to the four-flux form by applying the condition of axial symmetry to the six-flux model. A detailed derivation of the recommended equations is given in Appendix C.

8. Optical and Radiative Properties of Radiating Components

8.1. Optical Properties

In order to use the above model in a heat transfer situation, certain coefficients and parameters used in the equations are determined from the optical and physical properties of the components of the system.

One of the principal parameters to be determined is the angular distribution function or the phase function $P(\theta)$ for the char and ash particles that are scattering radiation. Also, the absorption and scattering cross sections are required to complete the analysis. The refractive indices of particles are needed, in addition to the size of particles and wavelength of radiation, in order to find the angular distribution and the cross sections.

If the refractive index of the particle is known, then the needed parameters can be calculated based on Mie theory.[32] These calculations can be carried out with the help of computer programs which are available to do such calculations. For example, the char particle is estimated to have an index of refraction of 1.93 $(1 - 0.53i)$ at a wavelength of $2\,\mu m$, based on Blokh's calculations.[10] Using the SMIE[33] program, the absorption, scattering, and extinction efficiencies of char particles, up to a size parameter of $a = 100$, have been calculated and plotted as shown in Figure 4. The scattering efficiency as obtained from these calculations includes a contribution of 1.0 due to diffraction which must be discounted in heat transfer calculations. This makes the modified scattering efficiency 1.0 less than that calculated by the above-mentioned program. Besides giving the efficiencies of char particles, above program also yields the angular distribution of scattered radiation for each particle size.

Use of such programs can be made only if the refractive index of the particle is known. Such is not the case for ash particles. There is a lack of knowledge about the radiative properties of ash particles that behave as opaque to radiation in the near-infrared region, but are transparent to radiation in the visible range of the spectrum. It is assumed that ash particles can be considered opaque and diffuse in pulverized-coal combustion, since the average wavelength of radiation remains around 2 μm in the near-infrared region. Based on this assumption[34] and calculations from Mie theory, the phase function for ash particles can be approximately represented by

$$P(\theta) = (8/3\pi)(\sin \theta - \theta \cos \theta) \tag{34}$$

Similar assumptions can be made for all char particles, which scatter weakly but cannot be neglected.

The overall extinction, absorption, and scattering coefficients for the system can be computed by adding the corresponding coefficients for gases and particles present in the medium. Although none of the constituents of the medium are gray when considered individually, their combined effect can be considered gray without introducing great error.[10,23,24]

The absorption coefficient for the medium is then given by

$$K_a = K_{ag} + K_{ap} \tag{35}$$

and the emissivity is related to the absorption coefficient by the equation

$$\varepsilon = 1 - \exp(-K_a l) \tag{36}$$

The absorption coefficient K_{ag} for gases can be determined by the equations

$$\varepsilon_g = \varepsilon_{CO_2} C_{CO_2} + \varepsilon_{H_2O} C_{H_2O} - \Delta\varepsilon \tag{37}$$

and

$$K_{ag} = 1/l[\ln(1 - \varepsilon_g)] \tag{38}$$

Only CO_2 and H_2O are considered to be absorbing gases, since other gases make relatively insignificant contributions to absorption. Detailed information concerning emissivisites of CO_2 and H_2O at different temperatures and pressures is available in reference 21.

The absorption and scattering coefficients for the char and ash particles can be calculated from the equations[10]

$$K_a = \sum_j K_{aj} = \sum_j (\pi/4) Q_a N_j d_j^2 \tag{39}$$

and

$$K_s = \sum_j K_{sj} = \sum_j (\pi/4) Q_s N_j d_j^2 \tag{40}$$

where Q_a and Q_s are the absorption and scattering efficiencies of particles, respectively. The soot particles absorb highly, but scatter negligibly due to their microscopic size. The absorption coefficient for soot must be determined experimentally.

Summing up all the above absorption coefficients, the overall absorption coefficient for a small volume is

$$K_a = K_{ag} + K_{ap} \tag{41}$$

and

$$K_a = 1/l \big[\ln(1 - \varepsilon_g) \big] + (\pi/4) \sum_j \big[(Q_a N_j d_j^2)_{\text{char}} + (Q_a N_j d_j^2)_{\text{ash}} \big] + (K_a)_{\text{soot}} \tag{42}$$

The overall scattering coefficient is

$$K_s = (\pi/4) \sum_j \big[(Q_s N_j d_j^2)_{\text{char}} + (Q_s N_j d_j^2)_{\text{ash}} \big] \tag{43}$$

8.2. Equilibrium Temperature

In order to define the emissivity, etc., of a small control volume it is important to define an equilibrium temperature of the control volume. Due to the presence of a number of solid and gaseous components, the temperature within the control volume changes from point to point. The mean temperature or the equilibrium temperature can be defined by[10]

$$T^4 = \left(\sum_i K_{ai} T_i^4 \right) \Big/ \left(\sum_i K_{ai} \right) \tag{44}$$

where the subscript i is for different components such as char, ash, soot, CO_2, H_2O, etc.

8.3. Mean Beam Length

The emissivity of a radiating component is defined as

$$\varepsilon_\lambda = 1 - \exp(K_\lambda l) \tag{45}$$

where l is the mean beam length of the control volume. The value of mean beam length depends on the shape and size of the radiating medium, and these different values are given in reference 6. For our purposes, a single value is used:

$$l = 3.5(V/A) \qquad (46)$$

8.4. Boundary Conditions

The following assumptions are made in order to define the boundary conditions[25]:

(1) The boundary is opaque to radiation (transmissivity $= 0$).
(2) The boundary is gray.
(3) The boundary surface is a diffuse emitter/reflector (Lambert surface).

The boundary surfaces in a pulverized-coal combustor will, in general, consist of the combustor walls and the cold surfaces of tubes carrying water and steam. Since the boundary surfaces are opaque to radiation, all the thermal flux coming from the surface will be due to either emission or reflection, and may be written in terms of surface properties as

$$I(\Omega^-) = (1/\pi)\varepsilon_w \sigma T_w^4 + r_w \mathbf{F}^+ \cdot \mathbf{n} \qquad (47)$$

where Ω^- represents the ray direction from the surface and $\mathbf{F}^+$ represents flux incident on the surface as shown in Figure 5. Similar equations can be written for cold surfaces.

Since the transmissivity of the surfaces is zero, it follows that

$$\varepsilon_w + r_w = 1 \qquad (48)$$

or

$$\varepsilon_w = 1 - r_w \qquad (49)$$

where ε_w and r_w are the total hemispherical emissivity, and reflectivity, respectively. This is the simplest term of boundary condition suited to the present problem and is consistent with the simplicity desired in formulation of the entire radiative transfer problem in pulverized-coal flames.

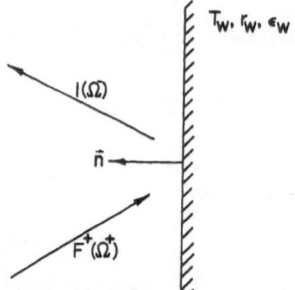

Figure 5. Wall boundary conditions for incident and reflected flux.

As the level of complexity in formulating the radiative transfer model is increased, the boundary conditions can also be modified to include effects such as Fresnel reflection at the boundary, curved boundary surfaces, and conduction due to the boundary layer.

The problem of radiative transfer is not completely solved merely by developing a simplified radiative transfer equation. The radiative properties of the particles and combustor walls are equally important in obtaining satisfactory results.[35] Little information is available in this area. A great deal of experimental and analytical work is needed to determine properties such as the optical efficiencies of char and ash particles. The nature of these properties depends largely on the refractive indices of these particles, which are functions of temperature, wavelength, and composition.[23] There are some data on the refractive index of char particles, but there is a serious lack of such information for ash particles.[36] Ash is a significant component in radiative heat transfer in pulverized-coal combustors, even though it has been largely overlooked. Some suggestions have been made in this work that can help in using the model in the absence of required information about these properties.

9. Notation

A	Area (m^2)	K_{ev}	Volumetric emission coefficient (m^{-1})
b	Backward scattering factor		
C	Velocity of light ($m\ s^{-1}$)	K_a	Absorption coefficient (m^{-1})
C_0	Velocity of light in a vacuum ($m\ s^{-1}$)	K_s	Scattering coefficient (m^{-1})
		K_t	Extinction or attenuation coefficient (m^{-1})
C_{CO_2}, C_{H_2O}	Pressure correction factor for CO_2 and H_2O, respectively	l	Mean beam length (m)
d	Diameter of solid particles (m)	n	Outward normal unit vector
e_v	Monochromatic radiation energy (J)	N	Number of solid particles per unit volume (m^{-3})
$E_{\lambda,g}$	Hemispherical monochromatic black emissive power ($W\ m^{-2}$)	P_c	Net emissivity of cold surface
		p	Proportion of total ash in suspension
f	Forward-scattering factor	P	Total pressure of gases (Pa)
F_v	Monochromatic radiation flux ($J\ m^{-2}\ s^{-1}\ \mu m^{-1}$)	$P(\theta)$	Phase function
		Q_a, Q_s	Absorption and scattering efficiencies, respectively
I	Specific intensity of radiation ($J\ m^{-2}\ s^{-1}\ \mu m^{-1}\ sr^{-1}$)	r_w, r_c	Total hemispherical reflectivity of furnace wall and cold surface, respectively
I_b	Blackbody emission intensity ($J\ m^{-2}\ s^{-1}\ \mu m^{-1}\ sr^{-1}$)	R	Radial distance (m)
K_1, K_2, K_3	Constants related to emissivity, surface area and the Stefan–Boltzmann constant, respectively	s	Sidewise scattering factor
		s	Unit vector in the direction of propagation

S	Furnace wall surface (m^2)	θ	Angle between the direction of beam and the outward normal $\bar{n}$
t	Time (s)		
T	Absolute temperature (K)		
$T_F, T_R,$ T_C	Absolute temperatures of flame, refractory, and cold surfaces respectively (K)	Ω	Solid angle about direction of propagation (sr)
ΔT	Temperature difference between wall and moving gases (K)	Ω'	Solid angle about the in-scattering beam (sr)
		Ω^+	Direction of ray toward the surface
U_v	Radiant energy density ($J\ m^{-3}\ \mu m^{-1}\ K^{-1}$)	Ω^-	Direction of ray away from the surface
V	Volume (m^3)		

S Furnace wall surface (m^2)
t Time (s)
T Absolute temperature (K)
T_F, T_R, T_C Absolute temperatures of flame, refractory, and cold surfaces respectively (K)
ΔT Temperature difference between wall and moving gases (K)
U_v Radiant energy density ($J\ m^{-3}\ \mu m^{-1}\ K^{-1}$)
V Volume (m^3)
W_0 Albedo for single scattering
X_a, X_s Absorption and scattering cross sections, respectively (m^2)
Z Coefficient for heat transfer by convection ($J\ m^{-2}\ s^{-1}\ K^{-1}$)
α Absorptivity
ε Emissivity
$\varepsilon_w, \varepsilon_c$ Total hemispherical emissivity of furnace wall and cold surface, respectively
ρ Mass density of emitting matter ($kg\ m^{-3}$)
λ Wavelength of radiation (μm)
ν Frequency of radiation (s^{-1})
σ Stefan–Boltzmann constant ($J\ m^{-2}\ s^{-1}\ K^{-4}$)
Θ Angle between direction of propagation and direction of scattered flux

θ Angle between the direction of beam and the outward normal $\bar{n}$
Ω Solid angle about direction of propagation (sr)
Ω' Solid angle about the in-scattering beam (sr)
Ω^+ Direction of ray toward the surface
Ω^- Direction of ray away from the surface

Subscripts

a absorption
g gas
j particle size
p particle
r radial direction
s scattering
t extinction
z axial direction
λ wavelength
ν frequency
θ angular direction

Superscripts

$+$ In the positive direction of a major axis
$-$ Opposite to the positive direction of a major axis

10. References

1. W. J. Wohlenberg and D. G. Marrow, Radiation in the pulverized-fuel furnace, *Trans. ASME* **47**, 127–176 (1925).
2. W. J. Wohlenberg and E. L. Lindseth, The influence of radiation in coal-fired furnaces on boiler-surface requirements, and a simplified method for its calculations, *Trans. ASME* **48**, 849–937 (1926).
3. R. T. Haslam and H. C. Hottel, Combustion and heat transfer, *Trans. ASME* **FSP50-3**, 9–22 (1928).
4. H. C. Hottel and E. S. Cohen, Radiant heat exchange in a gas-filled enclosure: Allowance for nonuniformity of gas temperature, *AIChE J.* **4**, 3–14, (1958).
5. F. C. Lockwood, Imperial College, London, U.K., personal communication at the University of Utah, Salt Lake City, Utah (1977).
6. W. A. Gray, J. K. Kilham, and R. Muller, *Heat Transfer from Flames*, Eleck Science, London, England (1976).
7. H. C. Hottel and A. F. Sarofim, *Radiative Transfer*, McGraw–Hill Book Company, New York (1976).

8. S. Chandrasekhar, *Radiative Transfer*, Dover Publications, Inc., New York (1960).
9. R. Viskanta, Radiation transfer and interaction of convection with radiation heat transfer, in *Advances in Heat Transfer* (T. F. Irvine, Jr., and J. P. Harnett, ed.), Vol. 3, 175–251, Academic Press, New York (1966).
10. M. A. Field, D. W. Gill, B. B. Morgan, and P. G. W. Hawksley, *Combustion and Pulverized Coal*, The British Coal Utilization Research Association, Leatherhead, England (1967).
11. T. J. Love and R. J. Grosh, Radiative heat transfer in absorbing, emitting and scattering media, *J. Heat Transfer* 7, 161–166 (1965).
12. G. N. Hatsopoulos and J. H. Keenan, *Principles of General Thermodynamics*, John Wiley and Sons, Inc., New York (1965).
13. C. M. Chu and S. W. Churchill, Numerical solution of problems in multiple scattering of electromagnetic radiation, *J. Phys. Chem.* 59, 955–863 (1955).
14. G. R. Whitacre and R. A. McCann, Comparison of Methods for the Prediction of Radiant Flux Distribution and Temperature, ASME paper No. 75–HT-9 (1975).
15. A. D. Gosman, W. M. Pun, A. K. Runchal, D. B. Spalding, and K. Wolfshtein, *Heat and Mass Transfer in Recirculating Flows*, Academic Press, New York (1969).
16. J. Gibb, Central Electricity Generation Board, U.K., personal communication at the University of Utah, Salt Lake City, Utah (1977).
17. R. G. Siddal, Flux methods for the analysis of radiant heat transfer, in *Institute of Fuel, 4th Symposium on Flames and Industry*, 169–179 (1972).
18. A. D. Gosman and F. C. Lockwood, Incorporation of a flux model for radiation into a finite-difference procedure for surface calculations, in *Fourteenth Symposium (International) on Combustion*, The Combustion Institute, Pittsburgh Pa., 661–671 (1972).
19. F. C. Lockwood and N. G. Shah, An Improved Flux Model for the Calculation of Radiation Heat Transfer in Combustion Chambers, ASME paper No. 76–HT-55 (1976).
20. R. Siegel and J. R. Howell, *Thermal Radiation Heat Transfer*, McGraw–Hill Book Co., New York (1972).
21. H. C. Hottel, Radiant-heat transmission, in *Heat Transmission* (3rd ed.), McGraw–Hill Book Co., New York (1954).
22. A. Lowe, T. F. Wall, and I. M. C. Stewart, A zoned heat transfer model of a large tangentially fired pulverized coal boiler, in *Fifteenth Combustion Symposium*, The Combustion Institute, Pittsburgh, Pa., 1261–1270 (1974).
23. A. Lowe, Ash and gas radiative data for mathematical models of pulverized fuel flames, in *Furnace and High Temperature Heat Transfer*, The University of New Castle, Department of Chemical Engineering, Paper No. 1, 1–4 (1977).
24. P. J. Street and C. S. Twamley, Fuel particle emissivities, *J. Inst. Fuel* 44, 477–478 (1971).
25. J. Gibbs, Notes on Monte-Carlo method for radiative transfer calculations, in *Furnace and High Temperature Heat Transfer*, University of New Castle, N.S.W. 2308 (1977).
26. E. J. McCartney, *Optics of the Atmosphere*, John Wiley and Sons, New York (1976).
27. M. C. Van De Hulst, *Light Scattering by Small Particles*, John Wiley and Sons, New York (1957)
28. M. Kerker, *The Scattering of Light and Other Electromagnetic Radiation*, Academic Press, New York (1969).
29. H. H. Theising, Macrodistribution of light scattered by dispersions of spherical dielectric particles, *J. Opt. Soc. AME* 59, 232–243 (1950).
30. S. Fritze, Ph.D. Thesis, M.I.T., Cambridge, Mass. (1953).
31. C. M. Chu and S. W. Churchill, Numerical solution of problems in multiple scattering of electromagnetic radiation, *J. Phys. Chem.* 59, 855–863 (1955).
32. G. Mie, *Ann. Physik* 25, 377 (1908).
33. F. V. Dave, Subrouting for Computing the Parameters of the Electromagnetic Radiation Scattered by a Sphere, 360 D-17, 4.00Z, IBM Corporation.

34. A. F. Sarofim, M.I.T., Cambridge Mass, personal communication, at the University of Utah, Salt Lake City, Utah (1977).
35. R. H. Essenhigh, D. T. Pratt, H. E. Shull, and L. D. Smoot, Current status of fluid mechanics in practical heterogeneous combustors, in *Central States Section*, The Combustion Institute, Pittsburgh, Pa. (1977).
36. S. A. Varma and D. T. Pratt, An improved model for anisotropic and multiple scattering of thermal radiation in pulverized coal combustion, in *Western States Section*, Pittsburgh, Pa. (1977).
37. G. N. Plass, Mie scattering and absorption cross sections for aluminium oxide and magnesium oxide, *Applied Optics* **3**(7), 867–872 (1964).

6

Gas–Particle Flow

Clayton T. Crowe

The parameters relating to the interaction of particles and gas in a dispersed-phase flow are presented in this chapter. These parameters are essential to the development of numerical and analytic submodels of pulverized-coal combustion.

1. Particle Diffusion

Particle diffusion can occur in a gas due to Brownian motion of the molecules colliding with the particles, or due to the aerodynamic forces created by turbulent velocities. Typically, the latter mechanism is dominant for the dispersed-phase flow of pulverized coal in a gas stream because the particles are much larger than the gas-phase molecules, so that Brownian motion is of no consequence.

The motion of particles in turbulent flows has attracted the interest of mathematicians and scientists for several years. One of the earliest analytical studies was carried out by Tchen,[1] who assumed that the particles were subjected to harmonic gas-phase oscillations, in which the same fluid element always surrounded the particle. Tchen's original approach was modified and extended by others[2,3] to predict the ratio of particle diffusivity to mass diffusivity as a function of particle–gas density ratio, particle relaxation time, and frequency. In general, the results predict decreasing particle diffusivities with increasing frequency or particle size. Several of the assumptions needed

Clayton T. Crowe • Professor of Mechanical Engineering, Washington State University, Pullman, Washington

to make the problem mathematically tractable shed doubt on the quantitative reliability of the predictions.

Experimental studies of particle dispersion due to turbulence have led to conflicting results. Much of the experimental data are presented in terms of the Schmidt number for particle diffusion, defined by

$$Sc_p = v/\varepsilon_p \tag{1}$$

where v is the effective (turbulent) kinematic viscosity of the gas and ε_p is the diffusivity of the particles due to turbulence. The particle mass flux due to turbulence is related to the diffusivity by

$$\mathbf{j}_p = -\rho \varepsilon_p \nabla x_p \tag{2}$$

where x_p is the mass concentration of particles and ρ is the mixture density.

Goldschmidt and Eskanazi,[4] using hot-wire anemometry, measured the concentration of liquid aerosol in a plane turbulent air jet. They measured a Schmidt number slightly larger than unity, indicating a lower spread rate of the droplet concentration profile than that of the velocity profile. Further studies[5] by the same researchers indicated that increasing droplet size leads to Schmidt numbers below unity, implying that increasing droplet size leads to increased droplet dispersion due to turbulence. This is contrary to what might be expected, suggesting that a critical particle size may exist for which the dispersion due to turbulence is a maximum.

Lilly[6] reports an extensive study of particle dispersion in a turbulent air jet. He also found that the particle diffusivity increased with particle size. He attempted to correlate his results with the ratio of particle relaxation time to the Lagrangian time macroscale of turbulence. He expected to observe a decrease in particle dispersion rate due to turbulence as the time ratio increased. However, he discovered the opposite trend, which was similar to the later work of Goldschmidt *et al.*[5] Lilly measured a particle Schmidt number of 0.17 for the particles when the time ratio was somewhat less than unity, which corresponds to a particle diffusivity more than five times the momentum diffusivity in the turbulent jet.

Some simple theories[7] suggest that *slight* increases in particle diffusivity might be realized with increasing particle size, but *not* of the magnitude measured. The argument revolves around a prediction that, even though the turbulence intensity of the particle motion will decrease with increasing particle size, the turbulence scale increases, thereby effecting an increasing particle diffusion. No theory developed to date satisfactorily predicts the magnitude of the trend.

Hedman and Smoot[8] report experiments on small- and large-particle dispersion in coaxial jets confined in a tube. The small-particle data indicate a turbulent Schmidt number near unity. It was noted that the mixing rate of the large particles was slower than that of the small particles, which does not agree with the trend observed by some other experimenters.

More recently Smoot and co-workers[9,10] reported a series of cold-flow experiments wherein the turbulent mixing rates of gas and silicon particles were measured in a laboratory-scale test chamber. All initial test conditions (except scale size), including primary and secondary velocities, densities, particle mass percentage, and particle size distribution, were typical of pulverized-coal furnaces. In all measurements, gases mixed at several times the rate of the particles. Further, in tests where particle size was varied, the smaller particles dispersed more rapidly than the larger particles, especially when the secondary air stream was injected on an angle (30°) to the primary jet, or when the flow was recirculating. However, even the smallest particles (19 μm) dispersed more slowly than the gas. It was also observed that mixing rates of gases and particles were significantly greater when the mixing chamber was enlarged to promote recirculating flows. These results are the most specific data available for evaluating pulverized-coal mixing rates, and provide a useful basis for determining gas and particle mixing coefficients for furnace submodels.

The increasing capability and availability of high-speed computers will ultimately lead to reasonably accurate descriptions of turbulent flow fields and, in turn, a predictive capability for the dispersion of particles due to turbulence. Peskin and Kau[11] have recently developed a numerical model for particle dispersion in a turbulent flow field. This is an extension of the numerical model of Deardorff[12] for turbulent flow in a duct. The model is complicated and not readily usable at the present time.

A reasonably useful and pragmatic description of a turbulent flow field is the two-equation model developed by Launder and Spalding,[13] and described in Chapter 3, in which the local turbulent field is characterized by the turbulence energy and dissipation rate. It seems that this turbulence model may provide a basis for modeling particle dispersion due to turbulence. A development of this potentially useful approach follows.

The equation of motion of a single particle is given by

$$d\mathbf{v}_p/dt = (\mathbf{v}_g - \mathbf{v}_p)/\tau \tag{3}$$

where t is the particle relaxation time, or the time required for a particle to reach dynamic equilibrium with the flow. Other forces, such as buoyancy, virtual mass, and the Basset force are neglected since they are unimportant when the ratio of gas density to particle material density is low, as for pulverized-coal combustion and gasification processes. Integration of Eq. (3) yields the particle velocity, and a second integration yields the particle trajectory. Taking k as the kinetic energy of turbulence (see Section 3.2 of

Chapter 3), the local gas velocity can be modeled as

$$v_g = \mathbf{v}_{g0} + c_k \mathbf{r} k^{1/2} \tag{4}$$

where $\mathbf{v}_{g0}$ is the time average velocity, c_k is an empirical constant, and $\mathbf{r}$ is a randomly oriented vector whose magnitude varies between 0 and 1.

The local time scale of turbulence, or the time an eddy persists, is given by

$$T = k/\varepsilon \tag{5}$$

where ε is the dissipation rate. It is assumed that the local gas velocity persists for a period given by the turbulence time scale, after which a new random vector and local velocity provide a new local gas velocity according to Eq. (4). This model is attractive because of its computational simplicity. It is currently under development.

2. Particle Drag

The equation of motion for a reacting particle, assuming a uniform mass efflux rate from the surface, is

$$\mathbf{F} = m(d\mathbf{v}_p/dt) \tag{6}$$

where m is the instantaneous mass of the particle. If the burning is not uniform over the surface, a thrustlike term has to be included. However, for purposes of modeling the combustion of pulverized coal, it is reasonable to assume uniform burning. The purpose of this section is to propose the most appropriate form of the equations for aerodynamic drag.

2.1. Steady-State Aerodynamic Drag

Often, the steady-state aerodynamic drag is the largest force acting on a particle in the drag direction (parallel to the relative gas–particle velocity vector) and is quantified by

$$\mathbf{F}_D = (\rho_g A_p C_D/2)|\mathbf{v}_g - \mathbf{v}_p|(\mathbf{v}_g - \mathbf{v}_p) \tag{7}$$

where C_D is the drag coefficient, ρ_g is the gas density, A_p is the particle projected area and $\mathbf{v}_g$ and $\mathbf{v}_p$ are the gas and particle velocities, respectively. Note that the force is in the direction $(\mathbf{v}_g - \mathbf{v}_p)$. The drag coefficient is primarily a function of the Reynolds number based on the relative velocity but may also depend on Mach number, particle shape, and burning rate.

Mach number (relative Mach number) effects depend on the magnitude of the Reynolds number. If the Reynolds number is greater than 50, an increasing Mach number tends to increase the drag coefficient while the opposite trend is observed for Reynolds numbers less than 50. The magnitude of the Mach number effect on drag coefficient is small[14] provided $M < 0.4$, unless the Reynolds number is also small, in which case rarefied flow conditions are approached. The degree of rarefaction is quantified by the Knudsen number, or the ratio of mean free path to particle diameter:

$$Kn = \lambda/d_0 \tag{8}$$

If the Knudsen number is greater than unity, the flow is identified as free-molecule flow, and the drag coefficient for continuum flow is no longer valid. Rarefaction effects can still be significant for Knudsen numbers $\frac{1}{10}$ or less when the particle lies in the slip or transition flow regimes. The mean free path for combustion gases at typical furnace conditions is less than 0.5 μm. Since the coal and char particles are significantly larger than this, the effect of Knudsen number is unimportant.

The drag coefficient of a spherical particle is most strongly influenced by Reynolds number. If the Reynolds number is less than unity, the drag coefficient is given by Stokes law:

$$C_D = 24/Re \tag{9}$$

For Reynolds numbers of 1 or more, Stokes law underpredicts the drag coefficient, as evidenced by the standard drag coefficient curve for a sphere. An empirical equation for C_D, valid up to Reynolds number of 1000, is[15]

$$C_D Re/24 = 1 + 0.15\, Re^{0.687} \tag{10}$$

As the Reynolds number approaches 0, this result approaches the Stokes law value. This expression is valid over the Reynolds number range encountered by particles in a coal-fired furnace. By defining f as

$$f = C_D Re/24 \tag{11}$$

the aerodynamic drag force on a particle is

$$\mathbf{F}_d = 3\pi\mu\, df(\mathbf{v}_g - \mathbf{v}_p) \tag{12}$$

which again reduces to Stokes drag as f approaches unity.

Particle burning tends to reduce the drag coefficient.[16] No information is available, however, on the drag coefficient of rapidly burning particles or particles giving off large mass flows of gas, such as coal particles during volatilization. Studies of burning droplets[17] indicate that the drag coefficient is correlated by

$$C_{Db} = C_{D0m}/(1 + B_m) \tag{13}$$

where C_{D0m} is the drag coefficient with no mass transfer and B_m is the mass transfer number. Because there can be a large variation in temperature-dependent properties between the particle surface and free stream, it is necessary to define the temperature at which the properties are evaluated. The subscript m refers to properties evaluated at the arithmetic mean. Thus, the gas density in Eq. (7) as well as the Reynolds number in Eq. (9) are evaluated at the mean temperature. The mass transfer number, B_m, for a droplet is defined by

$$B_m = C_{pm} \Delta T / L \qquad (14)$$

where C_{pm} is the mean specific heat, ΔT is the droplet-gas temperature difference and L is the latent heat of vaporization. The appropriate transfer number for a burning particle is

$$\phi_m = 2\dot{m} C_{pm} / k_m d_p \qquad (15)$$

where $\dot{m}$ is the burning rate and k_m is the mean thermal conductivity of the gases in the boundary layer. It can be shown that d_m is related to B_m by

$$B_m = \exp(\phi_m) - 1 \qquad (16)$$

Thus, extending the data for burning droplets to burning or volatilizing coal particles, a plausible correlation for drag coefficient is

$$C_{Db} = C_{D0m} \exp(-\phi_m) \qquad (17)$$

It is unlikely that burning effects on the drag coefficient are important during burning of the char particle.

Another important parameter affecting the aerodynamic drag is the particle shape. The drag coefficient of nonspherical particles has been the subject of several studies.[18] One approach is to define the sphericity of a particle in the following manner:

$$\theta = S_s / S_p \qquad (18)$$

where S_s is the surface area of a sphere of the same volume as the particle and S_p is the surface area of the particle. Waddell[19] correlated data obtained by various investigators and found that the sphericity serves reasonably well as a parametric variable for particle shape. A study by Robins[20] indicates that a reasonable value for the sphericity of coal particles less than 100 μm in diameter is 0.7. By interpolating the data assembled by Wadell, the following empirical equation for nonspherical particles with $\phi_m = 0.7$ results:

$$C_D/C_{Ds} = 1.7 Re^{0.23}, \qquad Re > 1 \qquad (19a)$$

$$= 1.7, \qquad Re < 1 \qquad (19b)$$

where C_{Ds} is the drag coefficient of the spherical particle.

Incorporating the effects of particle asphericity and burning on the drag factor f, defined by Eq. (11), the following correlation is recommended for a typical coal particle:

$$f = 1.7 Re^{0.23}(1 + 0.15 Re^{0.687}) \exp(-\phi_m), \qquad Re > 1 \qquad (20a)$$

$$f = 1.7(1 + 0.15 Re^{0.687}) \exp(-\phi_m), \qquad Re < 1 \qquad (20b)$$

2.2. Buoyancy

The buoyant force on a particle due to pressure gradient in the gas phase acts in the same manner as the buoyant force in a quiescent fluid resulting from the pressure gradient due to gravity. The magnitude and direction of the force is

$$F_p = -V_p(\partial p/\partial x) \tag{21}$$

where V_p is the volume of the particle. The relative importance of the buoyant force can be assessed by comparing its magnitude to the mass–acceleration product for the particle:

$$F_p/m\alpha_p \sim (1/\rho_p\alpha_p)(\partial p/\partial x) \tag{22}$$

The pressure gradient in the gas can be approximated by

$$\partial p/\partial x \sim \rho_g\alpha_g \tag{23a}$$

so that

$$F_b/m\alpha_p = (\rho_g/\rho_p)(\alpha_g/\alpha_p) \tag{23b}$$

If the acceleration of the particle and gas are comparable, then the (buoyant force)/(inertial force) ratio is approximately equal to the density ratio, which is generally 10^{-3}. Under such conditions, the buoyancy force can be neglected. Thus the buoyancy force term is only significant in regions of large gas accelerations, such as the throat region of a nozzle or through a shock wave, but typically insignificant in pulverized-coal combustion applications.

2.3. Virtual Mass Effect

When the particle accelerates with respect to the gas, a force is developed on the droplet which is proportional to the relative acceleration. This force, called the virtual or apparent mass effect, is given by

$$\mathbf{F}_{vm} = K_m\rho_p V_p(d\mathbf{v}_g/dt - d\mathbf{v}_p/dt) \tag{24}$$

where K_m is an empirical constant. The virtual mass effect arises because of the relative acceleration of the surrounding gas with respect to the particle, which gives rise to an additional force on the particle. For a sphere, the theoretically derived value for K_m is 0.5, so the virtual mass effect can be thought of as the force necessary to accelerate a mass of vapor in a volume equal to half the volume displaced by the particle. Further experiments by Odar[21] suggest that the empirical constant depends on the acceleration modulus, Ac, in the following way:

$$K_m = 1.05 - 0.066/(Ac^2 + 0.12) \tag{25}$$

where Ac is defined as the ratio of dynamic to acceleration-induced forces:

$$\text{Ac} = |\mathbf{v}_g - \mathbf{v}_p|^2 / \alpha_r d_p \tag{26}$$

The virtual mass effect is negligible for pulverized-coal combustion applications when compared to the mass–acceleration product of the particle.

2.4. Basset Force

Whereas the virtual mass effect is operative in a viscous or inviscid fluid, the Basset force occurs only in a viscous fluid and relates to the unsteadiness of the boundary layer. For example, consider a flat plate which is impulsively started in a quiescent fluid. As the momentum diffuses from the plate and a boundary layer develops, the shear stress produced by the boundary layer continually changes with time until the steady-state condition is achieved. The deviation of the shear stress during the transient period from the steady-state value is the Basset force. Thus, the Basset force depends on the history of the relative acceleration and is given by

$$\mathbf{F}_B = [K_B d_p^2 (\pi \rho_g \mu)^{1/2}/4] \int_0^t (d\mathbf{v}_g/dt - d\mathbf{v}_p/dt)\, dt'/(t-t')^{1/2} \tag{27}$$

Basset,[22] by a theoretical calculation, showed that $K_B = 6$. Experimental studies by Odar[21] indicate that K_B depends on the acceleration modulus and is represented empirically by

$$K_B = 2.88 + 3.12/(\text{Ac}+1)^3 \tag{28}$$

Dividing the Basset force by the mass–acceleration product of the particle, and assuming a constant relative acceleration, there results

$$F_B/m\alpha_p \sim 20[(\rho_g/\rho_p)(\text{Ac}/\text{Re})^{1/2}] \tag{29}$$

This term is generally small, since the density ratio is of the order of 10^{-3}, but may become significant in regions where Ac/Re is large, such as for small particles in shock waves. Once again, however, it is not significant in modeling the dynamics of a coal particle in a combustor, unless it is significant for particle accelerations created by turbulence fluctuations.

2.5. Magnus Effect

The Magnus effect is the lift force due to rotation of the particle and is given by

$$F_r = (\pi/8)[\rho_g d_p^3 (v_g - v_p)\omega] \tag{30}$$

where ω is the rate of rotation of the particle. It is difficult to assess the rotational rate of the particle. If the particles are small, the rotational rate is likely to be equal to that of the fluid, in which case the Magnus effect becomes significant only in high-shear regions such as in a boundary layer.

2.6. Saffman Lift Force

The Saffman lift force is due to a velocity gradient in the fluid as shown in Figure 1. The velocity gradient causes a higher velocity at A than at B and produces a lift force on the particle. The particle need not be rotating and is thereby distinct from the Magnus effect. The Saffman lift force is given by[23]

$$F_s = 1.61(\mu\rho_g)^{1/2}d_p^2(v_g - v_p)|dv_g/dy|^{1/2} \tag{31}$$

which is valid only for low Reynolds numbers (<1). There appears to be no data for the Saffman lift force at higher Reynolds numbers. Once again, this force becomes significant only in high-shear regions.

To summarize, the largest aerodynamic force acting on a particle in a pulverized-coal combustor is the conventional aerodynamic drag, modified by nonspherical and mass-transfer transpiration effects.

3. Gas–Particle Heat Transfer

The energy equation for an individual particle was given in Chapter 2, Eq. (92), as

$$\dot{Q}_p + (dm/dt)(h_s - h_p) = m(di_p/dt) \tag{32}$$

where the kinetic energy associated with the efflux velocity was neglected

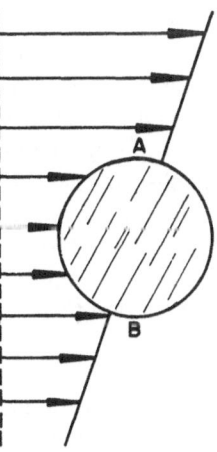

Figure 1. Schematic diagram of a particle in a velocity gradient field.

compared to the thermal energy. The purpose of this section is to investigate the nature of the term $\dot{Q}_p$, the heat transfer rate to the particle.

3.1. Convective Heat Transfer

The convective heat transfer rate to a spherical particle is quantified by the Nusselt number, defined as

$$\mathrm{Nu} = \dot{Q}_p d_p / S_p k (T_g - T_p) \tag{33}$$

where d_p is the particle diameter, S_p is the surface area of the particle, k is the thermal conductivity of the gas, and $(T_g - T_p)$ is the gas–particle temperature difference. If the particle is not spherical, it is suggested that the diameter of a sphere of equivalent surface area be used to represent the particle.

The Nusselt number is primarily a function of Mach number and Reynolds number, based on the relative (particle to gas) velocity. For incompressible flow, the Nusselt number data of a sphere are well correlated by the empirical relation[24]

$$\mathrm{Nu}_0 = 2 + 0.654 \mathrm{Re}^{0.5} \mathrm{Pr}^{1/3} \tag{34}$$

where the subscript 0 signifies incompressible flow, and Pr is the Prandtl number of the gas. In the subsonic flow regime, Kavanau[25] has obtained experimental results for the Nusselt number of a spherical particle for Mach numbers less than 0.7 and Reynolds numbers from 2 to 1000. Results indicate a variation of Nusselt number which can be expressed empirically as

$$\mathrm{Nu} = \mathrm{Nu}_0 / [1 + 3.42 (M \mathrm{Nu}_0 / \mathrm{Re} \mathrm{Pr})] \tag{35}$$

where M is the Mach number. For small Reynolds numbers, this equation approaches

$$\mathrm{Nu} \simeq 0.292 (\mathrm{Re} \mathrm{Pr} / M) \tag{36}$$

which agrees closely with the limit for free molecular flow and indicates the validity of Eq. (35) for all Reynolds numbers.

Other factors which can influence the Nusselt number are mass transfer from the surface and particle rotation. No data are available on the effect of mass transfer on the convective heat transfer to a coal particle. The burning of a coal particle occurs in two stages: rapid release of the volatiles followed by a slower burning of the remaining char. Assuming spherical symmetry, the Nusselt number is affected by mass transfer from the surface according to the relation[26]

$$\mathrm{Nu}_b = \mathrm{Nu}_{0m} \phi_m / [\exp(\phi_m) - 1] \tag{37}$$

where ϕ_m is the mass flow parameter defined by Eq. (15) and Nu_{0m} is the Nusselt number with no mass transfer, and is based on properties corresponding to the mean temperature between the particle and the gas. Order-of-magnitude calculations indicate that during coal devolatilization, the mass flow parameter, ϕ_m, may be of the order of unity, signifying the importance of blowing on convective heat transfer to the particle. It is advisable, then, to use Eq. (37) to account for the effect of blowing at the particle's surface on convective heat transfer during devolatilization. During the char-burning period, ϕ_m is small and effects of blowing are likely to be insignificant.

Rotation causes the convective heat transfer to increase after the surface speed due to rotation exceeds the free stream velocity by approximately 50%.[27] It is unlikely that this phenomenon is important in coal-particle combustion.

3.2. Radiative Heat Transfer

Radiative heat transfer occurs by absorption and emission from the particle. Assuming the particle can be represented as a gray body, the heat emitted by the particle due to radiation is $A_s \varepsilon \sigma T_p^4$, where ε is the emissivity and σ is the Stefan–Boltzmann constant. The amount of energy absorbed by a particle due to radiation is $A_s \alpha I$, where I is the intensity of radiation averaged over the particle's surface and α is the absorptance. Equating the emittance and absorptance by Kirchhoff's law, the net heat transfer to a particle of diameter d_p due to radiation is given by

$$\dot{Q}_{pr} = \varepsilon \pi d_p^2 (I - \sigma T_p^4) \tag{38}$$

assuming the particle temperature is uniform. Because the particle sizes are very small in pulverized-coal applications, cloud effects are important. Radiative effects of these coal–char–ash clouds are treated in detail in Chapter 5.

4. Gas–Particle Mass Transfer

Computation of char oxidation rates requires oxidizer mass transfer rates from the bulk gas to the particle surface. Treatment of convective mass transfer is closely analogous to that of convective heat transfer presented in the preceding section, as demonstrated theoretically by Bird *et al.*[26] Thus, from Eq. (34), the mass-transfer coefficient to spheres in the absence of surface transpiration effects is

$$(Nu_0)_{AB} = 2 + 0.654 Re^{0.5} Sc^{1/3} \tag{39}$$

where $(Nu_0)_{AB}$ is the Nusselt number for mass transfer (or the Sherwood number), which is equal to $(k_x d_p / c D_{im})$, and Sc is the Schmidt number. Effects of surface transpiration are estimated from Eq. (37), where Nu is replaced by

$(\text{Nu})_{AB}$. Further discussion of the prediction of mass transfer to and from char particle surfaces is presented in Chapter 13.

5. Notation

Ac	Acceleration modulus	Sc	Schmidt number
A_p	Particle projected area (m²)	S_s	Surface area of equivalent sphere (m²)
B	Burning parameter	S_p	Surface area of particle (m²)
c	Molar concentration (kg mol m⁻³)	t	Time (s)
c_k	Empirical coefficient	T	Temperature (K)
C_p	Specific heat capacity (J kg⁻¹ K⁻¹)	T_k	Turbulence time scale (s)
C_D	Drag coefficient	v	Velocity (m s⁻¹)
d_p	Particle diameter (m)	V_p	Particle volume (m³)
D_{im}	Mass diffusivity (m² s⁻¹)	x	Mass concentration (kg m⁻³)
f	Particle drag factor	α	Acceleration (m s⁻²)
F	Force (N)	ε	Emissivity
h	Specific enthalpy (J kg⁻¹)	ε_p	Particle diffusivity (m² s⁻¹)
h_s	Specific enthalpy at particle surface (J kg⁻¹)	θ	Sphericity factor
i	Specific internal energy (J kg⁻¹)	λ	Mean free path (m)
I	Radiation intensity (W m⁻²)	μ	Viscosity (N-s m⁻²)
$\mathbf{j}_p$	Particle flux vector (kg m⁻² s⁻¹)	ν	Kinematic viscosity (m² s⁻¹)
k	Turbulence kinetic energy (m² s⁻²)	ρ	Density (kg m⁻³)
k_m	Mean thermal conductivity (W m⁻¹ K⁻¹)	σ	Stefan–Boltzmann constant (W m⁻² K⁻⁴)
k_x	Mass transfer coefficient (kg mol m⁻² s⁻¹)	τ	Particle relaxation time (s)
K_B	Basset coefficient	ϕ	Burning rate parameter
K_M	Virtual mass coefficient	ω	Rotation rate (s⁻¹)
L	Latent heat of fusion (J kg⁻¹)		
m	Particle mass (kg)		

Subscripts

$\dot{m}$	Particle burning rate (kg s⁻¹)	b	burning
Nu	Nusselt number	g	gas
p	Pressure (Pa)	i	ith gas species
Pr	Prandtl number	m	based on mean temperature
$\dot{Q}$	Heat transfer rate (W)	m	mixture
r	Random unit vector	p	particle
Re	Reynolds number	r	relative

6. References

1. C. M. Tchen, Mean Value and Correlation Problems Connected with the Motion of Small Particles Suspended in a Turbulent Fluid, Ph.D. Thesis, Delft, The Hague (1947).
2. S. K. Friedlander, Behavior of suspended particles in a turbulent fluid, *AIChE J.* **3**(3), 381–385 (1957).
3. S. L. Soo, *Fluid Dynamics of Multiphase Systems*, Blaisdell, Waltham, Mass. (1967).
4. V. W. Goldschmidt and S. Eskanazi, Two-phase turbulent flow in a plane jet, *J. App. Mech.* **33**(E) (4), 735–747 (1966).

5. V. W. Goldschmidt, M. Householder, G. Ahmadi, and S. C. Chuang, Turbulent diffusion of small particles suspended in turbulent jets, in *Progress in Heat and Mass Transfer*, Vol. 6 (G. Hetsroni, S. Sideman, and J. P. Hartnett, eds.), Pergamon Press, New York, 487–508 (1972).

6. G. P. Lilly, Effect of particle size on particle eddy diffusivity, *Ind. Eng. Chem. Fundam.* **12**(3), 269–275 (1973).

7. J. O. Hinze, Turbulent fluid and particle interaction, in *Progress in Heat and Mass Transfer, Vol. 6 (G. Hetsroni, S. Sideman, and J. P. Hartnett, eds.)*, Pergamon Press, New York, 433–452 (1972).

8. P. O. Hedman and L. D. Smoot, Particle-gas dispersion effects in confined coaxial jets, *AIChE J.* **21**(2), 372–379 (1975).

9. V. J. Memmott and L. D. Smoot, Cold flow mixing rate data for pulverized coal reactors, *AIChE J.* **24**(3), 466–474 (1978).

10. C. L. Tice and L. D. Smoot, Cold-flow mixing rates with recirculation for pulverized coal reactors, *AIChE J.* (1978), accepted for publication.

11. R. Peskin and C. J. Kau, Numerical Simulation of Particle Motion in Turbulent Gas-Solid Channel Flow, ASME Paper 76-WA/RE-37, ASME Winter Annual Meeting, New York (1976).

12. J. W. Deardorff, A numerical study of three-dimensional turbulent channel flow at large Reynolds numbers, *J. Fluid Mech.* **41**(2), 453–480 (1970).

13. B. E. Launder and D. B. Spalding, *Mathematical Models of Turbulence*, Academic Press, London (1972).

14. C. T. Crowe, W. Babcock, and P. G. Willoughby, Drag coefficient for particles in rarefied, low Mach number flows, in *Progress in Heat and Mass Transfer*, Vol. 6 (G. Hetsroni, S. Sideman, and J. P. Hartnett, eds.), Pergamon Press, New York, 419–431 (1972).

15. G. B. Wallis, *One-Dimensional Two-Phase Flow*, McGraw–Hill Book Co., New York (1969).

16. C. T. Crowe, J. A. Nicholls, and R. B. Morrison, Drag coefficients of inert and burning particle accelerating in gas streams, in *Ninth International Symposium on Combustion*, Academic Press, New York, 395–406 (1963).

17. P. Eisenklam, S. A. Arunachalam and J. A. Weston, Evaporation rate and drag resistance of burning drops, in *Eleventh International Symposium on Combustion*, The Combustion Institute, Pittsburgh Pa., 715–728 (1967).

18. R. G. Boothroyd, *Flow Gas-Solids Systems*, Chapman and Hall, London (1971).

19. H. Wadell, The coefficient of resistance as a function of Reynolds number for solids of various shapes, *J. Franklin Inst.* **217**(4), 459–490 (1934).

20. W. H. M. Robins, The significance and application of shape factors in particle size analysis, *Brit. J. Appl. Phys.*, Supplement No. 3, S82–S85 (1954).

21. F. Odar, Verification of proposed equation for calculation of forces on a sphere accelerating in a viscous fluid, *J. Fluid Mech.* **25**(3), 591–592 (1966).

22. A. B. Basset, *A Treatise on Hydrodynamics*, Deighton, Bell and Co., Cambridge (1888); republished by Dover Publications, New York (1961).

23. P. G. Saffman, The lift on a small sphere in a slow shear flow, *J. Fluid Mech.* **22**(3), 385–400 (1965). Correction appeared *J. Fluid Mech.* **31**(3), 624 (1968).

24. P. N. Rowe, K. T. Claxton, and J. B. Lewis, Heat and mass transfer from a single sphere in an extensive flowing fluid, *Trans. Inst. Chem. Eng.* **43**(1), T14–T31 (1965).

25. L. L. Kavanau, Heat transfer from spheres to a rarefied gas in subsonic flow, *Trans. ASME* **77**, 617–624 (1955).

26. R. B. Bird, W. E. Stewart, and E. N. Lightfoot, *Transport Phenomena*, John Wiley and Sons, Inc., New York (1960).

27. T. D. Eastop, The influence of rotation on the heat transfer from a sphere to an air stream, *Int. J. Heat and Mass Trans.* **16**, 1954–957 (1973).

Part III
Coal Characteristics and Rate Processes

7

General Characteristics of Coal

L. Douglas Smoot

1. Formation and Variation

Coal is a black, inhomogeneous, organic fuel, formed largely from partially decomposed and metamorphosed plant materials. Formation has occurred over long time periods, often under high pressures of overburden and at elevated temperatures. Differences in plant materials and in their extent of decay influences the components present in coals, such as vitrain (from "vitro" meaning glass), clarain (from "clare" meaning clear or bright), durain (from "dur" meaning hard or tough), and fusain (meaning charcoal). Descriptions of such coal components are part of the science of petrography.[1]

Coals vary greatly in their composition. Of 1200 coals categorized by the Bituminous Coal Research Institute, no two had the same composition.[1] Typical compositions (mass percentages) of coal include 65–95% carbon, 2–7% hydrogen, up to 25% oxygen and 10% sulfur, and 1–2% nitrogen.[2] Inorganic mineral matter (ash) as high as 50% has been observed, but 5–15% is more typical. Moisture levels commonly vary from 2 to 20%, but values as high as 70% have been observed.

The process of conversion of plant materials such as peat to coal is called "coalification" and takes place in stages, producing a variety of coal products. Hendrickson[1] provides the following description of some of these coal types:

L. Douglas Smoot • Professor of Chemical Engineering, Brigham Young University, Provo, Utah

Lignite, the lowest rank of coal, was formed from peat which was compacted and altered. Its color has become brown to black and it is composed of recognizable woody materials imbedded in pulverized (macerated) and partially decomposed vegetable matter. Lignite displays jointing, banding, a high moisture content, and a low heating value when compared with the higher ranks of coal.

Subbituminous coal is difficult to distinguish from bituminous and is dull, black colored, shows little woody material, is banded, and has developed bedding planes. The coal usually splits parallel to the bedding. It has lost some moisture content, but is still of relatively low heating value.

Bituminous coal is dense, compacted, banded, brittle, and displays columnar cleavage and a dark black color. It is more resistant to disintegration in air than are subbituminous and lignitic coals. Its moisture content is low, volatile matter content is variable from high to medium, and its heating value is high. Several varieties of bituminous coal are recognizable.

Anthracite is the highly metamorphosed coal, is jet black in color, is hard and brittle, breaks with a conchoidal fracture, and displays a high luster. Its moisture content is low and its carbon content is high.

Neither peat nor graphite are coal, but they are the initial and end products of the progressive coalification process.

Lowry,[3] Hendrickson,[1] and Given[4] discuss in some detail the origin and characteristics of coal. Pennsylvania State University has undertaken a study to determine, categorize, and store information for U.S. coals.[5]

2. Coal Classification

Efforts have been made to classify the almost limitless number of coals into broad classifications, and to relate similarities among coals to their potential behavior in coal conversion processes. Possibly, the most common of these is the ASTM Classification, which is based upon fixed carbon level and heating value. Figure 1 illustrates the general characteristics of 12 such coal groups, ranging from soft lignite to very hard meta-anthracite.

Classifications have also been based on petrographic parameters. For example, a correlation has been observed between chemical rank and maximum mean reflectance of incident light by coal samples.[1] Lowry[3] summarizes several other systems for general classification of coal including:

(a) English National Coal Board System, based upon percent of proximate volatiles and Gray–King coking properties.
(b) International System for hard coals and brown coals.
(c) Mott System, based upon volatile matter, heating values, and ultimate (O–H) analysis.

None of these approximate systems are able to deal with the complex structural and compositional differences in coals. The Pennsylvania State study[5] is attempting to predict the behavior of coals during conversion processes, from a knowledge of coal composition.

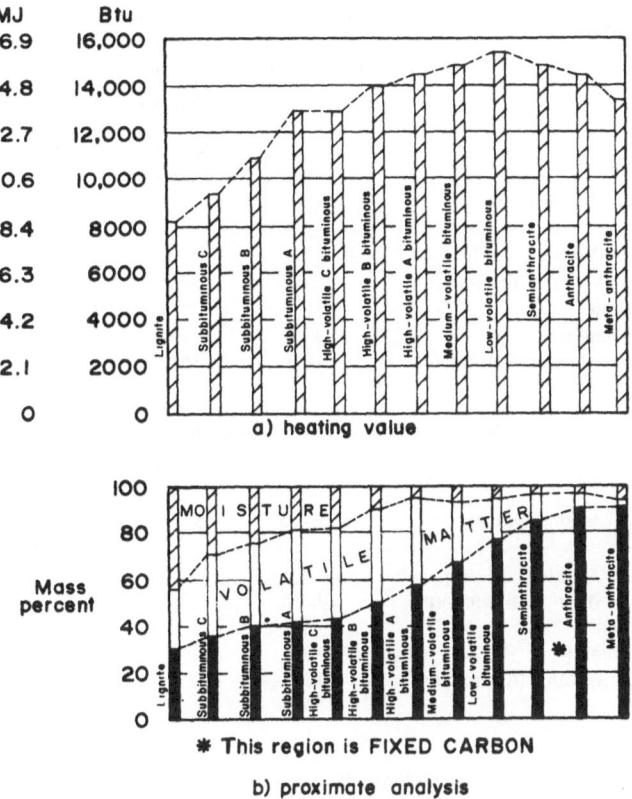

Figure 1. Energy content and composition of coals according to ASTM coal rank. (Figure used with permission from Hendrickson.[1])

3. Coal Physical and Chemical Properties

Modeling of coal conversion processes requires data for physical properties of coal, such as thermal conductivity, specific heat, density, etc. Table 1 summarizes selected values of common physical properties of coal, including specific heat, specific gravity, thermal conductivity, and swelling index. Heating value data are shown in Figure 1. These properties will vary among coals, even of common rank, and will often be related to temperature and moisture content. Lowry[3] and Hendrickson[1] discuss several other physical, mechanical, and thermal properties of coal, including grindability, friability, compressive strength, dustiness, electrical resistivity, plasticity, optical density, indices of refraction, reflection, and absorption, magnetic susceptibility, electrical conductivity, and dielectric constants.

Table 1. Typical Values for Selected Coal Physical Properties[a]

(a) Specific Heats of Air-Dried Coals

	Proximate analysis			
Coal sample source	Moisture (%)	Volatile matter (%)	Carbon (%)	Ash (%)
West Virginia	1.8	20.4	72.4	5.4
Pennsylvania (Bituminous)	1.2	34.5	58.4	5.9
Illinois	8.4	35.0	48.2	8.4
Wyoming	11.0	38.6	40.2	10.2
Pennsylvania (Anthracite)	0.0	16.0	79.3	4.7

Mean specific heat for °C temperature ranges

	Temperature range (°C)			
Coal sample source	28–65	25–130	25–177	25–227
West Virginia	0.261	0.288	0.301	0.314
Pennsylvania (Bituminous)	0.286	0.308	0.320	0.323
Illinois	0.334			
Wyoming	0.350			
Pennsylvania (Anthracite)	0.269			

(b) Specific Gravity[b] of Coal

Probable rank	Specific gravity
Anthracite	1.7
Semianthracite	1.6
Bituminous	1.4
Subbituminous	1.3
Lignite	1.2

(c) Thermal Conductivity[c]

Coal type	Temperature (K)	$kJ\ s^{-1}\ m^{-1}\ K^{-1}$
Monolithic anthracite	303	0.2 –0.4
Monolithic bituminous	303	0.17–0.3
Pulverized bituminous	Ambient	0.10–0.15

———— continued

Table 1 (Continued)

(d) *Average Free-Swelling Index Values for Illinois and Eastern Bituminous Coals*

Rank	Coals	ASTM free-swelling index
High-volatile C	Illinois No. 6	3.5
High-volatile B	Illinois No. 6	4.5
High-volatile B	Illinois No. 5	3.0
High-volatile A	Illinois No. 5	5.5
High-volatile A	Eastern	6.0–7.5
Medium-volatile	Eastern	8.5
Low-volatile	Eastern	8.5–9.0

[a]Table used with permission from Hendrickson.[1]
[b]Specific gravity has been shown by Lowry[3] to be a function of hydrogen content.
[c]Thermal conductivity usually increases with increasing apparent density, volatile matter content, ash content, temperature, and probably with moisture content; coal is thermally anisotropic with k greater perpendicular to bed.

Properties of char are even more variable, since such properties are a function of the nature of the conversion process from which they were produced, in addition to being related to the coal from which they were derived. Chars are generally richer in carbon and leaner in hydrogen than the parent coal and are often porous and more regular in shape, having softened during the conversion process.

The composition of coal is traditionally characterized by ASTM proximate analysis or ASTM ultimate analysis. The former determines only the moisture content (by drying), percent volatiles (from inert devolatilization at about 900°C), ash (residual after complete combustion in air), and fixed carbon (by difference). Coal rank vs. proximate analysis is shown in Figure 1. Proximate analysis for a selected variety of coals is shown in Table 2. In these coals alone, percent volatiles varies from 8.8 to 45.5% by weight. The char shown has only 2.4% proximate volatiles.

ASTM ultimate analysis gives elemental analyses for carbon, hydrogen, nitrogen, sulfur, and oxygen, the latter often determined by difference. The residual mineral matter is shown as ash. Ultimate analyses for selected coals and a char are also shown in Table 2. Ash in coals has been shown to contain significant amounts of several elements, together with trace amounts of several elements, as shown in Table 3 for selected coals. Padia *et al.*[6] discuss particle sizes of ash from pulverized coals and note that some ash components are volatile at higher temperatures. Further details of ash formation are discussed in Section 4 of Chapter 11.

Particle sizes of coal dust vary greatly, depending upon grinding technique and desired application. Typical size distributions for a fluid-bed gasifier application and a utility boiler application are shown in Table 4.

Table 2. Typical Proximate and Ultimate Analyses for Coals and Char[a]

Coal I.D. rank	Utah Church Mine, bituminous	Pittsburgh,[b] bituminous	Pittsburgh,[b] bituminous (high-volatile)	Sewell[b] bituminous (medium-volatile)	Anthracite[b] (low-volatile)	Illinois coal, bituminous	Illinois[c] coal char
Moisture %	2.5–2.7	2.0	1.0	1.9	1.3	10.1	0.9
Proximate %							
Volatiles	44.1–45.5	36.6	28.9	16.3	8.8	39.9	2.4
Fixed carbon	42.6–44.2	55.4	63.2	75.6	71.8	52.0	76.8
Ash	9.2–9.5	6.0	6.9	6.2	18.1	8.1	20.8
Ultimate %							
Carbon	69.8–71.5	77.5	80.6	84.2	73.2	68.3	74.0
Hydrogen	5.5–5.6	5.3	4.9	4.3	3.1	5.0	0.7
Nitrogen	1.4–1.5	1.5	1.5	1.2	0.9	1.3	1.0
Sulfur	0.4–0.7	1.2	0.7	0.7	0.9	3.5	3.3
Oxygen	11.2–13.2	8.5	5.4	3.4	3.8	13.8	0.2
Ash	9.2–9.5	6.0	6.9	6.2	18.1	8.1	20.8

[a]Data courtesy of Bureau of Mines, Brigham Young University, and ERDA.
[b]From U.S. Bureau of Mines, Pittsburgh, Pa.
[c]From COED Gasification, supplied by ERDA, Pittsburgh, Pa. (4th stage product).

Table 3. *Compositions of Typical Ashes*[a]

(a) Variations in Coal Ash Compositions with Rank

Rank	SiO_2 (%)	Al_2O_3 (%)	Fe_2O_3 (%)	TiO_2 (%)	CaO (%)	MgO (%)	Na_2O (%)	K_2O (%)	SO_3 (%)	P_2O_5 (%)
Anthracite	48–68	25–44	2–10	1.0–2	0.2–4	0.2–1	—	—	0.1–1	—
Bitumirous	7–68	4–39	2–44	0.5–4	0.7–36	0.1–4	0.2–3	0.2–4	0.1–32	—
Subbituminous	17–58	4–35	3–19	0.6–2	2.2–52	0.5–8	—	—	3.0–16	—
Lignite	6–40	4–26	1–34	0.0–0.8	12.4–52	2.8–14	0.2–28	0.1–1.3	8.3–32	—
Utah bituminous[b]	43–48	16–19	3.8–4.2	0.0–1.0	6.5–8.1	0.9–1.1	4.3–4.9	0.4–0.7	3.5–4.1	<1.0

continued overleaf

Table 3 (Continued)

(b) Range of Amount of Trace Elements Present in Coal Ashes (ppm on Ash Basis)

Element	Anthracites	High volatiles, bituminous	Low volatiles, bituminous	Medium volatiles, bituminous	Lignites and subbituminous	Utah[b] bituminous
Ag	1	1–3	1–1.4	1	1–50	—
B	63–130	90–2800	76–180	74–780	320–1900	700–1500
Ba	540–1340	210–4660	96–2700	230–1800	550–13900	700–1500
Be	6–11	4–60	6–40	4–31	1–28	5–7
Co	10–165	12–305	26–440	10–290	11–310	7–15
Cr	210–395	74–315	120–490	36–230	11–140	70–100
Cu	96–540	30–770	76–850	130–560	58–3020	62–68
Ga	30–71	17–98	10–135	10–52	10–30	30–70
Ge	20	20–285	20	20	20–100	—
La	115–220	29–270	56–180	19–140	34–90	70–100
Mn	58–220	31–700	40–780	125–4400	310–1030	400
Ni	125–320	45–610	61–350	20–440	20–420	15–30
Pb	41–120	32–1500	23–170	52–210	20–165	35–45
Sc	50–82	7–78	15–155	7–110	2–58	15
Sn	19–4250	10–825	10–230	29–160	10–660	—
Sr	80–340	170–9600	66–2500	40–1600	230–8000	1000–1500
V	210–310	60–840	115–480	170–860	20–250	150
Y	70–120	29–285	37–460	37–340	21–120	50–70
Yb	5–12	3–15	4–23	4–13	2–10	7
Zn	155–350	50–1200	62–550	50–460	50–320	58–64
Zr	370–1200	115–1450	220–620	180–540	100–490	200–300
Cd	—	—	—	—	—	<1
Li	—	—	—	—	—	133–155
Nb	—	—	—	—	—	20–30

[a]Table used with permission from Hendrickson.[1]

[b]Utah Power and Light Coal, Church Mine, Analyses by U.S. Geological Survey, Denver, Colorado.

Table 4.   Typical Size Distributions for Pulverized Coal

(a) Fluid-Bed Gasifier[a]

Tyler screen	Illinois bituminous	Illinois char
14	15.2	3.3
28	47.0	28.7
48	69.0	54.0
100	81.9	72.1
200	90.2	84.2
325	94.2	91.2
Pan	99.7	99.8

(b) Utility Boiler[b] (Utah Bituminous[c])

Increment size (μm)	Percentage in increment[d]	Increment size (μm)	Percentage in increment[d]
2.85	0.3–0.5	22.80	7.1–9.0
3.59	0.4–0.6	28.70	9.0–11.1
4.52	0.5–0.7	36.15	10.9–13.3
5.70	0.7–1.0	45.55	10.7–12.4
7.18	1.2–1.6	57.40	11.5–12.1
9.04	2.0–2.5	72.30	9.2–12.4
11.39	3.0–3.7	85.30	5.3–8.3
14.35	4.1–5.1	90+	7.4–12.9
18.10	5.5–6.9		

[a]Courtesy of ERDA.
[b]Termed 70% through 200 mesh.
[c]Church Mine, BYU data.
[d]Mass mean diameter is about 50 μm (Coulter counter measurement).

The coal dust size is much smaller in the latter case. Size distribution from one specific char is also shown in Table 4.

It is thought that coal structure is highly planar and layered with pore volume of 8–20%. Figure 2 shows one model for the detailed molecular structure of a typical coal. Hendrickson,[1] Anthony and Howard,[7] Lowry,[3] and Given[8] discuss additional chemical properties of coal, including details of proximate and ultimate analyses, plastic properties of coal, coal hydrogenation and halogenation, and solvent extraction of coal components, including properties of minerals and coal structure.

Presently, design and analysis of existing coal-conversion processes do not reflect significant use of available information concerning the structure of coals. However, with increased interest in the use of coal in the United States, it is anticipated that more information on coal structure will be forthcoming.

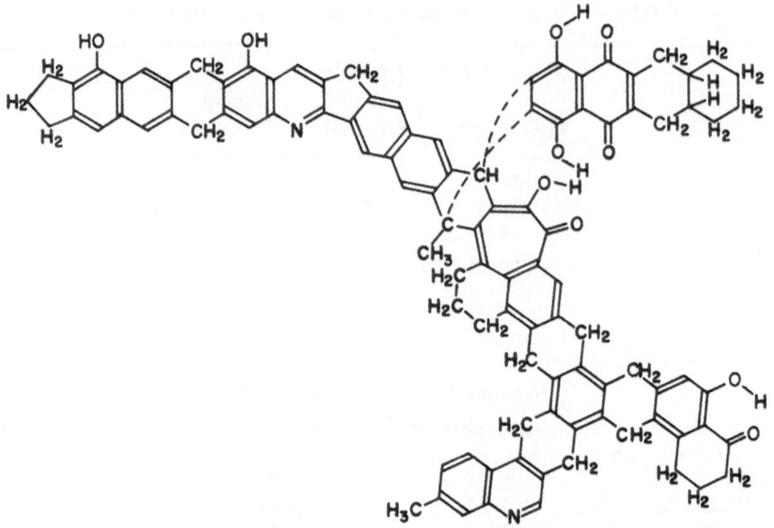

Figure 2. Proposed model for molecular structure of coal. Based upon a vitrinite of 82% C (dry, mineral-matter-free basis): $C_{102}H_{78}O_{10}N_2$. Figure used with permission from Given.[8]

4. References

1. T. A. Hendrickson (ed.), *Synthetic Fuels Data Handbook*, Cameron Engineers, Inc., Denver, Colo. (1975).
2. R. H. Essenhigh, Combustion and flame propagation in coal systems, in *Sixteenth Symposium (International) on Combustion*, pp. 353–374, Combustion Institute, Pittsburgh, Pa. (1977).
3. H. H. Lowry (ed.), *Chemistry of Coal Utilization*, Supplementary Volume, John Wiley and Sons, New York (1963).
4. P. H. Given (ed.), *Coal Science, American Conference on Coal Science*, Advances in Chemistry, Series 55, American Chemical Society, Washington, D.C. (1964).
5. W. Spackman, A. Davis, P. L. Walker, H. L. Lovell, R. H. Essenhigh, F. J. Vastola, and P. H. Given, *Characteristics of American Coals in Relation to Their Conversion into Clean Energy Fuels*, Quarterly Progress Report Fe-2030-3, U.S. ERDA Contract FE-2030-3, Pennsylvania State University, University Park, Pa. (1976).
6. A. S. Padia, A. F. Sarofim, and J. B. Howard, Behavior of Ash in Pulverized Coal under Simulated Combustion Conditions, paper given at the Spring Meeting of the Central States Section, The Combustion Institute, Pittsburgh, Pa. (April 1976).
7. D. B. Anthony and J. B. Howard, Coal devolatilization and hydrogasification, *AIChE J.* **22**, 625 (1976).
8. P. H. Given, The distribution of hydrogen in coals and its relation to coal structure, *Fuel* **39**, 147 (1960).

Fast Pyrolysis

M. Duane Horton

1. Introduction

In this chapter, coal pyrolysis is reviewed and then the modeling of this phenomenon is discussed. The intent is not to describe the entire subject of coal pyrolysis, which is reviewed elsewhere,[1-7] but to consider only rapid pyrolysis relevant to the combustion of pulverized coal. Within this context, the parameters of interest include particle sizes of 100 μm or less, heating rates of at least 10^4 K s^{-1}, final temperatures of 1000 K or greater, and residence times no.greater than a few seconds.

Pyrolysis (or devolatilization) as used in this chapter, refers to the destructive distillation of coal. *Volatiles* are the primary gaseous decomposition products, some of which will be liquids or even solids at ambient temperature and pressure. *Char* is the residual solid matter following fast pyrolysis, which may or may not yield further volatiles if the heating process is continued. Should the char be heated in the presence of air, all of the combustible material will be consumed, leaving a residual mineral matter called *ash*.

Even under the restrictions of small particles with high heating rates, the pyrolysis of coal is a very complex process. This is partly due to the fact that coal is not a homogeneous material; different portions (both microscopic and macroscopic) of a single coal sample exhibit widely differing chemical compositions and physical properties, as discussed in Chapter 7. In addition, samples from different portions of a mine will not be identical.

M. Duane Horton • Professor of Chemical Engineering, Brigham Young University, Provo, Utah

Furthermore, there is a broad range of coal types (bituminous, anthracite, etc.), each of which decomposes in a slightly different manner. Thus, it is possible to characterize only the "average" pyrolysis of coal particles of a given type in terms of the products produced as a function of the experimental variables.

In the following sections, the experiments used to study the pyrolysis of pulverized coal will be described. Next, the resulting data will be discussed. Finally, existing theoretical models of the process will be reviewed, with one selected that is sufficiently general to describe many aspects of coal pyrolysis, but simple enough for practical usage.

2. Experimental Results

Whether or not they are controlled in a given experiment, variables known to be important to the pyrolysis process are coal dust type, concentration, source, particle size, and size distribution. Also, the thermal history of the particles, including heating rate, final temperature, duration of heating, and the quench process, are influential. In addition, the composition and pressure of the ambient gas are significant.

To date, almost all experimenters have relied upon the collection and analysis of quenched samples of the gas and/or char from a pyrolysis experiment. The composition of the char has usually been determined by proximate or ultimate analysis. The latter yields elemental compositions and the former a determination of volatiles, fixed carbon and ash, as described in Chapter 7. The volatiles can be analyzed for molecular content but most often investigators have been content to simply evaluate the quantity of gasified material. This is because the primary volatiles often condense or react further so that existing species are only indirectly indicative of the primary products.

In this section, the experiments used to study the pyrolysis of pulverized coal, as well as the results obtained, will be described.

2.1. Experimental Methods

Primarily, detailed measurements relevant to pulverized-coal pyrolysis have been obtained from three types of experiments. In the first type, the coal dust is embedded in the pores of a wire screen that is heated electrically. This technique is particularly advantageous in that both the heating rate and ambient atmosphere can be controlled and are not dependent on the pyrolysis process. Probably the largest uncertainty in the experiment is that the coal particles must be heated by the screen and there is an unknown lag time and temperature difference involved. Earlier, it was believed that the screen might catalyze secondary reactions[7] but more recent evidence[8] suggests that such is not the case. References 8–11 report results from this type of experiment. The authors of references 8 and 11 refined the experiment so that the heating rate and final temperature could be independently controlled.

The second type of experiment[12–15] involves injecting the coal particles into a preheated gas. Thus the ambient atmosphere, final temperature, and coal-dust concentration are readily controlled. However, the heating rate of the particles is not precisely known because it is dependent on the ambient conditions experienced by the particle.

In the third type of experiment,[16–23] coal dust is burned in a flame. The primary advantage of this experiment is that it most nearly resembles the commercial processes using coal. The principal disadvantage, aside from the difficulty in designing the burner, is that the only independent variables are those of the feedstream. Heating rate, final temperature, and ambient atmosphere are then all determined by the resultant flame.

2.2. Data

Loison and Chauvin[9] reported the important result that a higher heating rate and final temperature produced a greater yield of volatile products. Because the heating rate and final temperature of the wire grid were not independently varied, it is not clear which variable was responsible for producing the high yield of volatiles.

Subsequent work has been directed toward obtaining a better understanding of this phenomenon, both because of its fundamental importance and because of the potential application to gasification processes. Cumulative results now indicate that dispersed coal particles pyrolyze at atmospheric pressure by at least a two-step process that is best understood by considering the experiments involved. Generally, the coal is heated rapidly to a high temperature which is maintained for a short period of time. Then the particles are cooled and collected, and the residual volatile matter is determined by proximate analysis which involves a relatively slow heating to a temperature of 1223 K and a total pyrolysis time of 7 min.

Often, a quasi-equilibrium state is observed before the particles are quenched for subsequent proximate analysis. Judged by the reaction occurring in the experiment, the pyrolysis process appears complete. However, additional devolatilization occurs when the quenched char particles are reheated in the proximate furnace. The implication is that if the particles had been maintained longer at the elevated temperature in the primary reactor (wire screen, hot gas, or flame), the additional pyrolysis would have occurred there. Thus, there is a second, slower step to the pyrolysis that continues during the proximate analysis.

Most of the attention has been focused on the first, relatively fast step in the reaction. This is because it is difficult to achieve residence times in the

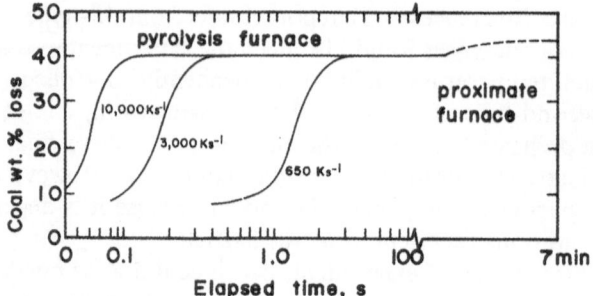

Figure 1. Devolatilization histories observed at different heating rates to a final temperature of 1273 K. The pyrolysis furnace data are from reference 11. Proximate furnace data are a qualitative representation of results from reheating the quenching char to 1223 K.

reactor long enough for the second reaction to take place. This may be illustrated by data summarized in Figure 1, which were obtained with a heated wire grid.[11] These data show the second reaction to be so slow that it is usually not significant during fast pyrolysis. The experimental data points result from different heating rates and reaction times ranging up to 20 s. At 1273 K, the temperature at which the experiment was conducted, the coal showed a 41% loss of volatiles, which was apparently independent of heating rate or residence time. However, also shown is the 45% volatiles yield observed in the proximate analysis, which is conducted at the slightly lower temperature of 1223 K. The slow, second step in the pyrolysis would almost surely yield additional volatiles ($\simeq 4\%$) if the heating time were increased from 20 s to 7 min. This conclusion is supported by other experiments, and applies to other temperatures as well.

The most comprehensive data describing the first, fast step in the pyrolysis were obtained at BCURA[3,13] with a hot-gas particle heater. Subsequent experiments[16-22] with small flames have supported these results. Figure 2, taken from reference 3, illustrates the general pyrolysis behavior observed in the fast step. The residual volatiles in the char, as determined by proximate analysis, decrease exponentially with residence time in the reactor. However, as already mentioned, there are residual volatiles in the char, even though the devolatilization in the reactor has apparently reached completion. The rate of pyrolysis is less at lower temperatures and/or heating rates, while the amount of residual volatiles decreases with increasing reactor temperature. Both the rates of pyrolysis and the residual limit also depend on coal type.

Results obtained by Anthony et al.,[11] with coal particles suspended in a wire grid, did not entirely agree with the conclusions of the preceding paragraph. In Anthony's experiments, the degree of pyrolysis for the fast step was the same, regardless of the time required for the coal particles to reach a given temperature. This discrepancy has not yet been explained. However, the time dependence has been observed by most researchers, and from at least two types of experiments.

The preceding discussion was concerned with the residual volatile matter in the char. Of equal or even greater importance is the liberated volatile matter. It is generally agreed that rapid heating influences not only the amount of volatiles generated, but also the product composition. References 3, 9, 10, 11, and 24 show that high heating rates and final temperatures yield an increasing amount of volatiles that can considerably exceed, and in some cases double, the amount observed during proximate analysis. References 3 and 11 present the most extensive set of data, and both conclude that the maximum yield of volatile matter increases with heating rate and carbon content of the feed coal. It is significant that the same result was obtained from experiments with flames, hot inert gases, and heated wire grids. As mentioned earlier, Anthony *et al.*[11] concluded that the volatiles yield depends only on final temperature and that an increase in final temperature produces an increased yield. Later work in the same program[8] supported this result.

There is one experimental finding that appears to further conflict with the preceding description. Howard and Essenhigh[16] concluded that the ignition and early reaction of coal particles in a flame involves a heterogeneous reaction occurring at the solid surface. However, this conclusion was reached by assuming that the amount of volatiles reacted could be calculated as the difference between the proximate volatiles of the feed coal and the char. This is not a good assumption because, as has been shown, the proximate volatiles of the feed coal does not characterize the potential amount of the volatiles that may be obtained.

An interesting correlation presented in reference 3 relates the total weight loss to the residual volatiles in partially reacted char. Figure 3 presents

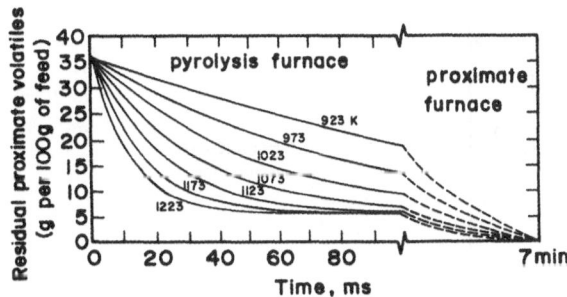

Figure 2. Typical devolatilization curves for different heating rates and final temperatures. The pyrolysis furnace data are from reference 3. The proximate furnace data are a qualitative representation.

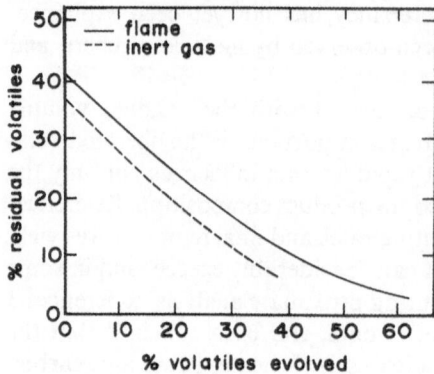

Figure 3. Comparison between residual volatiles in char and the volatiles evolved before quenching with data from inert coal pyrolysis experiments and coal–air flames.

the same data with modified coordinates. Included in Figure 3 are data from references 17 and 21. The correlation is broadly applicable, as both hot-gas and flame pyrolysis data are represented.

The volatile matter liberated in fast pyrolysis characteristically has a higher C/H ratio than that observed in proximate analysis. This is illustrated by Figure 4, which also shows that the volatiles first released are richer in carbon than those liberated later, with all being richer in carbon than the proximate volatiles.

Examination of the data in Figure 3 leads to the same interpretation even in the absence of detailed chemical analysis. Essentially all of the hydro-

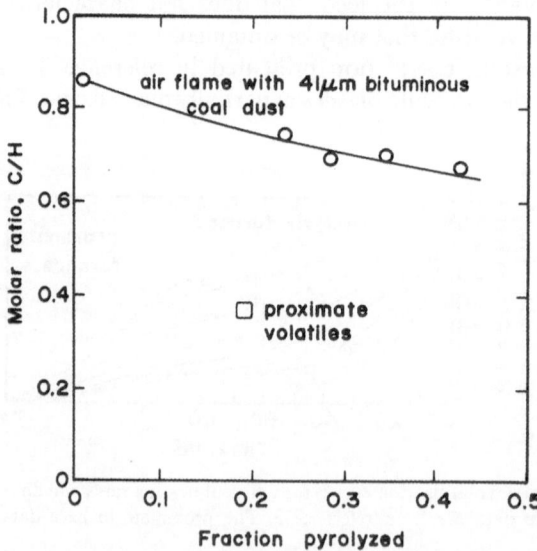

Figure 4. Composition of volatile matter as calculated from residual char composition.

gen is driven from the coal when the volatile matter is determined by proximate analysis. Thus, when the high-rate pyrolysis produces more volatiles than the proximate analysis, it means that the evolving hydrogen carries with it a larger amount of carbon so that the C/H ratio must be higher than the proximate value. Excess volatiles may be liberated even when there is residual hydrogen in the char, which also supports the conclusion and suggests an even higher C/H ratio.

Since the sort of data shown in Figure 3 emerges from flame experiments as well as those using a hot, inert gas as a heater, the composition of the emerging pyrolysis products is not strongly affected by the composition of the ambient gas. At least this is so over the range of conditions thus far studied, as judged by the residual char composition.[12,15,21] However, in some coal gasification experiments[5] the composition of the gaseous products varied greatly with the feed-gas composition. It is probable that the emerging volatiles react with the ambient gas so that the ultimate gas composition depends upon ambient-gas composition even though the primary volatile products do not.

Most of the experimental data cited above depend upon the use of ash as a tracer. This is satisfactory as long as the amount of residual ash is independent of the degree of pyrolysis or experimental procedure. The results of Padia *et al.*[25] suggest that this is not always the case but that it is a satisfactory approximation at lower temperatures.

In many experiments, the pyrolysis products have been sampled and then analyzed with a gas chromatograph. Measured compositions may not be a true indication of the primary pyrolysis products because secondary reactions may occur. Nevertheless, such results are the best information available. Table 1 shows that the products consist primarily of H_2O, CO_2, and CO, as well as light hydrocarbons that are primarily CH_4 and H_2, and heavy hydrocarbons called *tars*.

To some extent, the distribution of pyrolysis products may depend upon particle size or proximity. This conclusion is not supported by the data of references 12 and 21 (some of which are shown in Figure 3), which indicate that the pyrolysis products are independent of particle size. However, references 11 and 15 do report a particle size effect, and it is generally accepted

Table 1.   Apparent Products of Fast Pyrolysis at 1300 K

Reference	Coal dust	Ambient gas	Mole percent of product gas				
			Tar	CH_4	H_2	C_2-C_4	CO and CO_2
9	Bituminous	Vacuum	<9.9	50.3	13.1	—	26.7
5	Bituminous	O_2	—	8	59	—	26.7
10	Bituminous	Vacuum	53	<6	<31	<3	<8
8	Lignite	He	13	3	1	≃3	38

that very large particles behave differently than pulverized coal. The accepted explanation[7,8,11,15,26] is that the char surface provides a site where secondary reactions occur. Pyrolysis products generated near the center of a particle must migrate to the outside to escape. During this migration, they may crack, condense, or polymerize, with some carbon deposition taking place. The larger the particle, the greater the amount of deposition, and hence the smaller the volatiles yield. A high particle density would similarly provide a greater reactive surface which explains why volatiles yield also decreases as particle proximity increases.[15]

Anthony *et al.*[11] reported that a lower ambient pressure favors the liberation of a greater mass of volatiles. This appears to be consistent with other results, since a decrease in pressure also decreases the transit time of volatiles within the particle. A decrease in pressure thus has an effect analogous to a decrease in particle size. This explanation may explain the otherwise anomalous low-pressure results of Mentser *et al.*[10] that show a complex relation between temperature and total yield of volatiles.

As might be expected, the coal particles change not only in composition but also in appearance as the volatiles are driven off. Extensive data are available which describe the changes resulting as all of the combustibles are consumed, but comparatively less is known about the changes that happen early in the process while volatiles are still evolving. Based on samples obtained from flames,[17,23] there appears to be a systematic change in the coal particles. During early phases of pyrolysis, the particles become plastic,[23,27] lose their angularity, and become more spherical in shape. The size does not change greatly, with neither a large swelling nor shrinking occurring. As Figure 5 illustrates, a large fraction of the particles show the formation of "blow holes," presumably due to the escape of volatile matter from the interior of the plastic particles. This description results from a limited number of studies, and it is possible that other coal types, ambient atmospheres, particle diameters, or heating histories could show different results.

3. Pyrolysis Models

A complete pyrolysis model would describe the composition and physical state of the coal dust or char particle at all stages of pyrolysis. Also, the composition and evolution rate of the volatiles would be described. In case the solid phase reacted with the volatile products, the reaction and reaction rate would also be defined. Since these parameters depend on the thermal history of the particles, it is necessary to utilize the appropriate transport equations to calculate the temperature of the particles. The theoretical formulation must be sufficiently versatile to accommodate the

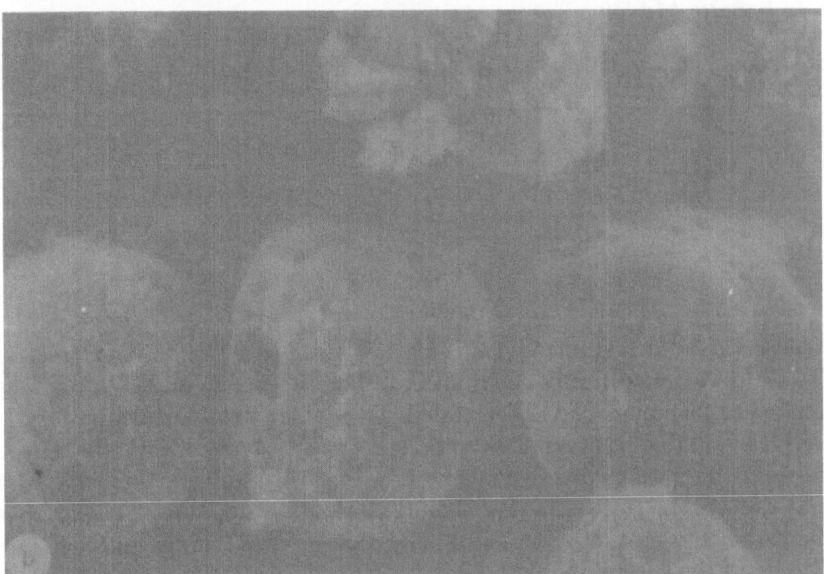

Figure 5. Selected scanning electron microscope pictures (33 μm Pittsburgh coal dust, 0.50 kg m^{-3}, 0.21 ms^{-1} flame). (a) Feed coal ($\times 200$); (b) 10 mm behind flame front ($\times 1000$). (Data from reference 20.)

different coal types, particle sizes, ambient atmospheres, and experimental techniques utilized. The relatively meager experimental data available, as well as the inherent difficulty in developing a comprehensive model, have required theoreticians to make the many simplifying assumptions to be described.

3.1. General Simplifications

It is customary to use several simplifying assumptions in describing the state of the coal, char, and volatiles. The variably sized, irregular fragments of coal dust are approximated as spheres having equivalent masses, and usually only one average particle size is used. The internal structure of the char particle is not defined, although the char particle is treated as a sphere whose size is the same as or is related to that of the raw coal dust. The complex chemistry of the coal dust and char[26] is avoided by defining the composition in terms of volatile matter, fixed carbon, and ash. As will be seen from the following section, more rigor is applied to the description of the gaseous products produced and the kinetics of their formation.

3.2. Kinetic Models

Badzioch *et al.*[12] use the equations

$$dv/dt = k(v_\infty - v) \tag{1}$$

and

$$v_\infty = Q(1 - v_c)v_p \tag{2}$$

with

$$k = A \exp\left(-E/RT\right) \tag{3}$$

to describe the formation of volatile matter from a coal dust. The parameters Q and v_c were empirically determined, although a value of 0.15 for v_c was applied to the nonswelling coals tested. This is the most satisfactory single-step reaction that has been used to describe pyrolysis. However, it lacks the flexibility required to describe much of the experimental data available, and may even be inadequate to describe nonisothermal pyrolysis. The fact that the parameters v_p, k, Q, v_c, A, and E may depend upon the specific coal dust also limits the generality of this model.

Stickler *et al.*[28] suggested that pyrolysis could be modeled with the following pair of parallel, first-order, irreversible reactions,

$$\begin{aligned} C &\xrightarrow{k_1} (1-\alpha_1)S_1 + \alpha_1 V_1 \\ C &\xrightarrow{k_2} (1-\alpha_2)S_2 + \alpha_2 V_2 \end{aligned} \tag{4}$$

with the rate equations

$$dc/dt = -(k_1 + k_2)c \tag{5}$$

and

$$dv/dt = (dv_1 + dv_2)/dt = -(\alpha_1 k_1 + \alpha_2 k_2)c \tag{6}$$

Again, k_1 and k_2 are Arrhenius-type constant, but an important feature of the model is that $E_1 < E_2$. This approach will satisfactorily correlate the data of references 12 and 13, and in addition will correlate the more recent data of reference 15 obtained under conditions of transient temperature. The model is conceptually sound in that the variation in volatiles yield with temperature is explained by a second reaction rather than by a correlating parameter. As before, the general utility may be limited because the parameters α_1, α_2, A_1, A_2, E_1, and E_2 depend on the specific coal dust.

A similar, but slightly different approach is presented by Nsakala et al.[29] These authors propose two different reactive species in the coal, C_1 and C_2, each decomposing via a different first-order reaction. Their reaction scheme is

$$C_1 \xrightarrow{k_1} V_i \rightarrow V_1 + S_1$$
$$C_2 \xrightarrow{k_2} V_2 + S_2 \tag{7}$$

The advantages and disadvantages of this treatment are much the same as those of Eqs. (4)–(6).

Perhaps next in order of complexity is the model of Anthony et al.,[11] which postulates that pyrolysis occurs through an infinite series of parallel reactions. A continuous Gaussian distribution of activation energies is assumed, along with a common value for the frequency factor so that

$$(v_\infty - v)/v_\infty = [\sigma(2\pi)^{1/2}]^{-1} \left\{ \int_0^\infty \exp\left[\left(-\int_0^t k\, dt \right) f(E)\, dE \right] \right\} \tag{8}$$

with

$$f(E) = [\sigma(2\pi)^{1/2}]^{-1} \exp\left[-(E - E_0)^2/2\sigma^2 \right] \tag{9}$$

This approach provides an excellent correlation of the data from Anthony et al.,[11] as well as the more recent experimental results of reference 8. However, minimal comparison has been made with the data of other investigators, and it appears likely that the model cannot correlate data of the sort shown in Figure 2. In this case also, the utility of the model may be restricted by the need to determine the parameters v_∞, A, E_0, and σ for the specific coal dust of interest.

A considerably more complex reaction mechanism was postulated by

Reidelbach and Summerfield.[30] A set of 10 reactions were used to describe the process of pyrolysis. Further refinements were later added by Antal and co-workers[31] and also by Reidelbach and Algermissen.[32] An interesting facet of one model variation[32] is the inclusion of a reversible vaporization step, which reference 21 also suggests. As yet, only limited comparison has been made between this model and experimental data, but the results have been favorable. Perhaps this is to be expected since the Reidelbach model also contains the reaction steps shown by Eq. (4), which by themselves quite successfully explain much experimental data. Conceptually, the Reidelbach approach seems quite sound and it can probably explain results obtained over a wide range of experimental conditions. However, when only fast pyrolysis is considered, the additional complexity is probably not necessary and the Stickler approach[28] is adequate. Also, the Reidelbach model requires that an even larger number of constants be evaluated from experimental data.

A rather complex pyrolysis model recently presented by Suuberg et al.[8] includes a set of 15 first-order reactions that produce 8 specific volatile products. This set of reactions is remarkably successful in describing the detailed pyrolysis data for a lignite coal, including the measured composition of the pyrolysis products. However, the simpler approach of Anthony et al.[11] was adequate to explain the pyrolysis rate of the solid. Also, as Table 1 shows, the gaseous products of Suuberg et al.[8] are sufficiently unique that this set

Table 2. Comparison between Experimental Observations and Model Predictions

Experimental observation	Model reference[a]				
	12	28	11,29	30, 32	8
Yield of pyrolysis products increases "exponentially" with time	+	+	+	+	+
Maximum volatiles increase with heating rate and/or temperature	+	+	+	+	+
Pyrolysis is at least a two-step process	−	+	+	+	+
The average volatiles composition depends on heating rate and/or final temperature	−	+	+	+	+
The instantaneous volatiles composition changes with time, heating rate, degree of pyrolysis or temperature	−	+	+	+	+
Pyrolysis products are a complex mixture of species	−	−	−	−	+
Volatiles yield depends on pressure, particle size, and particle proximity[b]	−	−	+	+	+
The irregular coal dust particles transit to "sponges" in the pyrolysis process	−	−	−	−	−

[a] + means the model can predict the observation; − means it cannot, or else no comparison has been made.
[b] See the specific references for the cracking reactions believed to explain this observation.

of specific reactions is probably inadequate to predict the gaseous pyrolysis products obtained in other experiments with other coals. For practical use, the model is limited by the need to use a large number of rate constants which may vary depending on the specific coal dust used.

Table 2 presents a summary comparing key experimental observations with the various pyrolysis models.

4. Summary

Each of the models requires a significant amount of experimental data to determine values for parameters used in the computations. For example, the ultimate extent of pyrolysis must be determined experimentally, then used as input in making pyrolysis calculations. This means that no model can be confidently applied to a new and different coal dust.

Another common feature of the various models is that the temperature history must be specified before the temperature-dependent pyrolysis may be calculated. This means that the pyrolysis equations must be coupled to the various transport equations for general use. Then, the solution of the coupled equations will describe the interdependent histories of temperature and volatiles evolution.

The selection of the "best" model depends heavily upon the amount of experimental data available for the specific coal dust being considered. However, in most cases a set of first-order, irreversible, parallel reactions is appropriate. A "set" of 1 reduces the treatment to that of Badzioch et al.,[12] while a set of 15 expands the treatment to that of Suuberg et al.[8] For most usage a set of 2, as Stickler et al.[28] suggest, is sufficient to describe the observed pyrolysis without adding unnecessary complexity to the model,

5. Notation

A	Frequency factor in Arrhenius rate constant (s^{-1})	t	Time from onset of heating (s)
		T	Absolute temperature (K)
c	Fraction of unreacted coal remaining in char	v	Fraction of solid lost as volatile matter up to time t
C	Concentration of unreacted coal in solid particles $(kg\ m^{-3})$	v_∞	Maximum fraction of solid lost as volatile matter in test reactor
E	Activation energy $(J\ kmol^{-1})$	v_p	Fraction of solid that is volatile matter as determined by proximate analysis
k	Reaction rate constant (s^{-1})		
Q	Constant relating quantity of high-rate volatiles to proximate volatiles		
R	Gas constant $(J\ kmol^{-1}\ K^{-1})$	v_c	Minimum fraction of char from test reactor that is volatile matter as determined by proximate analysis
S	Concentration of unreactive char in solid particles produced by devolatilization reaction $(kg\ m^{-3})$	V	Concentration of volatiles $(kg\ m^{-3})$

α	Mass stoichiometric coefficient	π	3.1416
σ	Coefficient describing spread in	1	reaction "1"
	Gaussian distribution of activation	2	reaction "2"
	energies		

Subscripts

| 0 | central activation energy in assumed | i | intermediate volatiles that may |
| | distribution | | decompose |

6. References

1. W. I. Jones, The thermal decomposition of coal, *J. Inst. Fuel* **37**, 3–11 (1964).
2. P. C. Yellow, Kinetics of the thermal decomposition of coal, *Brit. Coal Utiliz. Res. Assoc. Monthly Bull.* **29**(9), 285–308 (1965).
3. M. A. Field, D. W. Gill, B. B. Morgan, and P. G. W. Hawksley, *Combustion of Pulverized Fuel*, pp. 155–173, British Coal Utilization Research Association, Leatherhead, Surrey, England (1967).
4. R. H. Essenhigh and J. B. Howard, *Combustion Phenomena in Coal Dusts and the Two-Component Hypothesis of Coal Combustion*, Pennsylvania State Studies No. 31, Pennsylvania State University (1971).
5. R. L. Coates and J. M. Glassett, *High Rate, High Temperature Pyrolysis of Coal*, Final Technical Report to Bituminous Coal Research, Inc., Brigham Young University, Provo, Utah (1974).
6. D. Gray, J. G. Cogoli, and R. H. Essenhigh, Problems in pulverized coal and char combustion, in *Coal Gasification*, Advances in Chemistry Series, No. 131, pp. 72–91, American Chemical Society, Washington, D.C. (1974).
7. D. B. Anthony and J. B. Howard, Coal devolatilization and hydrogasification, *AIChE J.* **22**, 625–626 (1976).
8. E. M. Suuberg, W. A. Peters, and J. B. Howard, Product composition and kinetics of lignite pyrolysis, in *Symposium of Coal Gasification Kinetics*, American Chemical Society Preprints, Vol. 22, pp. 112–136 American Chemical Society, Washington, D.C. (1977).
9. R. Loison and F. Chauvin, Pyrolyse rapide de charbon, *Chim. Ind.* (*Paris*) **91**, 269–275 (1964).
10. M. H. Mentser, J. O'Donnel, S. Ergin, and R. A. Friedell, Rapid thermal decomposition of bituminous coals, in *Coal Gasification*, Advances in Chemistry Series, No. 131, pp. 1–7, American Chemical Society, Washington, D.C. (1974).
11. D. B. Anthony, J. B. Howard, H. C. Hottel, and H. P. Meissner, Rapid devolatilization of pulverized coal, in *Fifteenth Symposium* (*International*) *on Combustion*, pp. 1303–1317, The Combustion Institute, Pittsburgh, Pa. (1975).
12. S. Badzioch, P. G. W. Hawksley, and C. W. Peter, Kinetics of thermal decomposition of pulverized coal particles, *Ind. Eng. Chem. Process Design and Develop.* **9**, 521–530 (1970).
13. G. M. Kimber and M. O. Gray, Rapid devolatilization of small coal particles, *Combustion and Flame* **11**, 360–362 (1967).
14. R. L. Coates, C. L. Chen, and B. J. Pope, Coal devolatilization in a low pressure, low residence time entrained flow reactor, in *Coal Gasification*, Advances in Chemistry Series, No. 131, pp. 92–98, American Chemical Society, Washington, D.C. (1974).
15. S. K. Ubhayakar, D. B. Stickler, C. W. von Rosenberg, Jr., and R. E. Gannon, Rapid devolatilization of pulverized coal in hot combustion gases, in *Sixteenth Symposium* (*International*) *on Combustion*, pp. 426–436, The Combustion Institute, Pittsburgh, Pa. (1976).

16. J. B. Howard and R. H. Essenhigh, Mechanism of solid-particle combustion with simultaneous gas-phase volatiles combustion, in *Eleventh Symposium (International) on Combustion*, pp. 399–408, The Combustion Institute, Pittsburgh, Pa. (1967).

17. T. A. Milne and J. E. Beachey, *Summary Technical Progress Report on Contract H01220127*, Midwest Research Institute (1976).

18. R. L. Coates, L. D. Smoot, and M. D. Horton, *Exploratory Studies of Flame and Explosion Quenching*, Interim Report, U.S. Bureau of Mines, Contract No. H0122050, Chemical Engineering Dept., Brigham Young University, Provo, Utah (September, 1973).

19. L. D. Smoot and M. D. Horton, *Exploratory Studies of Flame and Explosion Quenching*, Interim Report No. 2, U.S. Bureau of Mines, Contract No. H0122052, Chemical Engineering Dept., Brigham Young University, Provo, Utah (September 30, 1974).

20. M. D. Horton and L. D. Smoot, *Exploratory Studies of Flames and Explosion Quenching*, Interim Report No. 3, U.S. Bureau of Mines, Contract No. H0122052, Chemical Engineering Dept., Brigham Young University, Provo, Utah (September 30, 1975).

21. L. D. Smoot and M. D. Horton, *Exploratory Studies of Flame and Explosion Quenching*, Interim Report No. 4, U.S. Bureau of Mines, Contract No. H0122052, Chemical Engineering Dept., Brigham Young University, Provo, Utah (December 15, 1976).

22. L. D. Smoot and M. D. Horton, Propagation of laminar, pulverized coal–Air Flames, in *Sixteenth Symposium (International) on Combustion*, pp. 375–387, The Combustion Institute, Pittsburgh, Pa. (1976).

23. M. D. Horton, F. P. Goodson, and L. D. Smoot, Characteristics of flat, laminar coal dust flames, *Combustion and Flame* **23**, 187–195 (1977).

24. R. T. Eddinger, L. D. Friedman, and E. Rau, Devolatilization of coal in a transport reactor, *Fuel* **45**, 245–252 (1966).

25. A. S. Padia, A. F. Sarofim, and J. B. Howard, The behavior of ash in pulverized coal under simulated combustion conditions, in *Central States Section*, The Combustion Institute, New York, N.Y. (April 1976).

26. R. H. Essenhigh, Combustion and flame propagation in coal systems: A review, in *Sixteenth Symposium (International) on Combustion*, pp. 353–374, The combustion Institute, Pittsburgh, Pa. (1976).

27. D. W. Van Krevelen, F. J. Huntjens, and H. N. M. Dormans, Chemical structure and properties of coal XVI—Plastic behavior on heating, *Fuel* **35**, 462–475 (1956).

28. D. B. Stickler, R. E. Gannon, and H. Kobayashi, Modeling of coal, in *Eastern States Section*, The Combustion Institute, Silver Springs, Md. (1975).

29. N. Nsakala, R. H. Essenhigh, and P. L. Walker, *Studies on Coal Reactivity: Kinetics of Lignite Pyrolysis in Nitrogen at 808° C*, Pennsylvania State University (1977), to be published.

30. H. Reidelbach and M. Summerfield, *Kinetic Model for Coal Pyrolysis Optimization*, Princeton University, Princeton, New Jersey (1974).

31. M. J. Antal, E. G. Platt, T. P. Chung, and M. Summerfield, Recent progress in kinetic models for coal pyrolysis, in *Symposium on Coal Gasification Kinetics*, American Chemical Society Preprints, Vol. 22, pp. 137–148, American Chemical Society, Washington, D.C. (1977).

32. H. Reidelbach and J. Algermissen, *Formulation of a Kinetic Model of Coal Pyrolysis including Chemical and Physical Processes*, Annual Report of Institut Fur Thermodyamik der Ruft- und Raumfahrt, Universität Stuttgart (1976).

9

Heterogeneous Reactions of Char and Carbon

F. Douglas Skinner and L. Douglas Smoot

1. Introduction

The combustion or gasification of coal may be thought of as occurring in several steps. First, the coal is heated at high rates, and rapid devolatilization occurs. The volatiles then react in the gas phase. (These processes are discussed in detail in Chapters 8 and 10.) The char remaining after devolatilization consists primarily of carbon, together with the major part of the ash components and some volatile matter. The heterogeneous reactions of the char with oxidizing gases (oxygen, steam, carbon dioxide, etc.) account for the majority of time required for particle burnout. In addition, in the case of gasification, the reaction of char with hydrogen may also become important, especially for high hydrogen partial pressures.

Photomicrographs of devolatilized coal char by Anson et al.[1] and Smoot et al.[2] show that the particles become more spherical, with "blow holes" formed, presumably from the ejection of volatiles. In addition, some particles form hollow spheres, and others are almost C-shaped.

Heterogeneous reactions between gases and solid particles have been conventionally modeled using the following assumptions[3]:

(1) Char particles are considered to be carbon spheres in an essentially infinite gas stream.

F. Douglas Skinner • Ph. D. Candidate, Chemical Engineering, Brigham Young University, Provo, Utah. *L. Douglas Smoot* • Professor of Chemical Engineering, Brigham Young University, Provo, Utah

(2) The gas-phase reactant first diffuses to the solid surface or into the pores in the char, and then absorbs and reacts with the surface.

(3) The gas-phase products desorb and diffuse away from the particle.

The rate-limiting step in this model can be chemical (adsorption of reactant, reaction, desorption of products) or diffusional (bulk phase or pore diffusion of reactants or products). Several investigators, such as Walker et al.[4] and Gray et al.,[5] have postulated the existence of different temperature regimes which determine which resistance is controlling. In Zone I, which occurs at low temperatures, chemical reaction is the rate-determining step. Zone II is characterized by control due to both chemical reaction and pore diffusion. Zone III, which occurs at high temperatures, is characterized by bulk mass-transfer limitations. Transition regimes exist between the various zones, in which combined effects are important. Kinetic data obtained by any investigator must be interpreted in light of the conditions under which the data were obtained.

In Zone I, the experimentally observed activation energy will be the true activation energy, and the reaction order will be the true order. In Zone II, the observed activation energy will be roughly one-half the true value, while the observed or apparent reaction order, n_o, will be related to the true order n as follows:

$$n_o = (n+1)/2 \tag{1}$$

In Zone III, where bulk phase mass transfer is the limiting resistance, the apparent activation energy will be small, generally in the neighborhood of 20,000 kJ kmol^{-1}.[5]

In general, most experimenters have tried to obtain data on the intrinsic chemical reaction rates of carbon or char with the various gases. To do this, a variety of schemes have been tried to minimize the effects of diffusion: by conducting tests at low pressure or low temperature, by microsampling of gases very close to the solid surface, by use of high-velocity gas streams to decrease the boundary layer thickness, and by use of small particles.[3,6] Field et al.[6] have dismissed data obtained by the technique of reduced pressure because of difficulties in extrapolating the results thus obtained to atmospheric pressure and greater.

The rates of carbon oxidation by steam and carbon dioxide are of the same order of magnitude, and are generally much slower than the reaction of carbon with oxygen, while the hydrogenation reaction is several orders of magnitude slower than the steam–char and CO_2–char reactions. For example, Walker et al.[4] estimate the relative rates of the reactions at 1073 K and 10 kPa pressure as being 3×10^{-3} for H_2, 1 for CO_2, 3 for H_2O, and 10^5 for O_2. At higher temperatures, the differences in rates will not be so extreme, but those for H_2, CO_2, and H_2O will still remain relatively unimportant as long as comparable concentrations of oxygen are present. This means that the only heterogeneous carbon oxidation reaction which needs to be considered for most combustors is that due to oxygen.[3] In coal gasifiers, however, the reactions with steam and CO_2 can be quite important, especially after the rapid depletion of oxygen. Batchelder *et al.*[7] calculate that the oxygen is essentially consumed in about 10 ms at the gasification temperatures in a pulverized-coal reactor. They also found "significant" CO and H_2 concentrations as early as 3 ms, which may be due to nonoxidized pyrolysis products.

The hydrogasification reaction will, under some conditions, be as important as the carbon–steam reaction.[4] Von Fredersdorff and Elliot[8] cite experimental evidence that under the conditions of 3,000–20,000 kPa H_2 pressure and 750–1200 K, 200–400 μm char particles are completely reacted in about $\frac{1}{2}$ hr. Anthony and Howard[9] have found that nearly complete conversion of raw coal can be achieved in a matter of a few seconds using high H_2 pressure; however, no particle sizes were mentioned.

In examining rate data from various sources, it is apparent that temperature and reactant concentration are not the only variables which influence reaction rates. The composition of the feed material, its pretreatment, and its thermal history are also important. Some experimenters have used pure forms of carbon, while others have used "char." Char consists of the residue formed by devolatilizing such fuels as coconut shells, solid petroleum residuals, or coal. In general, it has been found that the purer forms of carbon are less reactive than the chars. It has also been found that chars from the lower-rank coals are more reactive than those from higher-rank coals. Differences in reactivity at different heating rates and temperatures have been noted in chars formed from the same parent coal, but some of the differences in reactivity noted here may be due to additional volatiles reactions from the low-temperature chars, which were not totally pyrolyzed.

Other important variables are the internal and external surface areas of the char. Walker *et al.*[10] have performed experiments with CO_2 and carbon which show that the higher the internal area, the higher the reaction rate. Thus, under some conditions, the oxidation reactions do not occur only on the outside surface of the char particle, but reactant gases also diffuse into the particles before reacting. The observed increase in reaction rate with increasing burnoff, followed by a subsequent decline in rate, has been explained by some investigators in terms of expanding of the pores to expose more and more internal area.[1,10,11] The surface area continues to increase up to a certain point, then, due to coalescence of pores and thermal annealing, the area decreases, and with it the reaction rate. Wen,[12] on the other hand, explains the observed decline in terms of different macerals (microscopic organic constituents in the char) having differing activities. He postulates that the initial high activity is due to the reaction with aliphatic hydrocarbon sites and oxygenated functional groups. Once these are depleted, the low activity is due to the residual carbonaceous coke.

Another factor which may affect the oxidation rates is the presence of catalytic impurities in the char. Batchelder et al.[7] cited studies which indicated that oxides and carbonate salts act as catalysts for all four of the reactions being considered: C^*–CO_2, C^*–H_2, C^*–H_2O and C^*–O_2. Tomita et al.[13] recently conducted experiments in which various minerals were added to high-purity carbon to test their effect on the rates of the various oxidation reactions. They found no catalytic effect for the minerals tested on the C^*–O_2 reaction, inhibition of the C^*–CO_2 reaction with included minerals such as pyrite, gypsum, kaolinite, and calcite, and both positive and negative catalytic effects on the hydrogasification reactions. As might be expected, catalytic effects, whether positive or negative, tend to be more important at low temperatures. At actual combustion temperatures their effect is probably small.

The remainder of this chapter includes discussions of the postulated mechanisms, rate equations, and experimental rate data for the reactions of char with oxygen, carbon dioxide, steam, and hydrogen, as well as recommended equations and constants for use in modeling heterogeneous reactions in pulverized-coal combustors and gasifiers.

2. Char–Oxygen Reaction

It might be assumed, in view of the great amount of effort expended over many years, that a reaction as apparently simple as

$$C^* + O_2 \rightarrow CO_2 \tag{2}$$

where C^* refers to condensed or solid carbon or carbonaceous substance, would by now be completely characterized and understood. This is, however, not the case. Disagreements still exist among investigators on almost every aspect of the problem. In this section, some of the differing opinions will be reviewed as well as those aspects for which there is some agreement.

Several different controlling resistances are possible in char oxidation. Important factors which determine the controlling resistance for a given situation are temperature, particle size, relative velocities of the gas and particles, and pressure. Gray et al.[5] pointed out that, for the normal conditions of pulverized-char combustion, the reaction will probably be controlled either by intrinsic chemical reaction kinetics, or by a combination of pore diffusion and reaction kinetics. This conclusion is in agreement with the findings of Sergeant and Smith,[14] Hamor et al.,[15] and Field.[16] It appears that bulk diffusion becomes the controlling factor at high temperature for chars from low rank, highly reactive coals, or for large particles.[16,17] Batchelder et al.[7] cite evidence that diffusion effects become important for temperatures in excess of 1000 K. In order to account for possible diffusion effects, Field[16] and Gray et al.[5] recommend the use of an overall rate constant k, which is found by combining the rate constants k_s due to chemical kinetics and k_d due to bulk diffusion:

$$k = 1/[(1/k_s) + (1/k_d)] \tag{3}$$

Field gives an expression for k_d

$$k_d = 2.37\phi D/dR'T_a \tag{4}$$

where ϕ, a stoichiometric factor, is 2 if CO is the main product and 1 if CO_2 is the main product, D is the diffusion coefficient of oxygen through the other gas species, d is the particle diameter, R' is the universal gas constant, and T_a is the average gas temperature in the boundary layer.

Walker et al.[4] and von Fredersdorff and Elliot[8] review the mechanisms which have been proposed for the carbon–oxygen reaction. One of the uncertainties of formulating a mechanism has been the identification of the oxidation product or products. It is now generally held that both carbon monoxide and carbon dioxide are primary reaction products.[4] The ratio of the primary products CO/CO_2 increases with increasing reaction temperature. Arthur[18] showed that this ratio could be expressed, for temperatures between 730 and 1170 K, and for two different carbons, as

$$CO/CO_2 = 2500 \exp(-6240/T) \tag{5}$$

Rossberg[19] also reported similar results for temperatures between 790 and 1690 K for two different carbons. Day[20] showed that the CO/CO_2 ratio may not always be independent of carbon type, but verified the predominance of CO at higher temperatures.

At ordinary combustion temperatures, the predominant oxidation product still appears to be carbon monoxide. The carbon monoxide further oxidizes in the gas phase to carbon dioxide. According to von Fredersdorff and Elliot,[8] this reaction is very rapid, and is further accelerated by the presence of water vapor. This effect may be due to the water–gas shift reaction, which is discussed more fully in Section 4.

In most cases, experimental rate equations have been of the form

$$r = k_s p^n \tag{6}$$

where p is the partial pressure of reactant gas and n the reaction order. The rate constant k_s can be expressed in modified Arrhenius form as

$$k_s = AT^N \exp(-E/RT) \tag{7}$$

where most investigators take the temperature exponent N as zero.

Experimentally determined values for reaction order n, preexponential factor A, and activation energy E are found to vary considerably from investigator to investigator. Table 1 contains a summary of the experimentally determined rate data from several recent studies. The value obtained for the reaction order n also varies between 0 and 1, with most investigators correlating their data on the basis of first-order oxygen concentration dependence. However, Smith and Tyler[26] and Tyler et al.[27] indicate reaction orders which are close to zero, especially at high temperatures. The controversy as to reaction order is clearly not yet resolved. Gray et al.[5] explain the variation in terms of the controlling chemical reaction step: If adsorption is controlling, $n = 1$, whereas if desorption controls, $n = 0$.

The large variation in activation energies can be explained in part by several factors already enumerated in Section 1. Lower activation energies are expected for coal chars relative to pure carbons, as chars are more reactive, and for temperature regimes where the combined resistances due to pore diffusion and chemical reaction kinetics are controlling. Decreased values of activation energy are also anticipated for experiments performed within the same regime as coal rank decreases.

The values for the preexponential factors given in Table 1 are all based upon the external surface area of the particles. As pointed out previously, internal surface area probably plays an increasingly greater role as the particle burns out. Rates based on external area may be adequate up to about 5% particle burnout, then may be high relative to experiment. Modeling this dependence is an ongoing subject of study. Thomas[11] has proposed a model which attempts to take internal area changes into account. In a recent, major review article on heterogeneous kinetics of coal-char reactions, Laurendeau[28] discusses in some detail the influence of porosity and internal pore structure of coal chars on char reactions. Pore diffusion, adsorption–desorption, and surface reaction processes are considered. Advanced treatments of char reaction processes will likely account for such internal structure effects.

For the combustion of bituminous and lower-rank chars, the coefficients of Field et al.[6] in Table 1 are recommended. They are based upon curve-fitting of data from a number of different investigators. These coefficients are applicable for particle temperatures up to 1650 K. For higher temperatures, the predicted rates are too high, and a straight-line fit of k_s as a function of temperature is suggested by Field[16]:

$$k_s = -4.84 \times 10^{-2} + 3.80 \times 10^{-5} T_s \tag{8}$$

This fits the data reasonably well over the temperature range of 1400–2200 K.

3. Char–Carbon Dioxide Reaction

The majority of the work done on the reaction

$$C^* + CO_2(g) \rightarrow 2CO(g) \tag{9}$$

Table 1. Rate Parameters for Char Reaction with Oxygen for Rate Equation $r = AT^N[\exp(-E/RT)]p_{O_2}^n$

Reference	A (kg m^{-2} s^{-1} kPa^{-n} K^{-N})	E/R (K)	N	Order n	Particle type	Size-graded	Sizes (μm)	Temperature range (K)
21	1.32×10^{-1}	16,400	0	0	Brown coal char	Yes	22, 49, 89	630–1812
6	8.6×10^2	18,000	0	1	Various fuels	Varied	Varied	950–1650
22	—	20,100	0	0,1[a]	Carbon	Yes	2.54×10^4	—
23	—	3,000–6,000 / 15,000–32,700	0	1[b] / 0	Bituminous char	No	0–200	—
24	—	6,500–25,000[c]	1.75–3.5	1	Various coals	Yes	420–1000	1100–1500
15	9.18×10^{-1}	8,200	0	0.5	Brown coal char	Yes	22, 49, 89	630–2200
25	2.013	9,600	0	1	Semianthracite	Yes	6, 22, 49, 78	1400–2200
26	5.428	20,100	0	1	Semianthracite	Yes	6, 22, 49, 78	1400–2200
14	2.902[d]	10,300[d]	0	1	Bituminous char	Yes	18, 35, 70	800–1700

[a]0 for $T < 1000$ K; 1 for $T > 1000$ K.
[b]1 indicates adsorption control before flame front; 0 indicates desorption control in tail of flame.
[c]Most values between 11,600–14,600 K.
[d]Calculated from plot in Sergeant and Smith.[14]

has emphasized the determination of the reaction mechanism and has been conducted using mostly pure carbons over rather limited temperature ranges. Good reviews of work done on this reaction are by Batchelder et al.,[7] Walker et al.,[4] von Fredersdorff and Elliot,[8] and Ergun and Mentser.[29] The rate equation most often fit by researchers has been of the form of a Langmuir adsorption isotherm,

$$\bar{r} = k_1 p_{CO_2}/(1 + k_2 p_{CO} + k_3 p_{CO_2}) \tag{10}$$

where the k's are generally functions of the rate constants for the elementary steps comprising the reaction mechanism. Several investigators have determined Arrhenius forms for the k's; some of the reported results are summarized in Table 2. These rates have most generally been expressed in terms of mass of carbon gasified per unit mass of carbon remaining, and not on a unit area basis, as has been done in the case of the $C^* + O_2$ reaction. Since the apparent activation energies cited are not for elementary reactions, but are curve-fit parameters, negative values sometimes result. The large variations in both preexponential factors and activation energies point up the dependence of the results on the nature of the feed material and other experimental variables.

In general, it has been found that the form of Eq. (10) can be simplified under some conditions of temperature and reactant partial pressure. Walker et al.[4] and Lewis et al.[30] have noted that under conditions of low temperature and/or high CO_2 concentration, $k_3 p_{CO_2} \gg 1$, and a zero-order reaction should be observed. At low temperatures and low CO_2 concentration, or high temperatures and high CO_2 concentrations, the reaction should approach first order. The general Langmuir equation correctly predicts the "poisoning" effect of the reaction product, CO, on the reaction rate. This effect has been observed by several investigators such as Lewis et al.,[30] Blakely and Overholser,[31] and Turkdogan and Vinters.[32] The effect diminishes rapidly at higher temperatures, and is more pronounced for purer carbons. Turkdogan and Vinters[32] propose that the "poisoning" may be due to several factors, such as dilution of the CO_2, equilibrium considerations, or actual adsorption of CO on reaction sites.

Reif[33] and Grabke[34] give one of the simplest proposed mechanisms which leads to an expression of the proper type:

(1) Reaction of carbon dioxide with an active site on the solid surface, yielding a molecule of CO and an adsorbed oxygen atom:

$$CO_2 + (\ ) \leftrightharpoons CO + (O) \tag{11}$$

(2) Reaction of the surface-bound oxygen with another surface carbon atom, followed by desorption of the CO, leaving a vacant active site:

$$(O) + C^* \rightarrow CO + (\ ) \tag{12}$$

Table 2. Arrhenius Constants for Carbon–Carbon Dioxide Reaction with Langmuir-Type Rate Expression[a,b,c]

Carbon type	Temperature range (K)	A_1 (kg kg^{-1} s^{-1} kPa^{-1})	E_1/R (K)	A_2 (kPa^{-1})	E_2/R (K)	A_3 (kPa^{-1})	E_3/R (K)
Charcoal	1007–1103	1.24×10^6	29,600	1.24×10^{-10}	−22,900	3.12×10^4	15,100
Coke	1073–1363	1.38×10^3	24,000	1.4×10^{-4}	−7,500	2.07×10^{-3}	−3,200
Coke	—	6.2×10^4	31,100	4.0×10^8	−20,300	3.12×10^{-4}	−3,100
Electrode carbon	—	2.0×10^3	25,200	3.12×10^{-11}	−30,500	1.58×10^{-3}	−3,300
Coke	1199–1423	2.1×10^4	20,200	2.0×10^{-11}	−27,700	—	—
Anthracite	1073–1363	4.3	16,400	4.5×10^{-5}	−8,500	1.78×10^{-5}	−8,400

[a] Table used with permission from von Fredersdorff and Elliot.[8]
[b] This table with different units appeared in von Fredersdorff and Elliot,[8] where references were shown on p. 928.
[c] Equation form: $\bar{r} = k_1 p_{CO_2}/(1 + k_2 p_{CO} + k_3 p_{CO_2})$; $k_i = A_i \exp(-E_i/RT)$.

Other mechanisms have been proposed which also lead to the desired rate expression, as discussed in von Fredersdorff and Elliot,[8] Grabke,[34] and others.

Due to uncertainties involved in extrapolating the Langmuir-form constants to other carbonaceous materials and temperatures, it is recommended that these constants not be used outside the ranges for which they were derived. Further work is needed in this area to extend the constants to temperatures of greater interest.

Other investigators have chosen to express the rate of C^*–CO_2 reaction in the form

$$r = kp^n \tag{13}$$

where

$$k = A \exp(-E/RT) \tag{14}$$

These can be viewed as a special case of the Langmuir equation form, Eq. (10). A summary of the kinetic parameters from some studies which use this form of Eqs. (13) and (14) are given in Table 3. The rate constant in all these cases is based on a unit external surface area of the char, except that of Yang and Steinberg,[35] whose area dimension refers to the cross-sectional area of the diffusion cell used in their experiments.

Several recent investigators[40–42] have treated the C–CO_2 reaction somewhat differently. The kinetic parameters were obtained using the general reaction

$$r = k(C^*)(p_{CO_2})^n \tag{15}$$

where (C^*) is the weight of char left unreacted. A first-order dependence on unreacted char was found in every case. Table 4 presents the kinetic parameters from these studies.

Table 3. Arrhenius Factors for Carbon–Carbon Dioxide Reaction

Reference	Carbon type	A (kg m^{-2} s^{-1} kPa^{-n})	E/R (K)	Order n	Temperature range (K)
36	Graphite	1.35×10^{-1}	16,300	1	1123–1223
	Graphite	6.35	19,500	1	1223–1673
37	—	1.2×10^{-6}	17,600	0	1013–1133
38	—	7.8×10^{-1}	16,500	1	1133–1211
39	Electrode carbon	7.3×10^{-5}	4,990	1	2200–3200
10	"Gas-baked carbon"	—	23,700	1	1173–1373
	Graphite	—	24,200	1	1173–1473
35	Bituminous coal char	—	13,000	1	1473–1673
	Anthracite char	—	13,900	1	1673–1773

Table 4. *Arrhenius Factors and Reaction Orders for Carbon–Carbon Dioxide Reaction*[a]

Reference	Carbon type	Size, μm	A	E/R, K	Order n	Temperature range, K
40	Bituminous coal char	Pulverized[b]	3.13×10^6 kPa^{-1}	23,600	1	—
41	Subbituminous coal char	6,000	7.0×10^5 kg kg^{-1} s^{-1}	22,600	0	873–1,063
42[c]	Bituminous coal char	Pulverized[b]	1.4×10^5 kg kg^{-1} s^{-1}	15,600	0	1,600–1,800

[a] Treatment assumes first-order char dependence.
[b] "Pulverized" refers to the power plant standard of 70% through 200 mesh.
[c] Coates' data are for the combined reaction, char + H_2O + CO_2 → CO + H_2.

Attempts have been made to define the controlling resistances for various temperature regimes. The region of chemical kinetic control has been variously reported as up to about 1400 K by Blakeley,[43] up to 1473 K for "graphite" by Walker et al.,[10] and below about 1620 K for spectrographic grade graphite by Gulbransen et al.[44] Peterson and Wright[45] found pore diffusion control for the temperature range 1173–1573 K for "graphite" and CO_2. External gas diffusion was found to be the controlling resistance above 1670 K by Gulbransen et al.[44] For the temperatures achieved in most gasifiers, it is probable that the combined effects of pore diffusion and chemical reaction will control the overall reaction rate.

The safest procedure would be to use data which are either derived by direct experimentation using the same char, temperature, and pressure conditions that will be found in the desired reaction system, or to use literature data which come closest to the desired conditions. In the absence of such direct data, the parameters of Mayers[36] can be used for temperatures between 1123 and 1673 K.

4. Char–Steam Reaction

The reaction of carbon with steam, which can be expressed by the stoichiometric equation

$$C^* + H_2O = H_2 + CO \qquad (16)$$

has been shown to be similar to the reaction with CO_2, both in relative rate and mechanism. As with the C^*–CO_2 studies, work with this reaction has emphasized the determination of reaction mechanisms, and has unfortunately been done with relatively pure carbons at low temperatures and over small temperature ranges.

The rate equation form most often used in the past has been the Langmuir adsorption isotherm

$$\bar{r} = k_1 p_{H_2O}/(1 + k_2 p_{H_2} + k_3 p_{H_2O}) \qquad (17)$$

where the k's are generally functions of the rate constants for the elementary steps. Note that this equation predicts the hydrogen will act as a "poison" in much the same way that carbon monoxide does in the C^*–CO_2 reaction. This has indeed been found to be the case. However, Eq. (17) does not predict the rate retardation due to the presence of CO, which has been observed by Ergun and Mentser[29] and Blakely and Overholser,[31] among others. Gadsby et al.[46] postulated that CO did not by itself inhibit gasification, but by shifting the water–gas equilibrium, more hydrogen was produced, which did inhibit the C^*–H_2O reaction. Hydrogen has actually been found to increase the rate of gasification for some chars, while decreasing it for others.[47]

This result was explained by postulating that the hydrogen served to activate some catalytic impurity in the char for the carbon–steam reaction. As before, the effect of hydrogen on the reaction decreases markedly as the temperature increases, and Batchelder *et al.*[7] found that for relatively pure carbon, the effect was negligible above 1650 K. The temperatures at which the effect can be ignored are probably lower for impure chars.

The postulated reaction mechanisms for C^*–H_2O reaction closely parallel those for C^*–CO_2. The subject has been reviewed by Walker *et al.*[4] and by von Fredersdorff and Elliot.[8] A simple three-reaction mechanism is proposed by Reif[33]:

$$H_2O + (\ ) \leftrightharpoons (O) + H_2 \tag{18}$$

$$CO + (O) \leftrightharpoons CO_2 + (\ ) \tag{19}$$

$$C^* + (O) \to CO \tag{20}$$

Arrhenius constants for the various rate constants in the Langmuir rate expression, Eq. (17), are found in Table 5. Note that these data are all for relatively low temperatures. The effect of particle burnout on the rate parameters can be seen, as well as the diverse values obtained for different carbons.

As in the case of the C^*–CO_2 reaction, the effect of pressure and temperature conditions on the observed reaction rate of C^*–H_2O are such that reaction orders from 0 to 1 may be observed. Table 6, which presents the experimental findings of several investigators, shows this result. At the conditions found in most gasifiers, Walker *et al.*[4] conclude that the reaction will probably be first order.

Among recent attempts to develop a rate equation for the carbon–steam reaction have been the following:

(1) Taylor and Bowman[41] have used the following equation for the gasification of a subbituminous char over the temperature range of 770–950 K:

$$\bar{r} = 10^6 \, C^* \exp\left(-22{,}100/T\right) \tag{21}$$

where the dependence on unreacted char is first order.

(2) Wen[12] has used an equation of the form

$$\bar{r} = k_v(C_{H_2O} - C_{H_2O}C_{CO}RT/K_c) \tag{22}$$

where $\bar{r}$ is the number of kilomoles of C^* reacted per second per kilomole of remaining C^*, and K_c is the equilibrium constant for the char–steam reaction given by Eq. (16). He has plotted values of the volumetric rate constant k_v as a function of temperature for various carbonaceous materials, based on reanalysis of the data of previous investigators.

Table 5. *Arrhenius Constants for Carbon–Steam Reaction Langmuir-Type Rate Expression*[a,b,c]

Carbon type	Temperature range (K)	A_1 (kg kg^{-1} s^{-1} kPa^{-1})	E_1/R (K)	A_2 (kPa^{-1})	E_2/R (K)	A_3 (kPa^{-1})	E_3/R (K)
Graphite, 0% burnoff	1135–1211	5.8×10^{-2}	16,500	9.30×10^{-13}	$-30,600$	6.98×10^{-18}	$-39,900$
Graphite, 1% burnoff	1135–1211	4.4×10^{-3}	13,100	6.40×10^{-13}	$-31,200$	6.11×10^{-18}	$-35,100$
Graphite, 2% burnoff	1135–1211	7.0×10^{-4}	10,700	4.15×10^{-13}	$-31,800$	4.7×10^{-18}	$-40,400$
Graphite, 5% burnoff	1135–1211	1.2×10^{-4}	8,000	9.9×10^{-14}	$-33,600$	2.10×10^{-18}	$-41,700$
Graphite, 7.5% burnoff	1135–1211	4.6×10^{-5}	6,600	3.72×10^{-14}	$-34,800$	1.93×10^{-18}	$-41,700$
Charcoal	957–1043	3.12×10^8	31,400	3.3×10^{-1}	0	3.12×10^2	10,100

[a]Table used with permission from von Fredersdorff and Elliot.[8]

[b]A portion of this table with different units appeared in von Fredersdorff and Elliot,[8] where references were shown on p. 934.

[c]Equation form: $\bar{r} = k_1 p_{H_2O}/(1 + k_2 p_{H_2} + k_3 p_{H_2O})$; $k_i = A_i \exp(-E_i/RT)$.

Table 6. Arrhenius Factors for Carbon–Steam Reaction

Reference	Carbon type	A (kg m^{-2} s^{-1} kPa^{-n})	E/R (K)	Temperature range (K)	Order n
48	Graphite	3.19×10^2	25,020	1133–1233	1
	Graphite	1.92	17,680	1273–1433	1
49	Activated carbon	—	32,000	1033–1089	0.58
50	Graphite	—	8,240	873–1273	1
51	Electrode graphite	1.95×10^8	41,600	1256–1589	(0)
	Petroleum coke	3.66×10^2	26,700	1256–1589	(0)

(3) Coates[42] has formulated a rate equation,

$$\bar{r} = 1.4 \times 10^5 \, C^* \exp\left(-15,600/T\right) \tag{23}$$

which is based on the combined reaction of steam and carbon dioxide with char.

According to Dotson et al.,[51] the carbon–steam reaction is chemically controlled at gasification conditions. Ergun and Mentser[29] found that bulk diffusion was not an important factor. An additional factor complicating the picture for the steam–carbon system is the possibility of the so-called "water-gas shift" reaction, which has been alluded to previously:

$$H_2O + CO \leftrightharpoons CO_2 + H_2 \tag{24}$$

Von Fredersdorff and Elliot[8] state that above about 1370 K, the equilibrium concentrations may be assumed without serious error, especially in pulverized-coal gasifiers with oxygen and steam at ash-slagging conditions. Ergun and Mentser[29] found that deviations from equilibrium are not great in flow systems, and that the overall rates are relatively insensitive to such deviations. Batchelder et al.[7] found that the approach to equilibrium is more rapid near a carbon surface than in the gas phase. For modeling purposes, it is therefore probably safe to assume that equilibrium is attained.

The favored approach for the char–steam reaction is to derive rate data from direct experiment with coals and chars of interest where possible. Failing this, the parameters of Mayers[48] can be used, since they cover much of the temperature range of interest. The limitations on the use of Mayers' parameters, as pointed out by Dotson et al.,[51] are the low maximum temperatures over which the parameters are experimentally valid, and the failure to account for the effect of burnoff on the reaction rate. Mayers' data were developed to describe initial rates of gasification. The work of Feldkirchner and Huebner[52] indicates that, for most temperatures, the initial rate may be low when compared to that for higher percentages of particle burnout.

5. Char–Hydrogen Reaction

The reaction of char with hydrogen is usually represented by the equation

$$C^* + 2H_2 \rightarrow CH_4 \tag{25}$$

While methane is the main product of C^*–H_2 reactions, a wide spectrum of hydrocarbons can be obtained.[9,53] At the present time no satisfactory method exists to predict the distribution of products of the reaction.[9]

The reaction of coal with hydrogen proceeds through several phases.[8,9,12] The first phase consists of pyrolysis, or devolatilization of the coal, followed by vapor-phase hydrogenation. The reaction rate is often limited by the rate of evolution of volatiles from the solid. The second phase is a short period of rapid attack by hydrogen on the char, which consequently becomes less and less reactive, leading to phase 3, the period of low reactivity of hydrogen with the residual char. Phases 1 and 2 may overlap to a great degree, especially under conditions of rapid heating to temperatures above 1000 K.[8,9] Anthony and Howard[9] found a difference of several orders of magnitude between conversion rates in the high- and low-reactivity phases.

As for the other reactions discussed in this chapter, widely varying values of the activation energy and reaction order for the reaction of Eq. (25) have been obtained by various investigators. Values for the activation energy range from 20 to 25 MJ kmol^{-1} for the rapid rate period[9] to 63 to 209 MJ kmol^{-1} for the low-rate period.[4,8] The value increases as the percentage burnoff increases. This would be expected, since the char becomes less and less reactive as the reaction progresses.

The form of the rate equation most often used[4,8,54,55] is

$$r = a p_{H_2}^2 / (1 + b p_{H_2}) \tag{26}$$

The form of Eq. (26) suggests reaction orders varying from 2 to 1 with increasing hydrogen partial pressure, and indeed such orders have been observed. Zielke and Gorin[55] found that up to 3000 kPa hydrogen pressure at 1200 K the reaction rate was well fit by Eq. (26). For temperatures below 1200 K, however, the apparent reaction order varied with the degree of carbon burnoff as well as with hydrogen pressure.

Blackwood and McGrory[56] found that the rate of hydrogasification of 1500–2500 μm coconut char in the temperature range 923–1143 K, for hydrogen pressures between 500 and 4000 kPa, was well represented by a first-order equation,

$$\bar{r} = k_m p_{H_2} \tag{27}$$

where $k_m = 0.035 \exp(-17,900/T)$, if the simplifying assumption is made that methane is the only product. Johnson[57] gives some recent experimental results for the rapid-rate phase of the C^*-H_2 reaction. Recent reviews of carbon–hydrogen gasification have been given by Anthony and Howard[9] and by Laurendeau.[28] The contribution by hydrogasification is likely to be small in steam–oxygen gasifiers, but will be a major contributor in steam–hydrogen or pure hydrogen gasifiers.

6. Notation

A	Arrhenius preexponential factor	R'	Universal gas constant (m^3 kPa $kmol^{-1}$ K^{-1})
D	Molecular diffusivity (m^2 s^{-1})		
d	Particle diameter (m)	T	Temperature (K)
E	Activation energy (J $kmol^{-1}$)	ϕ	Mechanism factor
k	Rate constant (kg m^{-2} s^{-1} kPa^{-n}) or (kg kg^{-1} s^{-1} kPa^{-n})		

Subscripts

N	Temperature-dependence exponent	d	diffusion
n	Reaction order	m	mass
p	Partial pressure (Pa)	o	observed
r	Reaction rate, surface area basis (kg m^{-2} s^{-1})	s	surface
$\bar{r}$	Reaction rate, mass basis (kg kg^{-1} s^{-1})	v	volume
R	Gas constant (J $kmol^{-1}$ K^{-1})		

7. References

1. D. Anson, F. D. Moles, and P. J. Street, Structure and surface area of pulverized coal during combustion, *Combust. Flame* **16**, 265–274 (1971).
2. L. D. Smoot, M. D. Horton, and G. Williams, Propagation of laminar pulverized coal–air flames, in *Sixteenth Symposium (International) on Combustion*, The Combustion Institute, Pittsburgh, Pa., pp. 375–387 (1977).
3. M. A. Field, Predicting the burning time of the coke residue of pulverized fuel, *Brit. Coal Utiliz. Res. Assoc. Mon. Bull.* **28**(2), 61–75 (1964).
4. P. L. Walker, Jr., F. Rusinko, Jr., and L. G. Austin, Gas reactions of carbon, *Advan. Catal.* **11**, 135–221 (1959).
5. D. Gray, J. G. Cogoli, and R. H. Essenhigh, Problems in pulverized coal and char combustion, *Advan. Chem. Ser.* **131**, 72–91 (1974).
6. M. A. Field, D. W. Gill, B. B. Morgan, and P. G. W. Hawksley, Combustion of pulverized fuel. Part 6. Reaction rate of carbon particles, *Brit. Coal Utiliz. Res. Assoc. Mon. Bull.* **31**(6), 285–345 (1967).
7. H. W. Batchelder, R. M. Busche, and W. P. Armstrong, Kinetics of coal gasification, *Ind. Eng. Chem.* **45**(9), 1856–1878 (1953).
8. C. G. von Fredersdorff and M. A. Elliot, in *Chemistry of Coal Utilization, Supplementary Volume* (H. H. Lowry, ed.) pp. 892–1022, John Wiley and Sons, New York (1963).
9. D. B. Anthony and J. B. Howard, Coal devolatilization and hydrogasification, *AIChE J.* **22**, 625–656 (1976).
10. P. L. Walker, Jr., R. J. Foresti, Jr., and C. C. Wright, Surface area studies of carbon–carbon dioxide reaction, *Ind. Eng. Chem.* **45**(8), 1703–1710 (1953).

11. W. J. Thomas, Effect of oxidation on the pore structure of some graphitized carbon blacks, *Carbon* **3**, 435–443 (1977).

12. C. Y. Wen, *Optimization of Coal Gasification Processes*, R & D Report No. 66, Interim Report No. 1, Office of Coal Research Contract No. 14-01-0001-497 (1972).

13. A. Tomita, O. P. Mahajan, and P. L. Walker, Jr., Catalysis of char gasification by minerals, *ACS Div. Fuel Chem. Preprints* **22**(1), 4–6 (1977).

14. G. D. Sergeant and I. W. Smith, Combustion rates of bituminous coal char in the temperature range 800 to 1700 K, *Fuel* **52**, 52–57 (1973).

15. R. J. Hamor, I. W. Smith, and R. J. Tyler, Kinetics of combustion of a pulverized brown coal char between 630 and 2200 K, *Combust. Flame* **21**, 153–162 (1973).

16. M. A. Field, Rate of combustion of size-graded fractions of char from a low-rank coal between 1200 K and 2000 K, *Combust. Flame* **13**, 237–252 (1969).

17. M. F. R. Mulcahy and I. W. Smith, Kinetics of combustion of pulverized fuel: A review of theory and experiment, *Rev. Pure and Appl. Chem.* **19**, 81–108 (1969).

18. J. A. Arthur, Reactions between carbon and oxygen, *Trans. Faraday Soc.* **47**, 164–178 (1951).

19. M. Rossberg, Experimental results concerning the primary reactions in the combustion of carbon, *Z. Elecktrochem.* **60**, 952–956 (1956).

20. R. J. Day, Kinetics of the Carbon–Oxygen Reaction at High Temperatures, Ph.D. Thesis, The Pennsylvania State University, University Park, Pa. (1949).

21. I. W. Smith and R. J. Tyler, The reactivity of a porous brown coal char to oxygen between 630 and 1812 K, *Combust. Sci. Technol.* **9**, 87–94 (1974).

22. R. H. Essenhigh, R. Froberg, and J. B. Howard, Predicted burning rates of single carbon particles, *Ind. Eng. Chem.* **57**(9), 33–43 (1965).

23. J. B. Howard and R. H. Essenhigh, Mechanism of solid particle combustion with simultaneous gas-phase volatiles combustion, in *Eleventh Symposium (International) on Combustion*, pp. 399–408, The Combustion Institute, Pittsburgh, Pa. (1967).

24. M. A. Nettleton, Burning rates of devolatilized coal particles, *Ind. Eng. Chem. Fundam.* **6**(1), 20–25 (1967).

25. I. W. Smith, Kinetics of combustion of pulverized semi-anthracite in the temperature range 1400–2200 K, *Combust. Flame* **17**, 421–428 (1971).

26. I. W. Smith and R. J. Tyler, Internal burning of pulverized semi-anthracite: The relation between particle structure and reactivity, *Fuel* **51**, 312–321 (1972).

27. R. J. Tyler, H. J. Wouterhood, and M. F. R. Mulchahy, Kinetics of the graphite–oxygen reaction near 1000 K, *Carbon* **14**, 271–278 (1976).

28. N. M. Laurendeau, Heterogeneous kinetics of coal char gasification and combustion, in *Progress in Energy and Combustion Science*, Pergamon Press, London, England (1978), in press.

29. S. Ergun and M. Mentser, in *The Chemistry and Physics of Carbon* (P. L. Walker, ed.), Vol. 1, pp. 203–263, Marcel Dekker, Inc., New York (1965).

30. W. K. Lewis, E. R. Gilliland, and G. T. McBride, Jr., Gasification of carbon by carbon dioxide in fluidized powder bed, *Ind. Eng. Chem.* **41**(6), 1213–1226 (1949).

31. J. P. Blakely and L. G. Overholser, Oxidation of ARJ graphite by low concentrations of water vapor and carbon dioxide in helium, *Carbon* **3**, 269–275 (1965).

32. E. T. Turkdogan and J. V. Vinters, Effect of carbon monoxide on the rate of oxidation of charcoal, graphite and coke in carbon dioxide, *Carbon* **8**, 39–53 (1970).

33. A. E. Reif, The Mechanism of the carbon dioxide–carbon reaction, *J. Phys. Chem.* **56**, 785–788 (1952).

34. H. J. Grabke, Oxygen transfer and carbon gasification in the reaction of different carbons with CO_2, *Carbon* **10**, 587–599 (1972).

35. R. T. Yang and M. Steinberg, The reactivity of coal chars with CO_2 at 1100–1600°C, *ACS Div. Fuel Chem. Preprints* **22**(1), 12–16 (1977).

36. A. M. Mayers, The rate of reduction of carbon dioxide by graphite, *J. Am. Chem. Soc.* **56**, 70–76 (1934).
37. S. A. Pursley, R. A. Matula, and O. W. Witzell, The kinetics of carbon dioxide and carbon formation from carbon monoxide, *J. Phys. Chem.* **70**, 3768–3770 (1966).
38. C. Y. Chen, Mechanism of the Steam-Carbon Reaction in a Flow System, Ph.D. Dissertation, University of Illinois, Urbana, Illinois (1951).
39. E. S. Golovina and G. P. Khaustovich, Interaction of carbon with carbon dioxide at high temperatures, *Teplofiz. Vys. Temp.* **2**, 267–273 (1974).
40. A. K. Mehta, *Mathematical Modeling of Chemical Processes for Low Btu Gasification of Coal for Electric Power Generation*, ERDA Report No. FE-1545-26 (August 1976).
41. R. W. Taylor and D. W. Bowman, *Rate of Reaction of Steam and Carbon Dioxide with Chars Produced from Subbituminous Coals*, Lawrence Livermore Laboratory, Report No. UCRL-52002 (1976).
42. R. L. Coates, Kinetic data from a high temperature entrained flow gasifier, *ACS Div. Fuel Chem. Preprints* **22**(1), 84–92 (1977).
43. T. H. Blakeley, Gasification of carbon in carbon dioxide and other gases at temperatures above 900 K, in *Proceedings of the 4th Conference on Carbon, Buffalo, New York, 1959*, 95–105 (1960).
44. E. A. Gulbransen, K. F. Andrew, and F. A. Brassart, Reaction of graphite with carbon dioxide at 1000–1600°C under flow conditions, *Carbon* **2**, 421–429 (1965).
45. E. E. Peterson and C. C. Wright, Reaction of artificial graphite with carbon dioxide, *Ind. Eng. Chem.* **47**(8), 1624–1634 (1955).
46. J. Gadsby, C. N. Hinshelwood, and K. W. Sykes, The kinetics of the reactions of the steam–carbon system, *Roy. Soc. London Proc.* **187 A**, 129–151 (1946).
47. A. Linares, O. P. Mahajan, and P. L. Walker, Jr., Reactivities of heat-treated coals in steam, *ACS Div. Fuel Chem. Preprints* **22**(1), 1–3 (1977).
48. A. M. Mayers, The rate of oxidation of graphite by steam, *J. Am. Chem. Soc.* **56**, 1879–1881 (1934).
49. H. E. Klei, J. Sahagian, and D. W. Sundstrom, Kinetics of the activated carbon–steam reaction, *Ind. Eng. Chem. Process Des. Dev.* **14**, 470–473 (1975).
50. B. E. Reide and D. Hanesian, Kinetic study of carbon–steam reaction, *Ind. Eng. Chem. Process Des. Dev.* **14**, 70–74 (1975).
51. J. A. Dotson, W. A. Koehler, and J. H. Holden, Rate of the steam–carbon reaction by a falling-particle method, *Ind. Eng. Chem.* **49**(1), 148–154 (1957).
52. H. L. Feldkirchner and J. Huebner, Reaction of coal with steam–hydrogen mixtures at high temperatures and pressures, *Ind. Eng. Chem. Process Des. Dev.* **4**, 134–142 (1965).
53. R. Piccirelli, *Survey of Surface Rates of Carbon-Gas Reactions and Carbon Vaporization: A Final Report*, submitted to KMS Industries, Contract No. F12593, Feb. 28, 1975.
54. P. P. Feistel, K. H. von Heek, H. Juntgen, and A. H. Pulsifer, Gasification of a German bituminous coal with H_2O, H_2, and H_2O-H_2 mixtures, *ACS Div. Fuel Chem. Preprints* **22**(1), 53–76 (1977).
55. C. W. Zielke and E. Gorin, Kinetics of carbon gasification, *Ind. Eng. Chem.* **47**, 820–825 (1955).
56. J. D. Blackwood and F. McGrory, The carbon–steam reaction at high pressure, *Austral. J. Chem.* **11**, 16–33 (1957).
57. J. L. Johnson, Kinetics of initial coal hydrogasification stages, *ACS Div. Fuel Chem. Preprints* **22**(1), 17–37 (1977).

10

Volatiles Combustion

J. Rand Thurgood and L. Douglas Smoot

1. Introduction

The combustion of devolatilization products is a subject which has received little attention. This is probably due to the complicated reaction mechanisms involved, and also to the prior lack of need for such information. Recent emphasis on theoretical modeling of the details of coal conversion processes has brought about increased attention on the combustion of coal volatiles. Several investigators have proposed reaction sequences which treat certain aspects of volatiles combustion, but further work will be necessary before a complete and accurate description is possible. The intent of this chapter is to review the pertinent work reported to date and to suggest a tentative course to follow using the available information. A discussion of hydrocarbon oxidation mechanisms is given, including global reaction schemes, which have been proposed to model the reactions of the higher-molecular-weight hydrocarbons.

Field et al.[1] provided one of the first reviews and discussions on the combustion of coal volatiles. A general description of volatiles composition was provided, as well as information on soot formation and burning. The recent reviews of Anthony and Howard[2] and Coates and Glassett[3] present summaries of the work done in determining the composition of the volatiles products. Suuberg et al.[4] have reported product composition resulting from lignite pyrolysis. Although several experimental methods have been em-

J. Rand Thurgood • Ph.D. Candidate, Chemical Engineering Department, Brigham Young University, Provo, Utah. *L. Douglas Smoot* • Professor of Chemical Engineering, Brigham Young University, Provo, Utah

ployed in determining pyrolyzate compositions, most of them can be placed into one of the two broad categories: (1) methods using relatively slow heating rates (on the order of 10^3 K s^{-1} or less) and (2) methods using rapid heating rates (on the order of 10^4 K s^{-1} or more). Both ranges of heating rates yield the same types of product components, which include tars, various gases (CO_2, CO, CH_4, H_2, and others), and water. Table 1 of Chapter 8 summarizes overall pyrolyzate composition, including tars, at 1300 K. Typical gaseous components are presented in Table 1 of this chapter. The concentration levels of each component differ significantly, depending upon the particular coal studied, the heating rate used, etc. For example, methane concentration varied from 0.7%, using a fast heating rate,[13] to 54.3%, using a slow heating rate.[6] Regardless of the method used, definitive description of the volatiles composition remains incomplete.

The modeling approach adopted by the authors is based on a reaction mechanism for methane oxidation. Methane, although reported in varying concentration levels, appears to be a product component common to all studies. Using methane oxidation as a basis, it is possible to expand the combustion description to include other hydrocarbons and pyrolyzates. A majority of the reactions involved in a complete methane oxidation scheme are also found in mechanisms for other volatiles. Oxidation reactions for carbon monoxide, carbon dioxide, and hydrogen are included in reaction sequences for methane systems. Any water vapor formed must also be included in an overall mechanism. A complete mechanism would necessarily treat reactions involving pollutant species as well. Examples of such mechanisms are given by Edelman et al.[25] and by Osgerby.[26] Reactions of pollutant species are considered in Chapter 11.

2. Methane Oxidation

Methane oxidation at combustion temperatures has been studied in more detail than has the oxidation of any other hydrocarbon. Nevertheless, there remain questions as to the importance of certain species and reactions. Acquisition of reliable kinetic rate data associated with the reactions has also been difficult and is not nearly as well defined as is the accompanying thermodynamic data. Several mechanisms have been proposed during the last decade; some of the earlier mechanisms such as those of Chinitz[27] and Bowman and Seery[28] consisted of relatively few reactions but gave reasonable results and provided the basis for recently proposed reaction mechanisms. Most methane oxidation mechanisms presented to date have had limited comparison to experimental data. Few studies have actually incorporated proposed mechanisms into descriptive computer codes for modeling

Table 1. Composition of Gaseous Coal Pyrolysis Products

	Slow Heating Methods[a,d] Gas composition, vol. %				
Reference and conditions	CO_2	CO	H_2	CH_4	Other
(5) Flow tube, 530–650°C	10.6	4.9	25	42.3	7.6[b] 5.6 C_2H_6 4.0 C_3H_8
(6) Thin layer of particles	4.1	15.8	16.9	54.3	8.9[b]
heated to 1050–1500°C	2.4	10.0	21.8	52.8	13.0[b]
	6.1	20.6	13.1	50.3	9.9[b]
(7), (8) 600°C fluidized bed	12.0	12.3	14.7	50.5	10.5[b]
(7), (8) 600°C Fischer assay	10.3	8.7	22.4	54.2	3.7[b]

Rapid Heating Methods[c]

Heating method and references	CO_2	CO	H_2	CH_4	C_2H_6	Other, C_2
Arc image						
(9)	5.7	14.1	62.0	7.0	7.6	—
(10)	—	—	2.8	—	14.6	0.8
Plasma jet						
(11)	0.0	16.9	62.2	1.3	19.6	—
(12)	0.3	18.4	48.0	0.8	30.2	0.3
(13)	—	32.6	60.3	0.7	6.0	0.3
(14)	0.7	37.2	47.5	3.0	11.6	—
Laser						
(15)	—	2.0	45.4	4.8	20.9	14.4
(16)	—	—	—	18	15	7
(17)	8.7	22.5	52.2	5.1	10.6	—
Flash						
(15)	—	0.5	69.2	4.1	7.8	16.1
(18)	0.1	10.2	67.0	1.3	15.5	4.6
(19)	0.8	24.5	36.8	29.1	1.5	1.4
(20)	—	6	69	19	3	4
Tube furnace						
(21)	7.5	35.4	22.5	18.2	4.3	10.5
Carbonization						
(15)	0.4	7.4	55.6	31.5	0.1	4.6
(19)	1.7	5.7	56.7	29.6	0.1	3.8
(22)	3.1	17.0	61.7	15.8	—	1.7
(23)	3.0	6.0	56.7	26.0	—	2.5
(24)	2.9	7.5	54.7	25.4	—	5.3
(17)	0.4	7.4	55.6	31.5	0.05	4.6

[a]Data used with permission from Anthony and Howard.[2]
[b]By difference, composition unknown.
[c]Data used with permission of the publishers, IPC Business Press Ltd.©, Bond *et al.*[12]
[d]Additional data not included were recently reported by Solomon.[47]

combustion processes, but rather have used only partial solutions for experimental comparisons.

A recent study by Hecker[29] did propose an integrated reaction model to describe the kinetics, propagation, and suppression of methane/air flames. Hecker proposed a methane oxidation mechanism which was used in a computer code developed by Smoot et al.[30] The predicted results compared very well with velocity values and with temperature and concentration profiles of most species, as measured by other investigators.[30] Jachimowski[31] proposed a mechanism which was compared to experimental results of oxygen atom formation during methane oxidation behind shock waves. Bowman[32] conducted similar experiments in shock-initiated methane oxidation and devised a reaction mechanism which compared reasonably well to experimental results. These three mechanisms and two others, one by Edelman et al.[25] and one by Tsuboi,[33] are summarized in Table 2. Most of the reactions shown in Table 2 are found in each of the mechanisms. However, the value of the rate constant differs greatly from one study to another in many cases. Although the kinetic data for the reactions are not as well determined as corresponding thermodynamic data, good results have been obtained using the information that is available.

Engleman[34] undertook an extensive study to determine which reactions (with accompanying kinetics) should be considered in methane/air combustion, including an evaluation of kinetic data for 322 reactions in methane/air combustion. Engleman's review includes an evaluation of the probability of occurrence of each individual reaction in addition to recommended kinetic rate constants. Table 2 also contains the 24 reactions (and their recommended reaction rates) which Engleman indicates to be probably important; these reactions are those indicated to be from reference 34. Additional reactions in Table 2 listed to be possibly important by Engleman include reactions 6, 18, 21, 25, 26, 28, and 33. All other reactions shown in Table 2 and not mentioned above are deemed to be probably unimportant. Parametric computations by Smoot et al.[30] showed that reactions 1, 2, 5, 6, 12, 14, 18, 24, 27, 28, 33, and 37 were of major importance and reactions 17, 20, 25, 31, and 41 were insignificant in the propagating laminar flames studied.

Each of the sets of reaction mechanisms presented in Table 2 is similar and compares well with Engleman's recommendations. Differences occur mainly in the numerical values of the kinetic rates. Hecker's mechanism aligns most closely with Engleman's recommended reactions and has been more extensively tested against experimental data. If reactions 8, 11, and 23 of Table 2 were added to Hecker's mechanism, it would then include all of Engleman's reactions listed as "probably important." It should be noted that for several of the recommended reactions the rate constants used by Hecker and those shown by Engleman differ significantly. Hecker's mechanism and accompanying rate constants have been tested against experimental results

Table 2. Methane Oxidation Mechanisms[a,b]

Reaction	A	N	E/R	Reference
1. $CH_4 + OH \rightarrow CH_3 + H_2O$	10.48*	0	2520	30, 34
	10.48	0	3020	31
	11.78	0	6290	32
	11.54	0	4500	25
	10.40	0	2530	33
2. $CH_4 + H \rightarrow CH_3 + H_2$	11.30*	0	5990	25, 30
	11.10	0	5990	31
	1.35	3.0	4400	32
	7.70	1.0	5040	33
	10.80	0	5990	34
3. $CH_4 + O \rightarrow CH_3 + OH$	10.30*	0	3470	25, 30
	10.30	0	4640	31
	10.30	0	4560	32, 34
	7.00	1.0	4030	33
4. $CH_4 + M \rightarrow CH_3 + H + M$	14.30	0	44500	31, 32
	10.40	0	48520	33
5. $CH_3 + O \rightarrow CH_2O + H$	10.85*	0	500	30
	11.11	0	1010	31
	11.00	0	0	32
	10.28	0	0	25
	10.78	0	0	33
	10.70	0	0	34
6. $CH_3 + O_2 \rightarrow CH_2O + OH$	10.48	0	8810	30
	11.00	0	0	31
	10.30	0	· 0	32
	11.00	0	750	25
	8.95	0	6010	33
7. $CH_3 + O_2 \rightarrow CHO + H_2O$	7.30	0	0	25
8. $CH_3 + OH \rightarrow CH_2 + H_2O$	7.80*	0.7	1010	34
9. $CH_3 + OH \rightarrow CH_2O + H_2$	9.60	0	0	32, 33
10. $CH_3 + O \rightarrow CHO + H_2$	11.00	0	0	25
11. $CH_3 + H \rightarrow CH_2 + H_2$	11.2*	−0.3	6260	34[c]
12. $CH_2O + M \rightarrow CO + H_2 + M$	13.30*	0	17620	30, 31
13. $CH_2O + M \rightarrow CHO + H + M$	9.60	0	18500	32
14. $CH_2O + OH \rightarrow CHO + H_2O$	10.40*	0	500	30
	10.30	0	0	31
	11.73	0	3170	32
	10.48	0	0	25
	11.00	0	2050	33
	7.50	1.0	0	34
15. $CH_2O + O \rightarrow CHO + OH$	10.48*	0	0	25, 30
	10.70	0	2300	32
	10.62	0	2050	33
	8.00	1.0	1760	34
16. $CH_2O + H \rightarrow CHO + H_2$	10.23*	0	1510	25, 30
	10.13	0	1890	32
	10.52	0	2170	33
	7.10	1.0	1610	34

continued overleaf

Table 2 (Continued)

Reaction	A	N	E/R	Reference
17. $CHO + M \rightarrow CO + H + M$	9.30*	0.5	14500	25, 30, 31
	9.70	0	9570	32
	13.02	0	14920	33
	18.17	−1.83	8400	34[c]
18. $CHO + O_2 \rightarrow CO + HO_2$	10.48*	0	0	30
	10.62	0	7200	32
	9.90	0	0	33
19. $CHO + O_2 \rightarrow CO_2 + OH$	8.87	0.5	0	25
20. $CHO + O \rightarrow CO + OH$	8.73*	0.5	0	25, 30
	11.00	0	0	32, 33
	8.50	1.0	250	34
21. $CHO + O \rightarrow CO_2 + H$	8.73	0.5	0	25
22. $CHO + OH \rightarrow CO + H_2O$	11.00*	0	0	30, 32, 33
	10.00	0	0	31
	10.48	0	0	25
	7.50	1.0	0	34
23. $CHO + H \rightarrow CO + H_2$	11.30	0	0	32
	11.00	0	0	33
	9.20*	0.5	0	34
24. $CO + OH \rightarrow CO_2 + H$	8.74*	0	540	25, 30
	9.60	0	4030	31, 32
	9.00	0	2290	33
	3.18	1.3	380	34
25. $CO + O + M \rightarrow CO_2 + M$	12.56	−1.0	1260	30
	7.60	0	0	31, 33
	9.77	0	2060	32
	13.26	−1.0	1010	25
26. $CO + O_2 \rightarrow CO_2 + O$	10.20	0	20630	31
	9.48	0	12590	25
27. $H + O_2 \rightarrow OH + O$	11.34*	0	8460	30, 32, 33
	14.08	−0.9	8370	31
	11.35	0	4250	25
	12.53	−0.3	8530	34[c]
28. $H + O_2 + M \rightarrow HO_2 + M$	13.15*	0	−500	30, 32
	10.00	0	0	25
	9.23	0	−480	33
29. $H + OH + H_2O \rightarrow H_2O + H_2O$	17.15	−2.0	0	32
30. $H + OH + M \rightarrow H_2O + M$	13.85*	−1.0	0	30
	15.92	−2.0	0	31, 32
	11.00	0	0	25
	9.73	0	0	33
	16.30	−2.0	0	34
31. $H + O + M \rightarrow OH + M$	12.60*	−1.0	0	30
	10.00	0	0	25
	10.73	0	0	33
	9.90	0	0	34

continued

Table 2 (Continued)

Reaction	A	N	E/R	Reference
32. $H + H + M \rightarrow H_2 + M$	13.30*	−1.0	0	30
	11.81	−1.0	0	31
	7.95	0	−4140	32[c]
	9.70	0	0	25
	7.34	0	0	33
	7.95	0	−4140	34[c]
33. $H + HO_2 \rightarrow OH + OH$	11.30*	0	1010	30
	11.40	0	950	32
	10.00	0	0	25
34. $H_2 + O_2 \rightarrow H + HO_2$	10.10*	0.2	29440	30[c]
	10.48	0	31740	33
35. $H_2 + O_2 \rightarrow OH + OH$	10.23	0	24230	31
	10.23	0	12440	25
36. $H_2 + O \rightarrow H + OH$	10.23*	0	4760	30
	11.32	0	6920	31
	11.50	0	7540	32
	10.23	0	2390	25
	10.10	0	4690	33
	7.18	1.0	4550	34[c]
37. $H_2 + OH \rightarrow H_2O + H$	10.34*	0	2620	30
	10.72	0	3270	31
	11.46	0	5530	32
	10.34	0	1300	25
	10.40	0	2620	34
38. $HO_2 + O \rightarrow O_2 + OH$	10.40*	0	0	30
	9.78	0	0	33
	10.70	0	500	34
39. $HO_2 + OH \rightarrow O_2 + H_2O$	10.40*	0	0	30
	9.78	0	0	33
	10.70	0	500	34
40. $OH + OH \rightarrow H_2O + O$	9.78*	0	340	30
	10.74	0	3520	31, 32
	9.76	0	200	25
	9.79	0	500	33, 34
41. $O + O + M \rightarrow O_2 + M$	12.60*	−1.0	0	30
	14.11	−1.0	170	31
	5.65	0.2	−7180	32[c]
	8.97	0	0	25
	10.15	−0.8	210	34[a]

[a]Forward rate constant: $k = 10^4 T^N \exp(-E/RT)$. SI units: kmol, m³, s, K.
[b]Of the 41 reactions shown, all but reactions 4, 7, 9, 10, 13, 19, 21, 26, 29, and 35 are recommended for inclusion in the CH_4 O_2 mechanism. An asterisk (in the first column of numbers) indicates the recommended rate value for the remaining 31 reactions.
[c]Calculated from reverse rate constant and equilibrium constant.

while Engleman's recommendations reflect an extensive review of the work reported in the literature. Where significant differences occur, the rate constants listed by Hecker are being used until further experimental comparisons can be made on those suggested by Engleman. The reactions listed by Hecker as insignificant should be retained in the mechanism until proven unnecessary in all flames, not just those specific to that study. The recommended reactions and accompanying kinetic rate values are identified in footnote *b* of Table 2.

3. Oxidation of Other Hydrocarbons

Table 1 summarized pyrolyzate gas compositions reported in studies using both slow and rapid coal heating rates. In addition to H_2 and CH_4, pyrolysis products include such components as C_2H_6, C_2H_4, C_2H_2, C_3H_8, and tars. High-temperature oxidation of these latter species is not yet well understood. Work has been reported on ethane, ethylene, acetylene, and propane. Acetylene oxidation seems to have received the most attention, and its mechanism is perhaps the best characterized of the four.

Williams and Smith[35] presented an extensive review of the work done on acetylene oxidation up to 1969. A discussion of the reactions involved was made and an 11-step oxidation mechanism was proposed as shown in Table 3. Recently Vandooren and Van Tiggelen[37] also discussed reaction mechanisms in acetylene–oxygen flames, and Shaviv and Bar-Nun[36] proposed a kinetic scheme for the oxidation of acetylene by water vapor. This mechanism, also given in Table 3, consists of 13 reactions involving the decomposition of acetylene.

Fenimore and Jones[38] investigated the decomposition of ethane and ethylene in premixed flames and provided impetus for further investigation. One of the best efforts of investigating oxidation of ethane was conducted by Bowman.[39] He studied ethane oxidation behind reflected shock waves with temperatures ranging from 1300 to 2000 K. The experimental results were compared with an analytical study which incorporated an 11-step reaction mechanism. The model neglected all of the ethane consumption reactions indicated by Fenimore and Jones and also the reactions involving ethylene suggested by Hoare and Patel[40] and by Baldwin *et al.*[41] Bowman assumed that ethane breaks down only by the reaction

$$C_2H_6 + M \leftrightarrows CH_3 + CH_3 + M \tag{1}$$

with the methyl radicals further reacting as in a methane oxidation scheme. The model, although simple, produced reasonably good agreement between calculated and measured reaction times. Tsuboi[33] included reaction (1),

Table 3. *Acetylene Oxidation Mechanism*[a]

Reaction	A	N	E/R	Reference
1. $C_2H_2 + C_2H_2 \rightarrow C_4H_2 + 2H$	13.60	0	20630	36
2. $C_2H_2 + C_2H_2 \rightarrow C_4H_3 + H$	13.98	0	22300	35
3. $C_2H_2 + C_4H_2 \rightarrow C_6H_2 + 2H$	14.04	0	17110	36
4. $C_2H_2 + C_6H_2 \rightarrow C_8H_2 + 2H$	14.18	0	15350	36
5. $C_8H_2 + M \rightarrow C_8 + 2H + M$	14.30	0	22650	36
6. $C_2H_2 + H \rightarrow C_2H + H_2$	14.30	0	9760	35, 36
7. $C_2H_2 + OH \rightarrow C_2H + H_2O$	12.78	0	3520	35, 36
8. $C_2H_2 + M \rightarrow C_2H + H + M$	14.78	0	40260	35
9. $C_2H_2 + O_2 \rightarrow 2CO + 2H$	14.00	0	19120	35
10. $C_2H_2 + O \rightarrow CO + CH_2$	12.70	0	1260	35, 36
11. $C_2H_2 + O \rightarrow C_2H + OH$	15.51	−0.6	8560	35
	14.30	0	10070	36
12. $C_2H + OH \rightarrow CO + CH_2$	12.78	0	0	36
13. $C_2H + O_2 \rightarrow CO + CHO$	13.00	0	3520	35, 36
14. $C_2H + O \rightarrow CH + CO$	13.70	0	0	35
15. $CH_2 + OH \rightarrow CHO + H_2$	13.85	0	0	35
	13.48	0	0	36
16. $CH_2 + O \rightarrow CHO + H$	13.48	0	0	35, 36
17. $CHO + M \rightarrow CO + H + M$	13.85	0	7550	36

[a]Forward rate constant: $k = 10^A T^N \exp(-E/RT)$. SI units: kmol, m³, s, K.

those shown by Fenimore and Jones, and several other reactions dealing with ethylene. His reaction scheme is shown in Table 4.

Propane oxidation at high temperatures has also been studied somewhat. Perhaps the most informative work on the propane oxidation mechanism is that of Chinitz and Baurer,[42] whose mechanism consists of 69 reactions. The scheme includes reactions for the oxidation of methane, ethane, ethylene, acetylene, and hydrogen, as well as for propane. Although many of the reactions involve hydrocarbon radicals, a full set of kinetic rate data is presented with the mechanism. Hammond and Mellor,[43] following work of earlier investigators, also presented a combustion mechanism for propane which consisted of 9 reactions, initiated by a global step which takes propane directly to carbon monoxide and water at an infinitely fast rate.

Other hydrocarbons may be present to some extent in the pyrolysis products; these will not be discussed. Oxidation mechanisms of such hydrocarbons rapidly become quite complex with many difficulties arising in the acquisition of experimental data. For this reason, most work has been centered on the mechanisms presented here and on global reaction schemes to accommodate the oxidation of the higher-molecular-weight hydrocarbons.

Table 4.   Ethane/Ethylene Oxidation Mechanism^{a,b}

Reaction	A	N	E/R
1. $CH_3 + CH_3 + M \rightarrow C_2H_6$	7.00	0	0
2. $CH_3 + C_2H_6 \rightarrow CH_4 + C_2H_5$	8.30	0	5290
3. $C_2H_5 + M \rightarrow C_2H_4 + H + M$	10.58	0	19120
4. $C_2H_6 + O \rightarrow C_2H_5 + OH$	10.95	0	3730
5. $C_2H_6 + H \rightarrow C_2H_5 + H_2$	11.30	0	5410
6. $C_2H_6 + OH \rightarrow C_2H_5 + H_2O$	11.19	0	2890
7. $C_2H_4 + O \rightarrow CH_3 + CHO$	9.70	0	810
8. $C_2H_4 + OH \rightarrow CH_3 + CH_2O$	10.00	0	480

^{a}Table used with permission from Tsuboi.[33]
^{b}Forward rate constant: $k = 10^4 T^N \exp(-E/RT)$. SI units: kmol, m^3, s, K.

4. Global Reactions

As has been shown, mechanisms for hydrocarbon oxidation are not exact, and those proposed thus far lack accurate kinetic data for many of the involved reactions. More information is now available about the nature of the tar products as they devolatilize from the coal,[47] and these products amount to a substantial part of the total pyrolysis products (see Table 1, Chapter 9). To quantify these processes, several investigators have described hydrocarbon combustion through global reactions which take the various hydrocarbons to carbon monoxide and other products and allow them to further react as described in the mechanisms previously presented. Two such global descriptions appear to have become most widely recognized. One of these was presented by Hammond and Mellor.[44] They suggested that the most common features of hydrocarbon combustion mechanisms are (1) rapid, partial oxidation of the hydrocarbon to carbon monoxide and water, (2) a rate-limiting step for oxidation of carbon monoxide as follows:

$$CO + OH \leftrightarrows CO_2 + H \tag{2}$$

and (3) a series of free radical production reactions not involving carbon. A global reaction was proposed to allow for the partial oxidation of gaseous hydrocarbons as follows:

$$C_nH_m + (n/2 + m/4)O_2 \rightarrow nCO + (m/2)H_2O \tag{3}$$

Reaction (3), which was assumed to have an infinitely fast forward rate, was coupled with nine other reactions to form a simple, but general, hydrocarbon combustion mechanism. It is generally agreed that the partial oxidation of the given hydrocarbon is relatively fast and also that the oxidation of carbon monoxide is the rate-limiting step.[1]

Table 5.   Constants from Reference 46

Hydrocarbon	A	E/R
Long-chain	59.8	12.20×10^3
Cyclic	2.07×10^4	9.65×10^3

Edelman and Fortune[45] and Siminski *et al.*[46] also proposed a global reaction where reaction products are CO and H_2 rather than CO and H_2O as proposed by Hammond and Mellor. Siminski *et al.*[46] have correlated kinetic rates for heavy hydrocarbon combustion for this global reaction:

$$C_nH_m + (n/2)O_2 \rightarrow (m/2)H_2 + nCO \qquad (4)$$

Different finite rates are specified for long-chained hydrocarbons and for cyclic hydrocarbons. The rate is given as

$$dC_H/dt = -ATp^{0.3}(C_H)^{0.5}(C_0) \exp(-E/RT) \qquad (5)$$

where T is the temperature in $^\circ$K, p is the pressure in pascals, C_H and C_0 are molar concentrations in kmol m^{-3}, t is the time in seconds, E is the activation energy in kcal/kmol, and the constants from reference 46, are as given in Table 5. Edelman and Fortune[45] coupled the global reaction with nine elementary reactions to form a "quasi-global" model. The model was compared to the detailed kinetic model of Chinitz and Baurer[42] for propane in determining temperature and pressure histories. The two models gave results which compared reasonably well. Edelman *et al.*[25] also include the global reaction in their methane/air mechanism, which takes into account the generation of NO_x.

Once a global scheme is selected, a determination of the carbon-to-hydrogen ratio to be used for the pseudomolecule C_nH_m in Eq. (4) must be made. One possible approach is based upon ultimate analysis data of the char. All coal components other than char can be grouped together, and a simple material balance can be made to determine the carbon-to-hydrogen ratio of the group. The group ratio can then be used as a rough estimate of stoichiometric coefficients m and n for the global reaction. Inasmuch as the tars usually represent the major fraction of the volatiles products, the estimate should give fairly good results.

5. Recommendations

In order to accurately describe complete combustion of the volatiles, the mechanisms for each of the volatiles components (including the pollutants) should be combined into an overall mechanism. At present, this poses three significant problems: (1) insufficient knowledge of the product composition, (2) questionable reliability of the mechanistic kinetic rate data, and (3) computational difficulties and costs associated with computer modeling of very large systems of reactions. Additional study and investigation may advance combustion technology enough to provide solutions to the first two problems. The third difficulty may also be reduced as computer technology continues to develop and provide more efficient and rapid computational ability.

It is suggested that the advantages of the mechanistic and global approaches be coupled to provide an approximate description of the combustion process. This may be done by first selecting a methane oxidation mechanism as mentioned. With methane oxidation as a base, the mechanisms for ethane, ethylene, and acetylene could be added as the tradeoff between computational costs and desired accuracy warrants. One of the global reactions should be included to allow for oxidation of the heavier hydrocarbon tars, whose detailed composition and reaction mechanisms are not known. This system could then be modified as advances in related technology are made.

6. Notation

A Exponent (see Table 2)
C Concentration (kmol m^{-3})
E Activation energy (J kmol^{-1})
k Reaction rate constant for:
  (a) bimolecular reaction
    (m^3 kmol^{-1} s^{-1})
  (b) termolecular reaction
    (m^6 kmol^{-2} s^{-1})
N Exponent (see Table 2)
p Pressure (Pa)

R Gas constant (J kmol^{-1} K^{-1})
r Volumetric reaction rate (kg m^{-3} s^{-1})
T Temperature (K)

Subscripts

H hydrocarbon, C_nH_m
O oxygen (O_2)
m hydrogen number
n carbon number

7. References

1. M. A. Field, D. W. Gill, B. B. Morgan, and P. G. W. Hawkesly, *Combustion of Pulverised Coal*, pp. 175–185, The British Coal Utilization Research Association, Leatherhead, Surrey, England (1967).
2. D. B. Anthony and J. B. Howard, Coal devolatilization and hydrogasification, *AICHE J.* **22**(4), 625–656 (1976).
3. R. L. Coates and J. M. Glassett, *High Rate, High Temperature Pyrolysis of Coal, Final Technical Report*, OCR Contract No. 14-32-0001-1207, Brigham Young University, Provo, Utah (June 1974).

4. E. M. Suuberg, W. A. Peters, and J. B. Howard, Product composition and kinetics of lignite pyrolysis, *ACS Div. Fuel Chem. Preprints* **22**(1), 112–136 (1977).
5. A. Sass, Garrett's Coal Pyrolysis Process, paper presented at the 65th Annual AIChE Meeting, New York (November 30, 1972).
6. R. Loison and F. Chauvin, Pyrolyse rapide du charbon, *Chem. Ind. (Paris)* **91**, 269 (1964).
7. W. Peters, Stoff-und warmeubergang bei der schnellentgasung feinkorniger brennstoffe, *Chem. Ing. Tech.* **32**, 178 (1960).
8. W. Peters and H. Bertling, Kinetics of the rapid degasification of coals, *Fuel* **44**, 317–331 (1965).
9. E. Rau and R. T. Eddinger, Decomposition of coal in arc-image furnace, *Fuel* **43**, 246 (1964).
10. K. Littlewood, I. A. McGrath, The Use of an Arc-Image Furnace for the Formation of Acetylene Directly from Coal—Some Preliminary Results, paper C9 presented at the Fifth International Conference on Coal Science, Cheltenham, U.K. (May 1963).
11. R. D. Graves, W. Kawa, and P. S. Lewis, Reactions of coal in a plasma jet, *Amer. Chem. Soc. Div. Fuel Chem. Preprints* **8**(3), 118 (1964).
12. R. L. Bond, W. R. Ladner, and G. I. T. McConnell, Reactions of coal in a plasma jet, *Fuel* **45**, 381–395 (1966).
13. T. Aust, W. R. Ladner, and G. I. T. McConnell, in Sixth International Conference On Coal Science, Seventh Session, Muenster, Germany (June 1965).
14. L. L. Anderson, G. R. Hill, E. H. McDonald, and M. J. McIntosh, Flash heating and plasma pyrolysis of coal, in *Chemical Engineering Progress Symposium, Series No. 85, Vol. 64*, pp. 81–88, American Institute of Chemical Engineers, New York (1968).
15. A. G. Sharkey, J. L. Schultz, and R. A. Friedel, Gases from flash and laser irradiation of coal, *Nature* **202**, 988–989 (1964).
16. W. K. Joy, W. R. Ladner, and E. Pritchard, Laser heating of pulverized coal in the source of a time of flight mass spectrometer, *Fuel* **49**, 26–38 (1970).
17. F. S. Karn, R. A. Friedel, and A. G. Sharkey, Coal pyrolysis using laser irradiation, *Fuel* **48**, 297–303 (1969).
18. E. Rau and L. Seglin, Heating of coal with light pulses, *Fuel* **43**, 147–157 (1964).
19. C. O. Hawk, M. D. Schlesinger, and R. W. Hiteshure, Flash irradiation of coal, *Rep. Invest. U.S. Bur. Min.*, No. 6264, 1–7 (1963).
20. A. F. Granger and W. R. Ladner, The flash heating of pulverized coal, *Fuel* **49**, 17–25 (1970).
21. R. J. Eddinger, L. D. Friedman, and E. Rau, Devolatilization of coal in a transport reactor, *Fuel* **45**, 245–252 (1966).
22. D. Fitzgerald and D. W. van Krevelen, Chemical structure and properties of coal XXI— The kinetics of coal carbonization, *Fuel* **38**, 17–36 (1970).
23. I. E. Korobchaniskii and M. D. Kuznetsov, *Design of By-Product Recovery Units of Coke Plants*, p. 9, Asia Publishing House, New York (1932).
24. V. B. Lewes, *The Carbonization of Coal*, p. 176, D. Van Nostrand, New York (1914).
25. R. B. Edelman, J. Boccio, and G. Weilerstein, The Roles of Mixing and Kinetics in Combustion Generated NO_x, General Applied Science Labroatories, Paper 227, AIChE Symposia, Detroit, Michigan (June 1973).
26. I. T. Osgerby, Literature review of turbine combustor modeling and emissions, *AIAA J.* **12**(6), 743–754 (1974).
27. W. Chinitz, A theoretical analysis of nonequilibrium methane/air combustion, *Pyrodynamics* **3**, 197–219 (1965).
28. C. T. Bowman, and D. J. Seery, Investigation of NO formation kinetics in combustion processes: The methane–oxygen–nitrogen reaction, in *Emissions from Continuous Combustion Systems* (W. Cornelius and W. G. Agnew, eds.), pp. 123–139, Plenum Press, New York (1971).
29. W. C. Hecker, A Theoretical Study of Kinetics, Propagation, and Suppression of Methane–

Air Flames, M.S. Thesis, Chemical Engineering Department, Brigham Young University, Provo, Utah (April 1975).

30. L. D. Smoot, W. C. Hecker, and G. A. Williams, Prediction of propagating methane–air flames, *Combust. Flame* **26**, 323–342 (1976).

31. C. J. Jachimowski, Kinetics of oxygen atom formation during the oxidation of methane behind shock waves, *Combust. Flame* **23**, 233–248 (1974).

32. C. T. Bowman, Non-equilibrium radical concentrations in shock-initiated methane oxidation, in *Fifteenth Symposium (International) on Combustion*, pp. 869–882, The Combustion Institute, Pittsburgh Pa. (1975).

33. T. Tsuboi, Mechanism for the homogeneous thermal oxidation of methane in the gas phase, *Japan. J. Appl. Phys.* **15**(1), 159–168 (1976).

34. V. S. Engleman, *Survey and Evaluation of Kinetic Data of Reactions in Methane/Air Combustion*, EPA-600/2-76-003, Exxon Research and Engineering Co., Linden, New Jersey (January 1976).

35. A. Williams and D. B. Smith, The combustion and oxidation of acetylene, *Chem. Rev.* **70**, 267–293 (1970).

36. A. Shaviv and A. Bar-Nun, The oxidation of hydrocarbons by water vapor behind high-temperature shock-waves, *Int. J. Chem. Kinetics* **7**, 661–677 (1975).

37. J. Vandooren and P. J. Van Tiggelen, Reaction mechanisms of combustion in low pressure acetylene–oxygen flames, in *Sixteenth Symposium (International) on Combustion*, pp. 1133–1144, The Combustion Institute, Pittsburgh Pa. (1977).

38. C. P. Fenimore and G. W. Jones, The decomposition of ethylene and ethane in premixed hydrocarbon–oxygen–hydrogen flames, in *Ninth Symposium (International) on Combustion*, pp. 597–606, The Combustion Institute, Pittsburgh, Pa. (1963).

39. C. T. Bowman, An experimental and analytical investigation of the high-temperature oxidation mechanisms of hydrocarbon fuels, *Combust. Sci. Technol.* **2**, 161–172 (1970).

40. D. E. Hoare and M. Patel, Role of OH and HO_2 radicals in the slow combustion of mixtures of methane, ethane, and ethylene, *Trans. Faraday Soc.* **65**, 1325–1333 (1969).

41. R. R. Baldwin, D. E. Hopkins, and R. W. Walker, Addition of ethane to slowly reacting mixtures of hydrogen and oxygen at 500° C, *Trans. Faraday Soc.* **66**, 189–203 (1970).

42. W. Chinitz and T. Baurer, An analysis of nonequilibrium hydrocarbon/air combustion, *Pyrodynamics* **4**, 119–154 (1966).

43. D. C. Hammond and A. M. Mellor, A preliminary investigation of gas turbine combustor modelling, *Combust. Sci. Technology* **2**, 67–80 (1970).

44. D. C. Hammond and A. M. Mellor, Analytical calculation for the performance and pollutant emissions of gas turbine combustors, *Combust. Sci. Technology* **4**, 101–112 (1971).

45. R. B. Edelman and O. F. Fortune, A Quasi-Global Chemical Kinetic Model for the Finite Rate Combustion of Hydrocarbon fuels with Application to Turbulent Burning and Mixing in Hypersonic Engines and Nozzles, AIAA Paper No. 69–86, New York (January 1969).

46. V. J. Siminski, F. J. Wright, R. B. Edelman, C. Economos, and O. F. Fortune, *Research on Methods of Improving the Combustion Characteristics of Liquid Hydrocarbon Fuels*, AFAPL TR 72-74, Vols. I and II, Air Force Aeropropulsion Lab, Wright Patterson Air Force Base, Ohio (February 1972).

47. P. R. Solomon, *The Evolution of Pollutants During the Devolatilization of Coal*, Report R77-952588-3, United Technologies Research Center, East Hartford, Connecticut (November, 1977).

11

Mechanisms and Kinetics of Pollutant Formation during Reaction of Pulverized Coal

Philip C. Malte and Dee P. Rees

1. Introduction

Pollutants arising from the reaction of pulverized coal can be divided into two classifications. The first classification includes pollutants common to all industrial combustion systems: CO, UHC (unburned hydrocarbons), soot, and NO_x due to the fixation of atmospheric N_2. The second classification includes pollutants formed from the impurities in coal. Sulfurous and nitrogenous pollutant gases and flyash dominate this classification. Other pollutant impurities are chlorine, fluorine, and traces of toxic metals.

A complete understanding of pulverized-coal pollutant formation must involve three physical regions: (1) the particle, (2) the particle boundary layer, and (3) the free stream. Both chemical and physical factors must be considered. Chemical factors are: the composition and kinetics of the gases and condensable vapors emitted from coal particles during devolatilization and char reaction, and the free-stream gas composition. Physical factors are: gas-phase mass transport and particle behavior. The latter item includes the ejection velocity of volatiles from coal particles, which is important since it

Philip C. Malte • Associate Professor, Mechanical Engineering, Washington State University, Pullman, Washington. Dee P. Rees • Ph.D. Candidate, Chemical Engineering, Brigham Young University, Provo, Utah

influences mixing with the free stream, and char reaction in pores and cavities.

The present chapter is devoted primarily to describing and defining the fundamental chemical mechanisms and kinetics of pollutant formation and destruction in pulverized-coal reaction. Related information is also given in Chapters 4 and 10. Reactions rates, r, are specified by rate constants, k, which have the form

$$k = 10^4 T^N \exp(-E/RT) \tag{1}$$

Units appropriate to gas-phase, solid-decomposition, solid-gas, and surface-gas reactions are given in the Notation section (Section 7). The following combustion-generated pollutants will be examined: soot, nitrogenous pollutants (NO, NO_2, N_2O, HCN, NH_3), flyash, and sulfurous pollutants (SO_2, SO_3, H_2S). Carbon monoxide and unburned hydrocarbons are not examined, since their behavior has been reviewed in Chapter 10 and by Field et al.[1] Furthermore, these pollutants have a history of causing only minor concern in pulverized-coal combustion. An anticipated exception, due to future expansion of coal utilization, is the trace levels of complex organics, such as polynuclear aromatics; however, a detailed understanding of these species in combustion is lacking at this time. This chapter is concluded with a section devoted to pollutants arising from pulverized-coal gasification.

2. Soot

Soot is formed from volatile hydrocarbons in the fuel-rich pockets that surround the coal particles. Appleton[2] has summarized soot properties and structure for gaseous and liquid-spray combustion. Soot is primarily carbon (>90% by weight), but may include other elements such as oxygen and hydrogen. Individual soot particles have diameters between 100 and 500 Å. There is a tendency for soot particles to agglomerate.[3,4] Much of the carbon in the individual soot particles is present as embedded graphitelike crystallites, each of which is composed of several lamellae.[2] The lamellae appear to be arranged randomly and it appears that this anisotropic structure is responsible for the high resistance of soot to surface oxidation.

The formation of soot in flames has been discussed by Foster,[4] and reviewed by Homann[5] and Palmer and Cullis.[6] A characteristic formation time of 10^{-4} s was noted by Foster.[4] This rate is quite rapid, and thus, for practical applications, it is convenient to assume instantaneous soot formation whenever the local fuel-air equivalence ratio is sufficiently high. Foster[4] stated the following rule-of-thumb: Soot will form whenever the local carbon-to-oxygen atom ratio exceeds unity. This is in agreement with the equilibrium calculations of Boehman and Davison[7] at gas temperatures above 1000 K. As the gas temperature decreased progressively below 1000 K, these equilibrium calculations showed a progressive increase in sooting tendency. Soot formation under conditions of flame nonequilibrium was considered by Homann.[5] The apparent effect is to lower the critical C/O atom ratio for incipient soot formation to about 0.5 for many fuels.

The oxidation or burnoff of soot particles is markedly slower than the formation kinetics. The specific surface oxidation rate due to Lee, Thring, and Beer[8] is

$$r = 1.07(P_{O_2}/T^{1/2}) \exp{(-19,800/T)} \qquad (\text{kg m}^{-2} \text{ s}^{-1}) \qquad (2)$$

This result is valid for flame temperatures in the vicinity of 1500 K, and for oxygen partial pressures of 4–12 kPa. The linear dependence on P_{O_2} has been noted elsewhere[2] for conditions appropriate to pulverized-coal firing at atmospheric pressure.

Appleton[2] found that the mechanisms and kinetics of soot oxidation were similar to those of commercial pyrolytic graphite. He concluded that the semiempirical expression of Nagle and Strickland-Constable[9] for pyrolytic graphite was appropriate for estimates of soot oxidation in particle combustors. Figure 1, from Park and Appleton,[3] gives the specific oxidation rate of soot.

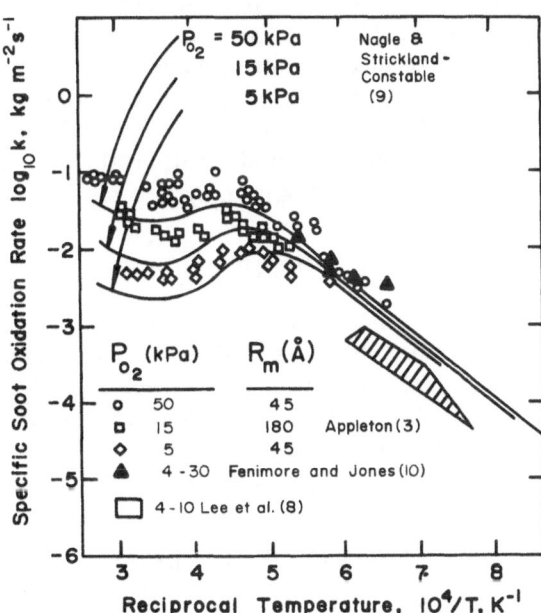

Figure 1. Measurement of the specific soot oxidation rate due to Park and Appleton,[3] Lee, Thring, and Beer,[8] and Fenimore and Jones[10] compared to the results of Nagle and Strickland-Constable[9] for pyrolytic graphite. (Figure used with permission from Park and Appleton.[3])

3. Nitrogenous Pollutants

The combustion of pulverized coal causes the formation of nitric oxide (NO), a major pollutant. Exhaust concentrations typical of pulverized-coal furnaces are 300–700 ppm by volume.[11,12] Nitric oxide results from (1) the fixation of atmospheric nitrogen and (2) the oxidation of the organically bound nitrogen contained in the coal (fuel-nitrogen). At moderate furnace gas temperatures (1700 K) the fuel-nitrogen was found to contribute about 75% of the total NO.[13]

Pollutants of secondary importance formed between N and O are nitrogen dioxide (NO_2) and nitrous oxide (N_2O). Nitrogen dioxide is apparently formed from the oxidation of a small fraction of the nitric oxide. The sum of NO and NO_2 is termed "NO_x." Nitrous oxide has been classified here as a pollutant because of its apparent reaction with singlet D-oxygen in the upper atmosphere, thereby forming two NO molecules which may react catalytically to deplete the earth's protective ozone layer. Pierotti and Rasmussen[14] and Weiss and Craig[15] have reported the measurement of exhaust levels of about 30-ppm N_2O from coal-fired power plants. Nitrous oxide may form from both atmospheric N_2 and from fuel-nitrogen.

Under reducing conditions, various cyanide and amine species can form in pulverized-coal combustion. The light molecules, hydrogen cyanide (HCN) and ammonia (NH_3), have been measured recently in several fuel-rich laboratory combustors burning fuel-nitrogen additives. At gas temperatures and fuel-nitrogen contents typical of pulverized-coal firing, but at enriched conditions of 40% excess fuel, HCN concentrations were about 100 ppm.[16,17] The principal source of the cyanides and amines is the fuel-nitrogen. However, small concentrations (≈ 1–10 ppm) may also result from N_2 fixation.

3.1. Fixation of Atmospheric Nitrogen

The fixation of atmospheric nitrogen to NO is associated principally with the high-temperature, fuel-lean regions of combustion chambers. Table 1 gives the established Zeldovich mechanism,[18] extended by reaction (3). A rough estimate of Zeldovich-NO in pulverized-coal firing can be obtained by neglecting reaction (3) and the reverse rates of reactions (1) and (2), and assuming a steady-state N-atom concentration. The resultant NO formation rate is simply

$$d[NO]/dt = 2k_{1f}[N_2][O] \tag{3}$$

At combustion temperatures below approximately 1600 K, the Zeldovich-NO formation rate is low due to the high activation energy of reaction (1). In fact, it may be overtaken by mechanisms which exhibit less temperature dependence, e.g., the nitrous oxide mechanism listed second in Table 1 and

Table 1. NO_x Kinetics Due to the Hydrogen–Oxygen Subsystem[a]

Reaction	A	N	E/R (K)	Reference
High Temperature				
(1) $N_2 + O \rightarrow NO + N$	11.13	0	37,950	19
(2) $O_2 + N \rightarrow NO + O$	6.81	1.0	3,150	19
(3) $N + OH \rightarrow NO + H$	8.8	0.5	0	20
Low Temperature				
(4) $N_2 + O + M \rightarrow N_2O + M$	6.21	0	1,600	23
(5) $N_2O + O \rightarrow 2NO$	10.66	0	12,130	23
(6) $N_2O + O \rightarrow N_2 + O_2$	10.58	0	12,130	23
(7) $N_2O + H \rightarrow N_2 + OH$	10.47	0	5,420	23
Nitrogen Dioxide				
(8) $NO + O + M \rightarrow NO_2 + M$	4.59	1.0	−4,800	23
(9) $NO_2 + O \rightarrow NO + O_2$	10.00	0	300	23
(10) $NO_2 + H \rightarrow NO + OH$	11.86	0	970	23
(11) $NO + HO_2 \rightarrow NO_2 + OH$	10.00	0	1,510	20

[a]SI units: kmol, m³, s, K.

the "prompt" NO noted below. However, if the high activation energy of reaction (1) is considered in the context of turbulent (furnace) combustion, the importance of the Zeldovich mechanism is extended to lower mean furnace gas temperatures. This is because turbulent temperature fluctuations will nonlinearly reinforce the Zeldovich formation rate.[21,22]

The low-temperature nitric oxide formation mechanism in Table 1, which proceeds through intermediate N_2O, has been detailed by Malte and Pratt.[23] For combustion applications, the reverse rates of reactions (5), (6), and (7) can be neglected. Nitrous oxide emission levels due to this mechanism are estimated to be 1–10 ppm.

Nitric oxide formation due to both of the above mechanisms is directly affected by the concentrations of the free radicals O, OH, and H. In particular, super equilibrium O-atom concentrations cause an enhancement in NO formation.[24] Thus, knowledge of the free-radical concentrations is necessary for accurate NO predictions. Under conditions of chemical equilibrium for the H C O system, as apparent in high-temperature post-flame zones, the Zeldovich mechanism yields what is termed "thermal NO."

The fixation of atmospheric nitrogen can also proceed through reaction of N_2 with small hydrocarbon fragments. This has been termed "prompt NO" by Fenimore,[25] and has also been studied by Iverach et al.[26] A plausible reaction is

$$N_2 + CH \rightarrow HCN + N \qquad (4)$$

with subsequent oxidation of the products of this reaction to NO. Engleman et al.[20] give the rate data as $A = 10.3$ (SI units), $N = 0$, and $E/R = 24,000$ K. However, it is cautioned that direct experimental measurement of this rate constant is not yet available.

The oxidation of nitric oxide to nitrogen dioxide is the final mechanism listed in Table 1. At normal combustion temperatures, the steady-state $[NO_2]/[NO]$ ratio can be determined from reactions (8), (9), and (10). The reverse rates are neglected, giving the result

$$[NO_2]/[NO] = (k_{8f}[M]/k_{9f})/\{1 + (k_{10f}/k_{9f})([H]/[O])\} \qquad (5)$$

At atmospheric combustion, this ratio is less than 10^{-2}. Even though NO_2 forms rapidly from reaction (8), it is quickly reduced to NO by reactions (9) and (10).

Certain investigators postulate NO_2 formation via reaction (11) in regions where the hot combustion gases are rapidly cooled, as is the case with turbulent mixing with a cool secondary air stream.[27,28] The HO_2 radical in reaction (11) requires calculation by a low-temperature H_2–O_2 combustion mechanism (e.g., see Baulch et al.[19]).

3.2. Fuel-Nitrogen

3.2a. Structure and Release

We now turn to the topic of fuel-nitrogen, and first discuss its chemical structure in coal and its release during combustion. Most of the nitrogen in coal is believed[29] to occur in aromatic ring structures that contain

Tingey and Morrey[30] caution that the current knowledge of the nitrogen structure in coal is preliminary.

The origin of nitrogen in coal has been discussed by Hauck.[31] Coalification has resulted in a relatively constant coal-nitrogen content of 0.5–2.0% by mass. Bituminous coals exhibit somewhat higher nitrogen contents than other ranks, while anthracites exhibit the lowest nitrogen levels.[31]

The composition of the nitrogen volatiles and their rates of volatilization depend on the thermal environment of the coal particles. Industrial coking of coal yields nitrogen mainly in the form of ammonia (NH_3), and as residual coke-nitrogen.[32] Since, in this case, coal particles of about 3-mm diameter are heated at 1300 K in a reducing bed-retort for 18 hr, the results only provide an indirect guide to pulverized-coal reactions because of its markedly different heating rate (10^4–10^5 K s^{-1}) and gas–particle interaction.

Studies by Klein and Jüntgen[33] on the evolution of elemental nitrogen from German coals at low heating rates of 1 K/min showed a general trend of increasing nitrogen-release rate with increasing temperature. At temperatures of 970 and 1300 K, respectively, breakage of the aliphatic and heterocyclic bonds was indicated. With increasing coal rank (from V.M. = 40% to V.M. = 10%) the fraction of evolved coal nitrogen decreased (respectively, from 53 to 20%), and the temperature of apparent bond breakage increased. Evolved nitrogen was first detected at 750 K for high-volatile coal, and at 900 K for anthracite.

Blair et al.[34] studied the nitrogen pyrolysis of coals for final temperatures of 1150–1970 K. Coal particles were sprinkled on a preheated graphite ribbon, and although the coal heating rate was not measured, it was undoubtedly rapid. Weight-loss measurements showed that the evolution of nitrogen had a greater temperature dependence than did the total volatiles, as shown in Figure 2. The tendency of the nitrogen to remain in the char at low temperatures (i.e., below 1370–1700 K) has also been observed by Solomon,[35] as shown in Figure 7 (Section 5.1). At higher temperatures, Figure 2 shows that 70–90% of the coal-nitrogen has devolatilized.

The following correlation was offered by Pershing and Wendt[36]

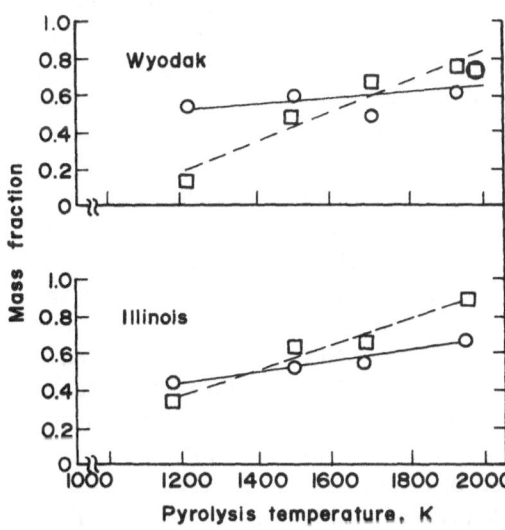

Figure 2. Release of volatile nitrogen from two kinds of coal particles vs. pyrolysis temperature; results compared to the release of total volatile matter. ○—— Mass fraction of coal evolved. □--- Mass fraction of coal nitrogen evolved. (Figure used with permission from Blair, Wendt, and Bartok[34].)

between nitrogen volatiles and total volatiles on a mass basis:

$$\text{(Volatilized N)/(Coal N)} = 1.95 \times \text{(Total mass fraction volatilized)} - 0.56$$

$$\tag{6}$$

As concluded by Blair et al.,[34] nitrogen species are evolved late in the volatilization process, and probably have low particle-escape velocities according to modeling studies. Chemical reduction in the fuel-rich particle vicinity is, therefore, a likely possibility. Likewise, a slow rate of mixing of the N-volatiles with air is important in influencing their reduction to N_2, compared to oxidation to NO.[37]

Blair et al.[34] examined the composition of the nitrogen volatiles. A controlled heating rate of 2×10^4 K s^{-1} and final temperatures of 1270–1770 K were used, along with a helium flow to immediately quench the evolved gases. At 1770 K, no more than 20% of the coal-nitrogen mass was converted to the light gases HCN, NH_3, and N_2, with HCN being the dominant species. Cyanogen and NO were not observed. The indicated low yield of nitrogen gases compared to nitrogen tars has also been observed by Solomon,[35] as shown in Figure 7 (Section 5.1).

Pohl and Sarofim[38] subjected pulverized-coal particles to heating rates of 10^4–10^5 K s^{-1}. The particles were conveyed in helium through an electrically heated furnace, and the approximate final particle temperatures were 1000–2100 K. Plots were constructed of the mass fraction of nitrogen evolved vs. the mass fraction of total material volatilized. No evolved nitrogen was observed until 10–15% of the total coal mass had been volatilized. For a furnace (wall) temperature of 1500 K, the nitrogen delay was about 80 ms. Thereafter, a linear dependence between nitrogen and total volatiles was observed. The yield of nitrogen was relatively greater than the yield of total volatiles; the constant of proportionality was between 1.25 and 1.50. These results further substantiate that nitrogen is evolved late in the particle heating sequence, indicating that most of the nitrogen is probably in the strongly bonded aromatic structures, and that at sufficiently high temperatures the relative yield of nitrogen exceeds the yield of total volatiles.

A rough, first-order fit of the data of Pohl and Sarofim[38] gave a rate constant of $9.3 \times 10^3 \exp(-11,400/T)$ s^{-1} for overall coal-nitrogen pyrolysis, with the reactant being the amount of the residual nitrogen. At 1500 K, this expression gives a characteristic pyrolysis time of 100 ms.

The oxidation of coal- and char-nitrogen was examined by Pohl and Sarofim.[38] For a furnace (wall) temperature of 1500 K, 60–80% of the NO was contributed by the volatile nitrogen, as shown in Figure 3. Char-nitrogen conversion was a factor of 2 to 3 less; however, trends with stoichiometry were the same as for the volatile nitrogen. It has been tentatively postulated by Pohl and Sarofim,[38] as well as by Pershing and Wendt[36] and Wendt and

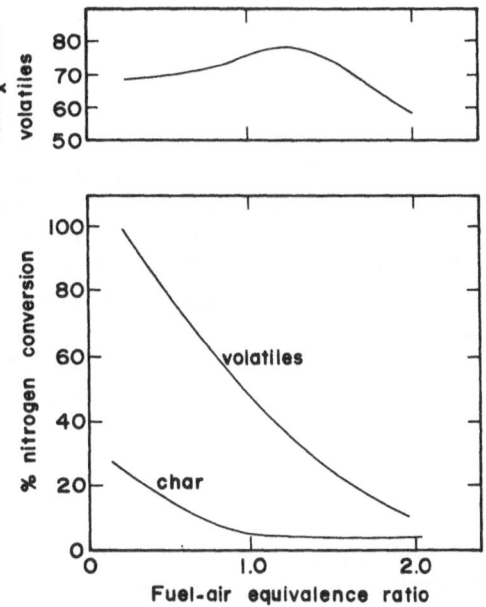

Figure 3. Contribution of volatile nitrogen and char-nitrogen to NO_x formation vs. fuel–air equivalence ratio. (Figure used with permission from Pohl and Sarofim.[38])

Schulze,[39] that char-nitrogen conversion may occur by surface oxidation plus reduction of resultant NO in the fuel-rich particle boundary layer.

Pershing and Wendt[13] measured the oxidation of coal- and char-nitrogen in a laboratory burner. At 15% excess air, between 17 and 38% of the coal-nitrogen was converted to NO_x; flames with enhanced mixing resulted in the higher conversion. For the same test conditions, approximately 13% of the char-nitrogen was converted to NO_x, with little apparent sensitivity to the mixing intensity. Conversion of coal-nitrogen was not sensitive to the combustion temperature, except at the highest adiabatic flame temperatures examined (>2500 K), where an increased conversion was recorded, possibly due to particle explosion. The NO derived from the volatiles appeared to be amenable to combustion control, while the char-NO did not.

Finally, the measurements of coal devolatilization at very high rates of heating with lasers by Sharkey *et al.*,[40] and with an argon plasma by Bond *et al.*,[41] are interesting since they indicate a high degree of conversion of the coal-nitrogen to HCN gas. No other nitrogen volatiles were reported.

These studies cited herein indicate that fuel-nitrogen contributes significantly to the NO_x formed during pulverized-coal combustion. Intermediates such as HCN and NH_i also appear to form, and evidence for a complex system of interconversion reactions is being accumulated. Variables such as mixing, heating rate, and stoichiometry are important and may provide methods for control of NO_x formation.

3.2b. Gas-Phase Kinetics

It is clear from the preceding section that the analysis of the gas-phase kinetics of fuel-nitrogen conversion must begin with the volatile nitrogenous material evolved from coal. It is reasonable to assume that these nitrogenous compounds involve pyridine and pyrrole ring structures. Tentative rate constants for the thermal (i.e., inert) decomposition of some model nitrogen compounds have been obtained by Axworthy *et al.*[42]: For example, pyridine $= 3.8 \times 10^{12}$ exp $(-35,200/T)$ s^{-1}; and pyrrole $= 7.5 \times 10^{15}$ exp $(-42,800/T)$ s^{-1}. (Quinoline and benzonitrile were also examined.) These

model compounds were thermally stable at $T < 1100$ K, while at higher temperatures, HCN was the dominant pyrolysis product. Its yield increased with pyrolysis temperature; for example, at 1375 K, pyridine was totally decomposed to HCN. In the lower temperature range, 1100–1375 K, pyridine decomposition yielded less than 10% NH_3, plus small amounts of benzonitrile, methyl cyanide (CH_3CN), vinyl cyanide ($CH_2 : CHCN$), quinoline, and residual pyridine. Pyrrole decomposition at $1100 < T < 1300$ K gave less HCN and more organic nitrogen, particularly methyl cyanide and vinyl cyanide.

High-temperature thermal soaking of the coal pyrolysis products in the fuel-rich environment of coal particles yields HCN as the dominant species, given sufficient temperature and residence time. For 90% pyridine decomposition, using the rates of Axworthy *et al.*,[42] 10 ms is estimated at 1500 K, 2 ms at 1600 K. Values of this magnitude were indicated in stirred reactor oxidation of pyridine.[17] This, then is the assumed entry point into the small-molecule fuel-nitrogen combustion mechanism.

A plausible mechanism for gas-phase fuel-nitrogen conversion is outlined in Figure 4, from the work of Haynes.[43,44] The mechanism connects the four major nitrogenous reactant systems: CN pool (largely HCN), NH pool, NO, and N_2. HCN would tend to equilibrate with CN, via the reaction $HCN + H \leftrightarrows CN + H_2$, having an equilibrium constant $K = 2.3$ exp $(-8430/T)$.[45] This equilibrium gives $[CN]/[HCN] < 0.01$. The NH pool (NH_3, NH_2, NH, and N) also tends toward internal equilibration, particularly among the higher-order species according to the basic mechanism given by Kaskan and Hughes[46] listed in Table 2; rate data are due to Sarofim *et al.*[47]

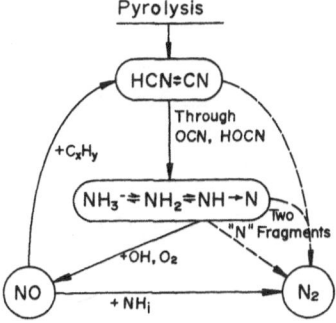

Figure 4. Small-molecule, fuel-nitrogen conversion mechanism at combustion temperatures. Major features due to Haynes.[43,44]

Depletion of the CN pool, leading to formation of species in the NH pool, may proceed through formation and subsequent reaction of the apparently bimodal free radical $\cdot O—C\equiv N \leftrightarrows O=C=N\cdot$. Possible reactions, due to Haynes[43] and Mulvihill and Phillips,[48] respectively, are

$$CN + OH \rightarrow OCN + H \tag{7}$$

$$OCN + H \rightarrow NH + CO \tag{8}$$

It has also been postulated by Haynes[43,49] and Fenimore[50] that intermediate HOCN could form from the CN pool and subsequently react to the NH pool. A first-order rate constant for HCN depletion in one-dimensional flames has been suggested by Fenimore[50]:

$$d(\log [\text{HCN}])/dt = -k = -6 \times 10^{13}([\text{H}_2\text{O}]^2/[\text{H}_2]) \exp(-34, 200/T), \quad \text{s}^{-1} \tag{9}$$

Reactions and rate constants for depletion of the NH pool are listed in

Table 2. Major Reactions between NH_i Species Mechanism from Kaskan and Hughes[46]a,b

Reaction	A	N	E/R (K)
(1) $NH_3 + H \rightarrow NH_2 + H_2$	8.28	0.67	1,710
(2) $NH_3 + OH \rightarrow NH_2 + H_2O$	7.60	0.68	550
(3) $NH_3 + O \rightarrow NH_2 + OH$	8.91	0.50	0
(4) $NH_2 + H \rightarrow NH + H_2$	8.15	0.67	2,165
(5) $NH_2 + OH \rightarrow NH + H_2O$	7.48	0.68	655
(6) $NH_2 + O \rightarrow NH + OH$	8.96	0.50	0
(7) $NH + H \rightarrow N + H_2$	9.00	0.68	960
(8) $NH + OH \rightarrow N + H_2O$	9.20	0.56	755
(9) $NH + O \rightarrow N + OH$	9.93	0.70	50

[a]Table used with permission from Sarofim *et al.*[47]
[b]SI units: kmol, m³, s, K.

Table 3. Mechanism and Rate Data for NO and N_2 Formation from NH_i Species[a]

Reaction	A	N	E/R (K)	Reference
(1) $N + OH \rightarrow NO + H$	8.8	0.5	0	20
(2) $N + O_2 \rightarrow NO + O$	6.8	1.0	3,150	19
(3) $NH + OH \rightarrow NO + H_2$	9.2	0.6	750	47
(4) $NH_2 + NO \rightarrow N_2 + H_2O$	9.9	0.0	0	50
(5) $NH + NO \rightarrow N_2 + OH$	9.4	0.0	0	47
(6) $N + NO \rightarrow N_2 + O$	10.3	0.0	0	19

[a]SI units: Kmol, m^3, S, K

Table 3. These reactions yield NO directly, or yield N_2 indirectly, through reaction with NO. This last statement is important since it suggests that N_2 can be formed only when sufficient NO exists. In principle, N_2 could also form through reaction of two amine or cyanide fragments; however, estimates indicate that this will be appreciable only at very fuel-rich conditions.[17]

Also, small concentrations of N_2O could arise from the oxidation of the NH pool by NO. Nitrous oxide has been observed to occur occasionally in low-pressure ammonia flames.[51]

Completion of the fuel-nitrogen loop in Figure 4 is facilitated by reaction between NO and hydrocarbons to form HCN. Possible reactions, listed by Haynes[43] and by Engleman et al.,[20] respectively, are

$$CH_3 + NO \rightarrow HCN + H_2O \tag{10}$$

$$CH + NO \rightarrow HCN + O \tag{11}$$

The stirred-reactor measurements of Malte et al.[17] indicate a first-order rate for the conversion of NO to the CN pool of magnitude 10^3 s^{-1} at conditions of atmospheric pressure, 1650 K, and fuel-air equivalence ratio equal to 1.4.

Global behavior of fuel-nitrogen conversion has been observed in stirred-reactor measurements.[52] Results due to Malte et al.[17] for combustion of propane in air doped with about 900-ppm pyridine are shown in Figure 5. Also shown are results with NH_3 and NO added to the reactor fuel inlet stream. The main conclusions are:

(1) Decreasing fuel-nitrogen conversion to NO occurred when the fuel-nitrogen concentration increased.

(2) Conversion of fuel-nitrogen to NO was similar for all additives.

(3) Above a fuel-air equivalence ratio of 1.3, significant HCN and NH_3 occurred. This corresponded with a marked increase in unburned hydrocarbon concentrations.

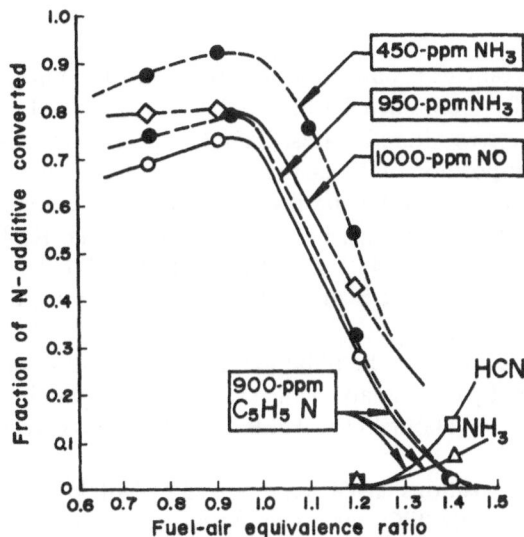

Figure 5. Conversion of model nitrogen additives (C_5H_5N, NH_3, and NO) to NO_x, HCN, and NH_3 as a function of fuel-air equivalence ratio. Jet-stirred reactor operated at atmospheric pressure on propane-air at ≈ 10 ms residence time and $1600 < T < 1800$ K. (Figure used with permission from Malte *et al.*[17])

4. Pollutants from Inorganic Matter

Ash, the inorganic mineral matter in coal, comprises about 5–20% of the mass of the coal and is a significant pollution source. Ash contains both pyritic and sulfate sulfur, some of which is evolved during combustion. Much of the mineral matter forms flyash particulates during combustion. The flyash size distribution is effectively bimodel.[53] There is the large size range which is closely related to the size distribution of the original pulverized coal; mean particle diameters of 5–20 μm appear to be typical for dense flyash.[54,55] There is also the submicron size range, which is significant in pollutant emissions because it escapes effective collection in flue gas cleaning equipment such as electrostatic precipitators,[54] and because it shows a propensity for concentrating heavy metals such as Cd, As, Se, Pb, Ni, Zn, Sb, as well as sulfur.[56,57] On a mass basis, the large, micron-size particles are dominant; however, on a number basis, the submicron size range has the markedly greater contribution.[53]

Mineral matter is distributed in coal more or less uniformly as small inclusions of variable composition of approximate 2-μm mean size.[58] However, pyritic particles of 20–60 μm are observed in x-ray scans of coal.[35] Large-sized pyrite particles are amenable to physical removal techniques.

4.1. Mineralogy and Behavior of Coal Ash

Several investigators have cataloged the elemental mineral composition of coal ash. Ode[59] conducted a comprehensive review of coals from West Virginia, the western United States, Canada, and Europe. Correlation with the nominal composition of the earth's crust was also shown. For the ash of a West Virginia coal, the following listing of elemental composition by mass was given: Si (44%), Al (30%), Fe (16%), Ca (3%), Na, K, Ti, Mg (1–2%); Sr, P, Ba, B, Li, As (0.08–0.4%); Cu, Zn, V, Pb, Mn, Ni, Zr, La, Rb, Cr, Ga, Sn, Mo, W, Hg Ge, Co, Cb (0.01–0.06%). Additional data are given in Chapter 7.

Of greater concern is the mineralogical composition of coal ash. This has been studied by x-ray analyses of ash samples prepared by low-temperature ashing of coal at about 450 K. Table 4 lists the mass percentages of major coal minerals.[58]

Similar compositions have been reported by Mitchell and Gluskoter,[60] who also listed quartz, SiO_2, as a significant ash component. Western U.S. coals were characterized as having high calcium and low pyrite contents compared to midwestern and eastern U.S. coals. Gypsum, $CaSO_4 \cdot 2H_2O$, is a source of calcium in western U.S. coals. Also, ferrous and ferric sulfates form in weathered coals, since pyritic sulfur readily oxidizes when exposed to moist air.[61]

Mitchell and Gluskoter[60] and Padia et al.[58] heated low-temperature ash samples slowly to high temperatures in an air environment, followed by x-ray analysis. The reader is referred to their observations of mineral transformations.[58,60] Such transformations, in toto, establish characteristic ash fusion (melting) temperatures of about 1100–1300 K,[55] depending on ash composition and combustion stoichiometry. Fuel-rich combustion appears to yield the lower ash fusion temperatures.

Table 4.   Major Minerals in Low-Temperature Ash[a]

Mineral	Montana lignite (%)	Pittsburgh bituminous (%)
Kaolinite (clay), $Al_2Si_2O_5(OH)_4$	50	50
Pyrite,[b] FeS_2	10	40
Calcium carbonate, $CaCO_3$; and Calcium sulfate, $CaSO_4$	40	10

[a]Table used with permission from Padia et al.[58]
[b]The term pyrite includes two structural forms: the dominant cubic FeS_2 (true pyrite) and the orthorhombic FeS_2 (marcasite): see Gluskoter.[61] While this convention may be unacceptable to mineralogists, it suffices for present purposes.

Table 5. Mechanism of Pyrite Decomposition at 750 K[a]

(1) $FeS_2 \rightarrow FeS + \frac{1}{2}S_2$ (gas)
(2) $FeS \rightarrow Fe + \frac{1}{2}S_2$
(3) $FeS_2 + H_2 \rightarrow FeS + H_2S$
(4) $FeS + H_2 \rightarrow Fe + H_2S$
(5) $FeS_2 + CO \rightarrow FeS + COS$
(6) $FeS + CO \rightarrow Fe + COS$

[a]Data from Given and Jones,[62] Fleming,[63] Attar *et al.*,[64] and Halstead and Raask.[65]

The behavior of the sulfur-containing minerals is most interesting from the pollution standpoint. Both Mitchell and Gluckoter[60] and Padia *et al.*[58] indicated that pyrite is quite unstable at high temperatures. At 750 K, the oxidation of pyrite to hematite occurs as follows:

$$2\,FeS_2 + 5.5O_2 \rightarrow Fe_2O_3 + 4SO_2 \tag{12}$$

In a reducing environment, FeS_2 is transformed to FeS by the reactions listed in Table 5. Ferrous sulfide (FeS) requires a higher temperature for decomposition, and thus tends to persist in the coal ash.[65] Halstead and Raask[65] have demonstrated the persistence of sulfur in the ash, as indicated in Table 6. Solomon[35] established the following experimental rate for the decomposition of FeS_2 to FeS

$$[\log(x-1)]/t = -480 \exp(-8400/T), \qquad s^{-1} \tag{13}$$

where x pertains to FeS_x.

Reactions involving calcium sulfate (or anhydrite) are also of interest. Mitchell and Gluskoter[60] indicate that during oxidation, FeS_2 can react

Table 6. Residual Sulfur in Pyrite Particles Heated for 0.5 s in $96\% \, N_2, 4\% \, O_2$[a]

Furnace temperature, K	Percent residual sulfur by mass
Unheated	53.1
1375	28.5
1475	22.7
1575	17.8
1675	15.1
1775	10.1

[a]Table used with permission from Halstead and Raask.[65]

with $CaCO_3$ to form $CaSO_4$. Also, kaolinite may react with $CaSO_4$ and $CaCO_3$ to form anorthite, $CaAl_2Si_2O_8$, above 1100 K.

A direct route to $CaSO_4$ formation is the dehydration of gypsum, originally present in western U.S. coals. Mitchell and Gluskoter[60] indicate complete dehydration at about 700 K. Calcium sulfate is a stable mineral and does not decompose to CaO (lime) until about 1400 K. A detailed study of $CaSO_4$ decomposition has been conducted by Swift et al.[66]

Calcium sulfate may be responsible, in part, for the ash sulfur retention observed during pulverized-coal firing of high-alkali western U.S. lignite coals[67]; see also Section 5.1. The affinity of calcium for sulfur, plus the thermal stability of $CaSO_4$, is the reason fluidized-bed coal combustors employ dolomite, $CaMg(CO_3)_2$, or limestone particles for sulfur-emission control. However, the injection of calcium particles into pulverized-coal furnaces[68] has not been nearly as effective for sulfur control, apparently because of poor calcium–sulfur contact and thermal decomposition of $CaSO_4$ at higher combustion temperatures.

4.2. Flyash Formation Mechanisms

Various aspects of flyash formation in pulverized-coal combustion have been examined by Ramsden,[69,70] Raask,[71,72] Street et al.,[73] and Ulrich et al.[74,75] Padia et al.[58] and Flagan[55] have presented overall models.

Figure 6, redrawn from Flagan,[55] schematically illustrates the mechanism of flyash formation. The model begins with the heating and devolatilization of the coal particles and the initial char burnout. For most pulverized-coal particles, this results in a fragile, skeletal char structure which continues to support the mineral inclusions. The exception is coal particles that become plastic and swell upon rapid heating (for example, vitrite particles). These tend to form hollow vesicular spheres.[70] The hot, molten ash inclusions tend to fuse together, rather than appreciably vaporize, as the char material burns away.[70] This agglomeration occurs principally during the last 25% of char burnout. At some point during char burnout (Flagan[55] suggests 90% char burnout), the weakened char structure fractures. The number of resultant fragments determines the number of large flyash particles formed per original coal particle. For nonswelling lignite, Padia et al.[58] indicate an average of three fragments per coal particle, while for bituminous coal, with some swelling, the yield is five fragments per coal particle.

The formation of hollow spherical flyash particles with coherent nonporous shells, called cenospheres, has been the subject of considerable attention, as well as uncertainty. The experimental results of Padia et al.,[58] for simulated furnace conditions, indicate substantial cenosphere formation during combustion of both pulverized lignite and bituminous coals. On the

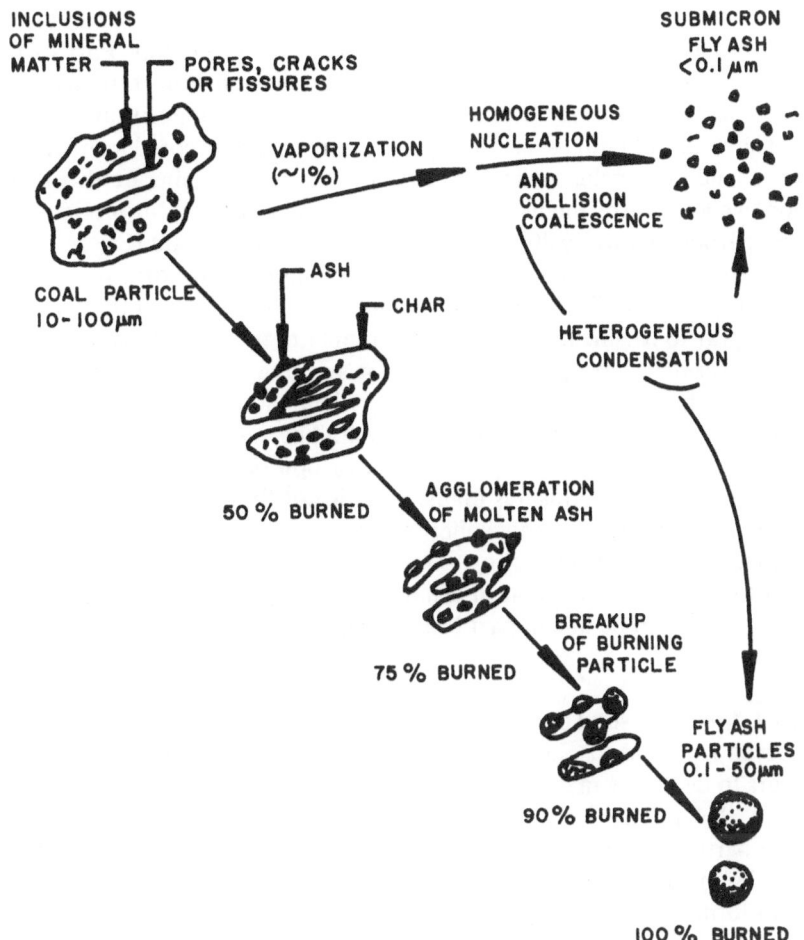

Figure 6. Schematic of flyash formation mechanism showing both dense, large-sized particles and submicron particles. (Figure used with permission from Flagan.[55])

other hand, Field *et al.*[1] indicate negligible cenosphere formation with anthracite coals. Raask[71] found as much as 20% by volume, or 5% by mass, of furnace flyash from British coals to be cenospheres; he indicated a size range of 20–200 μm, which is considerably above the size of the dense flyash particles.

Both Raask[71] and Padia *et al.*[58] indicate an optimum particle temperature for cenosphere formation. The temperature must be high enough so that the viscosity of the molten ash is low, yet at the same time, the temperature must be low enough so that the internal gas pressure, which facilitates cenosphere formation, is not so great as to cause the cenosphere shell to rupture. Padia et al.[58] suggest the range 1200–1400 K, while Raask[71] indicates a temperature of 1670 K for cenosphere formation. Part of this difference in reported ranges could be due to experimental uncertainty in the true particle temperature, or due to differences in ash composition.

The results of Padia et al.[58] imply that a cenosphere could form directly from a small ash inclusion in the coal, or from an agglomerate due to the fusion of a few nearby inclusions. They indicate a 30-ms formation time for a 12-μm particle to form a 40-μm-diameter cenosphere at 1400 K.

There appears to be a tendency for swelling coals to form more ceno-spheres than nonswelling coals. This could result from the fact that swelling coals appear to fragment into a greater number of total flyash particles.

Field et al.[1] have postulated that for swelling coals, cenospheres may form early in the heating sequence; i.e., when the coal particle is plastic during devolatilization, and thus can swell. The resultant hollow particle could then yield a large cenosphere as the carbon burned away and the ash fused. While this mechanism is plausible, it is not thought to be a dominant or general mechanism for cenosphere formation.

Finally, it is quite interesting to note that Raask[71] has shown a correla-tion between the iron content in ash and its tendency to form cenospheres. This is believed to be due in part to the reaction

$$2Fe_2O_3 + C^* \rightarrow 4FeO + CO_2 \tag{14}$$

where C^* refers to solid carbon. This reaction provides the CO_2 gas pressure (observed within cenospheres) for expansion. The critical amount of Fe_2O_3 appears to be 8% by mass in the ash.

To this point, the discussion has concerned the large-sized flyash. Flyash particles of submicron size appear to be a result of ash vaporization and subsequent condensation (see Figure 6). Ash vaporization occurs in several forms: (1) Above about 1350 K, alkali-metal salts vaporize[55]; (2) at high temperatures, compounds of heavy metals such as As, Cd, Cr, Ni, Pb, Se, Sb, Tl, and Zn will boil or sublime[56]; and (3) above 1900 K, silica volatilizes by first forming the more volatile SiO[58] according to

$$SiO_2 + C^* \rightarrow SiO + CO \tag{15}$$

As the furnace gases cool, the inorganic vapors undergo homogeneous nucleation and heterogeneous condensation on existing particles. Because small particles have greater specific surface areas, they tend to collect larger quantities of the condensed metallics, as well as condensable sulfates. Davison et al.[56] and Kaakinen et al.[57] have conducted measurements of flyash composition that confirm this behavior. Two toxic metals that escape from coal furnaces as a vapor are mercury, and to a lesser extent, Se.

Ulrich et al.[74,75] have proposed a mechanism to explain the growth of submicron particles. The theory relies on particle collision rates determined by free-molecule Brownian motion, followed by coalescence. The model appears to explain the measured growth in particle size for times of 30–100 ms following initial homogeneous nucleation.

4.3. Chlorine and the Alkali Metals

The volatilization and subsequent condensation of alkali metals on furnace surfaces have been examined by Halstead and Raask[65] and Bishop.[76] Sodium chloride, NaCl, in coal has received the most attention, followed by potassium salts. Coal-chlorine contents as high as 1% by mass have been reported by Bishop[76] for British coals.

While this subject relates directly to the corrosiveness of coal combustion products, it is discussed here because of its effect on pollutant emissions: (1) The presence of Na in coal causes some of the SO_x to be retained as Na_2SO_4 rather than emitted; (2) Cl in coal gives rise to pollutant emissions of HCl; and (3) fuel-rich regions, designed into burners for the control of NO_x formation, may enhance the corrosiveness of coal firing due to NaCl condensation.

The model of Halstead and Raask[65] involves the steady-state evaporation of NaCl from the coal particle surface. The evaporation temperature or wet-bulb temperature is 1300–1500 K for typically sized pulverized-coal particles. Evaporation times are about 0.5 s. The gas-phase sodium chloride will react with water vapor to form NaOH and HCl:

$$NaCl + H_2O \rightarrow NaOH + HCl \qquad (16)$$

Both NaCl and NaOH undergo overall reaction with SO_2, SO_3, H_2O, and O_2 to form gaseous sodium sulfate, Na_2SO_4. Reaction times are indicated to be less than 2 s. Fenimore[77] and Durie *et al.*[78] examined the kinetics of NaOH reaction with sulfur, but a complete mechanism is not yet available.

Equilibrium thermodynamic calculations by Halstead and Raask[65] for a high-chloride coal led to the following conclusions:

(1) With 3–5% excess O_2 (mole basis), typical of pulverized-coal firing, Na_2SO_4, rather than NaCl, is condensed. This assumes good air-fuel mixing and sufficient residence time for Na_2SO_4 formation. The condensation temperature is 1100–1300 K, and increases with coal-chloride content.

(2) In fuel-rich pockets, or for insufficient residence time, NaCl will also condense. The condensation temperature is indicated to be around 900 K. Increasing chlorine content increases the tendency for NaCl condensation.

5. Sulfurous Pollutants

The dominant sulfur pollutant formed due to the oxidation of fuel-bound sulfur in pulverized-coal combustion is sulfur dioxide (SO_2). For atmospheric

Table 7. *Organic and Inorganic Sulfur Levels in Some U.S. Coals*

Coal	Rank	Organic sulfur	Inorganic sulfur
		Mass % in coal	
Lyons (Va.)	HVA	0.95	0.75
Pittsburgh	HVA	1.89	1.46
Pittsburgh	HVA	0.65	0.19
Hazard # 7 (Ky.)	HVA	1.73	1.73
Ohio # 2	HVC	0.83	0.10
Ohio # 5	HVB	1.15	3.86
Montana	Lignite	0.64	0.54

*a*Table used with permission from Solomon.[35]

combustion with excess air, nominally 1% of the SO_2 is converted to sulfur trioxide (SO_3). It is convenient to term the sum ($SO_2 + SO_3$) as "SO_x." In fuel-rich combustion regions, hydrogen sulfide (H_2S) may form, as well as COS, CS_2, and sulfurous hydrocarbons.

Sulfur is bound in both the organic material and in the inorganic pyrites and sulfates. Inorganic forms and their behavior have been discussed in detail in Section 4. Total sulfur content can range from essentially zero up to about 10% of the coal mass.[79] Table 7, due to Solomon,[35] gives levels of organic and inorganic sulfur in some U.S. coals.

5.1. Structure and Release

The dominant form of organic sulfur in the coal molecule is thought to be thiophene.[29,30] Solomon[35] indicates that as much as 60% of the organic

sulfur resides in such heterocyclic rings. Other organic sulfur forms listed by Wiser[29] are mercaptans, disulfides, and sulfides.

The sulfurous composition of "coal gas" resulting from the coking of coal (typically, by heating at 1300 K in a reducing-bed retort for 18 hr) has been reviewed by Gollmar.[79] Hydrogen sulfide accounted for about 90% of the gaseous sulfur; carbon disulfide (CS_2) comprised much of the remainder. Traces of carbonyl sulfide (COS), thiophenes, and mercaptans were also detected. Approximately 60% of the sulfur remained in the residual coke. As elucidated in Section 3.2, these results only provide an indirect guide to sulfur reactions in pulverized coal.

During pulverized-coal pyrolysis, the sulfur side chains (—SH) and the linking sulfur chains (—S—) rupture first in the heating sequence, leading to early volatile sulfur. The thiophene structure, however, is more stable and does not decompose until temperatures of about 1200 K are attained.[64]

Solomon,[35] using a heating rate of 600 K s^{-1} to final temperatures of 700–1100 K, found that the gases evolved from coal particles consistently exhibited an enhanced sulfur content. Relative to the sulfur content of the original coal, the sulfur content of the volatile gas was greater by a factor of 2 (mass basis), with the dominant species being H_2S. Concomitantly, the sulfur content of the residual char tended to decrease, but not universally. Typical results due to Solomon[35] are redrawn in Figure 7.

Through an x-ray scanning technique sensitive to S and Fe, Solomon[35] was able to directly examine both inorganic and organic sulfur in the char.

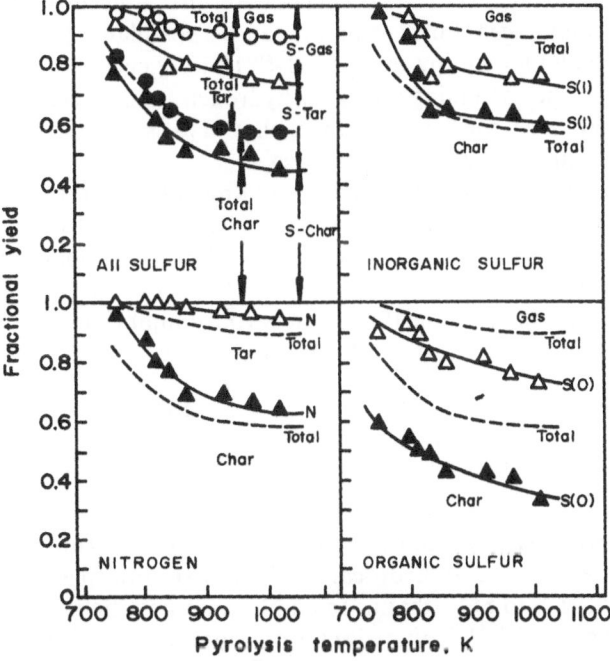

Figure 7. Fate of total sulfur, inorganic sulfur, organic sulfur, and fuel-bound nitrogen (clockwise from top left) on mass basis vs. pyrolysis temperature for coal from Lyons Seam (Virginia). For comparison, the fate of total coal matter as gas, tar, or char is also indicated. Vacuum heating at 6×10^2 K s^{-1}; final temperature held for 10 s. (Figure used with permission from Solomon.[35])

It was observed that the residual organic matter was depleted of sulfur, while the sulfur content of the inorganic ash was increased, as shown in Figure 7.

The axiom that "all of the sulfur is released from coal during pulverized firing" is not strictly correct. If this were true, then for every 1% sulfur by mass in coal, about 600 ppm SO_x would be emitted. For western U.S. lignites, however, an average of only 76% of the sulfur was observed to be released as SO_x during furnace combustion, and values as low as 48% were reported,[67] due to the high alkali content of the coal (20% Ca, 1–9% Na by mass in ash). Close contact between calcium molecules and the coal structure may promote retention of sulfur as $CaSO_4$, which has a high decomposition temperature (1400 K). In addition, as mentioned in Section 4.3, coals with high sodium content are prone to form Na_2SO_4, which will condense on furnace surfaces and possibly also on the flyash particulates.

5.2. Gas-Phase Kinetics

Modeling the gas-phase kinetics of sulfur conversion is a tentative procedure at present, because the composition of the sulfur volatiles is not precisely known. However, experimental results indicate that H_2S is the dominant volatile gas; furthermore, it can be postulated that the condensable sulfur volatiles will pyrolyze to H_2S. Heterogeneous reaction of char-sulfur is plausible, assuming attack by hydrogen to form H_2S, or by oxygen to form SO_2. Therefore, it is assumed that H_2S is the entry point for the purpose of evaluating gas-phase kinetics.

In the high-temperature reducing atmosphere of the coal-particle boundary layer, the pyrolysis of H_2S can be modeled by the shock-tube decomposition studies of Bowman and Dodge.[80] A summary of their work appears in Table 8. Cullis and Mulcahy[81] have suggested that the fast bimolecular reactions (2) and (5) exist in partial equilibrium. H_2S may also be decomposed by incorporation into the tarry volatiles[64] as

$$H_2S \; + \; \underset{/}{\overset{\backslash}{}}C{=}C\underset{\backslash}{\overset{/}{}} \; \longrightarrow \; \underset{|}{\overset{SH\;\;H}{-\underset{|}{C}-\underset{|}{C}-}} \qquad (17)$$

A proposed H_2S oxidation mechanism is shown in Table 9; most of the rate data are tentative. In hydrocarbon flames at atmospheric pressure, the oxidation of H_2S to SO_2 (that is, to equilibrium post-flame concentrations) occurs in about 1 ms.[81] The kinetic path of oxidation is $H_2S \rightarrow HS \rightarrow SO \rightarrow SO_2$.

Additional reactions could be included in Table 9. For one-dimensional flames, Levy et al.[83] included the initiation reaction $H_2S + O_2 \leftrightarrows HS + HO_2$, plus the reaction $SO + O \rightarrow SO_2 + h\nu$. Fenimore and Jones[84] considered the intermediate HSO_2 in the burnt gases of hydrogen-rich flames seeded with

Table 8. *Mechanism and Rate Data for H_2S Decomposition in a Shock Tube*[a,b]

Reaction	A	N	E/R (K)
(1) $H_2S + M \rightarrow HS + H + M^c$	11.3	0	37,300
(2) $H + H_2S \rightarrow HS + H_2$	9.9	0	860
(3) $HS + HS \rightarrow H_2S + S$	10.0	0	0
(4) $H + HS \rightarrow H_2 + S$	10.3	0	0
(5) $S + HS \rightarrow S_2 + H$	8.8	0.5	0
(6) $S + H + M \rightarrow HS + M$	9.0	0	0

[a]Table used with permission from Bowman and Dodge.[80]
[b]SI units: kmol, m³, s, K.
[c]M = argon.

additive sulfur. Finally, reactions of sulfur with hydrocarbon fragments, and reactions involving the oxidation of CS_2 and COS, are discussed by Cullis and Mulcahy.[81]

Furnace combustion modifications (in the sense of NO_x control) may be feasible for controlling the ratio of SO_2 to SO_3. Such control is important for a number of reasons. First, since SO_3 is corrosive in conjunction with other coal impurities,[86] exhaust temperatures must remain above 430 K to prevent H_2SO_4 condensation,[1] which leads to wasted furnace heat. Further, SO_3 is a visible and reactive pollutant due to its conversion to sulfuric acid and to other sulfates. In addition, SO_3 is beneficial for electrostatic precipitation of flyash due to its effect on particulate resistivity.

Table 9. *Mechanism and Rate Data for H_2S Oxidation*[a]

Reaction	A	N	E/R (K)	Reference
(1) $H_2S + H \rightarrow HS + H_2$	9.9	0	860	80
(2) $H_2S + OH \rightarrow H_2O + HS$	9.3	(at 300 K)		85
(3)[b] $H_2S + O \rightarrow OH + HS$	8.2	0	750	81
(4)[c] $HS + O \rightarrow SO + H$	11.0	(at 300 K)		85
(5) $HS + H \rightarrow H_2 + S$	10.3	0	0	80
(6) $HS + O_2 \rightarrow SO + OH$	<7.8	(at 300 K)		85
(7) $SO + O_2 \rightarrow SO_2 + O$	7.5	(assumed at 1250 K)		81
(8) $SO + OH \rightarrow SO_2 + H$	10.8	(at 300 K)		81
(9) $SO + O + M \rightarrow SO_2 + M$	11.5	(at 300 K)		81
(10) $S_2 + O \rightarrow SO + S$	9.6	(at 1050 K)		81
(11)[c] $S + O_2 \rightarrow SO + O$	9.1	(at 300 K)		81
(12) $SO + H \rightarrow S + OH$	[d]			81

[a]SI units: kmol, m³, s, K.
[b]Low-temperature data (<300 K).
[c]Weak temperature dependence.[82]
[d]No rate data are available.

Table 10.   Mechanism and Rate Data for SO_3 Formation[a,b]

Reaction	A	N	E/R (K)
(1) $SO_2 + O + M \rightarrow SO_3 + M$	11.38	0	1,260
(2) $SO_3 + O \rightarrow SO_2 + O_2$	11.46	0	6,040
(3) $SO_3 + H \rightarrow SO_2 + OH$	See discussion in text		

[a]Table used with permission from Merryman and Levy.[87]
[b]SI units: kmol. m³. s. K.

A mechanism for oxidation of SO_2 to SO_3 in combustion systems was deduced from the experiments of Merryman and Levy[87] for H_2S and COS flames. Their mechanism is shown in Table 10. This system is analogous to that for NO-to-NO_2 oxidation (Table 2). For completeness of this analog, the reaction $SO_2 + HO_2 \leftrightarrows SO_3 + OH$ could be included; at 300 K the rate constant is given by Hampson and Garvin[85] as $k = 10^{5.7}$ m³ kmol^{-1} s^{-1}.

Merryman and Levy[87] did not directly measure the rate for reaction (3); however, they did evaluate a rate constant for the combination of reactions (2) and (3):

$$k = k_2 + k_3[H]/[O]$$
$$= 10^{11.81} \exp(-5435/T) \quad (\text{m}^3 \text{ kmol}^{-1} \text{ s}^{-1}) \tag{18}$$

The steady-state $(SO_3)/(SO_2)$ ratio is easily evaluated if the reverse rates are neglected:

$$[SO_3]/[SO_2] = k_1[M]/k \tag{19}$$

This ratio is applicable to the fuel-lean regions of pulverized-coal furnaces. Nominally, the ratio is about 1%; however, because of the temperature dependence of k_1 and k, the ratio is enhanced somewhat in the cooler regions of the furnace. Also, combustion at elevated pressure will directly enhance the ratio because of the third-body influence. Without excess oxygen available, the $[SO_3]/[SO_2]$ ratio drops markedly, which is clearly observed in the flame measurements of Barrett et al.[88]

Interaction between the NO_x and SO_x kinetic subsystems may occur. Preliminary measurements by Wendt et al.[89] indicate the following:

1. Homogeneous catalytic recombination of O-atoms by reactions (1) and (2) in Table 10 may retard the fixation of atmospheric N_2 to NO.

2. The conversion of fuel-nitrogen to NO appears to be enhanced by SO_x. Interaction with the fuel-nitrogen mechanism shown in Fig. 4 is suspected.

6. Pollution in Gasifiers

The prospect that gasification of coal may become a major source of energy and feedstocks requires that the fate of the coal impurities during gasification be considered. Gasification has the advantage that the bound nitrogen and sulfur are converted to NH_3 and H_2S, which are gas-phase species that can be removed relatively easily from the product gas.

The present discussion is directed toward pollutant formation in entrained-flow, pulverized-coal gasifiers. In cases where specific information is lacking for entrained-flow systems, data to indicate potential trends are presented from fixed-bed and fluidized-bed systems.

An entrained-flow gasifier is fed by pulverized coal, 70% of which will pass through a 200-mesh screen. The reaction zone usually consists of an oxidizing combustion zone, followed by a reducing gasification zone. Gas temperatures are about 2200 K in the combustion zone and 1750 K in the gasification zone. Pollutant formation in these two areas is a complex process, about which relatively little detail is known. Many of the kinetic processes are similar to those occurring in combustion, and thus, some of the relevant fundamental aspects have already been discussed, particularly in Sections 3.2 and 5.1. In the following sections, observations specific to gasifiers will be discussed.

6.1. Gasification of Fuel-Nitrogen

Research into the fate of fuel-nitrogen during gasification has recently been initiated. A summary of process data by Becker and Murthy[90] on general gasification and by Gray et al.[91] on hydrogasification (free-fall dilute-phase reactor) indicate that nitrogen is gasified to a slightly greater extent than is the carbon. From these data, the fraction of nitrogen-to-carbon gasified is about 1.3 for hydrogasification, and about 2.0 for general gasification. These and other data[92] indicate that most of the gasified nitrogen exists as ammonia (NH_3).

The mechanism of nitrogen release from the coal is dependent upon several factors; among them are the nature of its existence in coal, mixing, the particle environment, pressure, and the time-temperature history of the particle. In an entrained-flow gasifier, nitrogen release is characterized by early pyrolysis in the oxidizing environment of the combustion zone, followed by char burnout in the reducing atmosphere of the gasification zone. Nitrogen compounds existing locally may be of considerably different nature than those exiting the reactor. For example, nitrogen oxides could form early in the combustion zone and then subsequently react to form ammonia (or other compounds) in the reduction zone. Possible reaction

Table 11. Distribution of Nitrogen in an
Entrained Koppers–Totzek Gasifier Process[90]

Form of N in product	Mass % of original N
NH_3	21.4
HCN	3.7
N_2	18.5
Particulates	?
Liquor	?
Unaccounted	54.7

mechanisms for such interconversion are shown in Tables 2 and 3 (Section 3.2).

In the reducing atmosphere of the gasification zone, nitrogen remaining in the char could react directly to form ammonia via a reaction proposed by Fleming,[63]

$$2\text{'N'} + 3H_2 \rightarrow 2NH_3 \qquad (20)$$

where 'N' represents the fuel-bound nitrogen. Subsequent reaction of ammonia with char to form hydrogen cyanide is also possible:

$$NH_3 + C^* \rightarrow HCN + H_2 \qquad (21)$$

Small amounts of HCN are found in most gasifiers.[92]

The distribution of nitrogen in the Koppers–Totzek entrained gasifier is shown in Table 11.[90] The molecular nitrogen yield in Table 11 probably results from the high-temperature decomposition of ammonia. This could occur by H-atom attack on NH_3 and NH_2, as shown by reactions (1) and (4) of Table 2 (Section 3.2b). The resultant NH_i fragments (and other N-fragments) could then combine to form N_2.

6.2. Inorganics and Particulate Matter

For an entrained-flow gasifier, char is the dominant particulate product in the exit stream, rather than flyash.[90] The char-to-ash ratio is a function of the extent of gasification.

The behavior and fate of trace inorganic metallic compounds in coal gasification is an important issue from both the process and environmental viewpoints. Recent studies[92,93,94] can be consulted in this regard, as well as the general discussion by Babu.[95]

6.3. Gasification of Fuel-Sulfur

The inorganic and organic forms of sulfur in coal have been described in Sections 4.1 and 5.1, respectively. In most gasification processes, the sulfur is released to a large extent with the major species being H_2S. In the Koppers–Totzek entrained-flow gasifier, 90% of the fuel-sulfur appears in the gas phase.[96] In the Synthane fluidized-bed reactor, 71% of the sulfur is gasified, 4.5% is found in the ash, and 1% remains in the char.[93]

Data on general gasification by Fleming[63] and on hydrogasification by Gray *et al.*[91] show a ratio of sulfur-to-carbon gasified of about 1.1–1.8. The extent of desulfurization is a function of the hydrogen concentration[97,98]; therefore, the value of 1.1 may be more applicable to entrained-flow gasifiers because of the lower hydrogen concentration.

Mechanistically, it is postulated that hydrogen attacks the sulfur binding sites—either pyrite or organic—and that H_2S is evolved.[99,100] Kinetic parameters of reactions suspected to occur under conditions of hydrogasification are shown in Table 12.[99] As the coal is heated, H_2S is observed to evolve from two apparently different types of organic sulfur (I and II). Then pyrite decomposes to FeS, after which the remaining sulfide and organic sulfur (III) react to form H_2S.

Since both H_2 and CO are present in general gasifiers, COS is also found in the product gas.[90] Thermodynamically, about 96% should be H_2S and 4% should be COS. The thermodynamic equilibrium assumption appears to be valid for entrained-flow gasifiers, as evidenced by the effluents listed by Fleming[63] and Becker and Murthy.[90] Heterogeneous reactions could also be postulated for the organic sulfur.

Other sulfur-containing compounds may also be formed during gasification. Sulfur is found in the tars, char, and ash as well as in other gas-phase species.[92,93,101,102] Under the conditions prevailing in the Koppers–Totzek entrained-flow gasifier,[90] however, only about 10% of the original feed sulfur was unaccounted for: 83% H_2S, 6% COS, and 1% SO_2.

Table 12. Kinetic Mechanism for Sulfur Release in Hydrogasification[a,b]

Reaction	A	N	E/R (K)
(1) $(Org{-}S)_I + H_2 \rightarrow H_2S$	3.71	0	17,360
(2) $(Org{-}S)_{II} + H_2 \rightarrow H_2S$	4.66	0	20,886
(3) $FeS_2 + H_2 \rightarrow H_2S + FeS$	5.66	0	23,654
(4) $FeS + H_2 \rightarrow H_2S + Fe$	6.54	0	27,680
(5) $(Org{-}S)_{III} + H_2 \rightarrow H_2S$	1.58	0	26,170

[a]Table used with permission from Vestal *et al.*[99]
[b]Units of k are $(Pa\ H_2)^{-1}\ s^{-1}$.

6.4. Summary

Complete nitrogen and sulfur mechanisms for entrained-flow gasifiers do not exist at this time. Engineering analyses, however, are possible employing the ratios of nitrogen-to-carbon and sulfur-to-carbon gasified, with thermodynamic calculations providing the gas-phase concentrations of NH_3, HCN, H_2S, COS, and other species.

7. Notation

A	Exponent in frequency factor of generalized rate constant, Eq. (1)	r	Reaction rate: Gas-phase (kmol m^{-3} s^{-1})
E	Activation energy (J kmol^{-1})		Surface-gas [kg(s) m^{-2} s^{-1}]
k	Generalized rate constant, Eq. (1):		Solid-gas [kg(s) s^{-1}]
	Gas-phase [(kmol m^{-3})$^{n-1}$ s^{-1}]		Solid [kg(s) s^{-1}]
	Solid-gas (Pa^{-1} s^{-1})	R	Universal gas constant (J kmol^{-1} K^{-1})
	Solid (s^{-1})		
K	Equilibrium constant	R_m	Mean particle radius (m)
$[M]$	"Third-body" concentration (kmol m^{-3})	t	Time (s)
		T	Temperature (K)
n	Order of gas-phase reaction	x	Pertains to FeS$_x$, Eq. (13)
N	Temperature exponent in generalized rate constant, Eq. (1)	$[\ ]$	Species molar concentration (kmol m^{-3})
P_{O_2}	Oxygen partial pressure (Pa)		

8. References

1. M. A. Field, D. W. Gill, B. B. Morgan, and P. G. W. Hawksley, *Combustion of Pulverized Coal*, The British Coal Utilisation Research Association, Leatherhead (1967). (Available from the Institute of Fuel, London.)
2. J. P. Appleton, Soot oxidation kinetics at combustion temperatures, in *Atmospheric Pollution by Aircraft Engines*, pp. 20/1–20/11, AGARD Conference Proceedings No. 125 (1973).
3. C. Park and J. P. Appleton, Shock-tube measurements of soot oxidation rates, *Comb. Flame* **20** 369–379 (1973).
4. P. J. Foster, Carbon in flames, *J. Inst. Fuel* **38**, 297–301 (1965).
5. K. H. Homann, Carbon formation in premixed flames, *Comb. Flame* **11**, 265–287 (1967).
6. H. B. Palmer and C. F. Cullis, The formation of carbon from gases, in *Chemistry and Physics of Carbon* (P. L. Walker, ed.), Vol. 1, pp. 265–325, Marcel Dekker, New York (1965).
7. L. I. Boehman and J. E. Davison, Refractory Metals for Advanced Gas Turbine Engines for Combined Cycle Power Plants, Paper presented at 2nd National Conference on Energy and the Environment, College Corner, Ohio (1974).
8. K. B. Lee, M. W. Thring, and J. M. Beer, On the rate of combustion of soot in a laminar soot flame, *Comb. Flame* **6**, 137–145 (1962).

9. J. Nagle and R. F. Strickland-Constable, Oxidation of carbon between 1000° and 2000°, in *Proceedings of Fifth Conference on Carbon*, Vol. 1, pp. 154–164, Macmillan, New York (1962).

10. C. P. Fenimore and G. W. Jones, Oxidation of soot by hydroxyl radicals, *J. Phys. Chem.* **71**, 593–597 (1967).

11. C. E. Blakeslee and H. C. Burbach, Controlling NO_x emissions from steam generators, *J. Air. Pollut. Control Assoc.* **23**, 37–42 (1973).

12. C. McCann, J. Demeter, R. Snedden, and D. Bienstock, *Combustion Control of Pollutants from Multi-Burner Coal-Fired Systems*, Report EPA-650/2-74038, U.S. Environmental Protection Agency, Washington, D.C. (1974).

13. D. W. Pershing and J. O. L. Wendt, Pulverized coal combustion: The influence of flame temperature and coal composition on thermal and fuel NO_x, in *Sixteenth Symposium (International) on Combustion*, pp. 389–436, The Combustion Institute, Pittsburgh, Pa. (1977).

14. D. Pierotti and R. A. Rasmussen, Combustion as a source of nitrous oxide in the atmosphere, *Geophys. Res. Lett.* **3**, 265–267 (1976).

15. F. Weiss and H. Craig, Production of atmospheric nitrous oxide by combustion, *Geophys. Res. Lett.* **3**, 751–753 (1976).

16. K. Yamagishi, M. Nozawa, T. Yoshie, T. Tokumoto, and Y. Kakegawa, A study of NO_x emission characteristics in two stage combustion, in *Fifteenth Symposium (International) on Combustion*, pp. 1157–1166, The Combustion Institute, Pittsburgh, Pa. (1975).

17. P. C. Malte, C. A. Halgren, L. E. Monteith, R. C. Corlett, and D. T. Pratt, The Fate of Organic Nitrogen in Jet-Stirred Combustion, Paper No. 76-31, Western States Section, Fall Meeting of the Combustion Institute, La Jolla, Ca. (1976).

18. Y. B. Zeldovich, The oxidation of nitrogen in combustion explosions, *Acta Physicochim. U.S.S.R.* **21** 577–628 (1946).

19. D. L. Baulch, D. D. Drysdale, D. G. Horne, and A. C. Floyd, *Evaluated Kinetic Data for High Temperature Reactions*, CRC Press, Cleveland, Ohio (1973). (Also, Reports: High Temperature Reaction Rate Data, Nos. 1, 2, and 3, Leeds University, England, 1968 and 1969.)

20. V. S. Engleman, V. J. Siminski, and W. Bartok, *Mechanisms and Kinetics of the Formation of NO_x and Other Combustion Pollutants, Phase II, Modified Combustion*, Report EPR-600/7-76-0096, U.S. Environmental Protection Agency, Washington, D.C. (1976).

21. F. Gouldin, Role of turbulent fluctuations in NO formation, *Combust. Sci. Technol* **9**, 17–23 (1974).

22. J. J. Wormeck and D. T. Pratt, Computer modeling of combustion in a Longwell jet-stirred reactor, in *Sixteenth Symposium (International) on Combustion*, pp. 1583–1592, The Combustion Institute, Pittsburgh, Pa. (1977).

23. P. C. Malte and D. T. Pratt, The role of energy-releasing kinetics in NO_x formation: Fuel-lean, jet-stirred CO–air combustion, *Combust. Sci. Technol.* **9**, 221–231 (1974).

24. P. C. Malte, and D. T. Pratt, Measurement of atomic oxygen and nitrogen oxides in jet-stirred combustion, in *Fifteenth Symposium (International) on Combustion*, pp. 1061–1070, The Combustion Institute, Pittsburgh, Pa. (1975).

25. C. P. Fenimore, Formation of nitric oxide in premixed hydrocarbon flames, in *Thirteenth Symposium (International) on Combustion*, pp. 373–380, The Combustion Institute, Pittsburgh, Pa. (1971).

26. D. Iverach, K. S. Basden, and N. Y. Kirov, Formation of nitric oxide in fuel-lean and fuel-rich flames, in *Fourteenth Symposium (International) on Combustion*, pp. 767–776, The Combustion Institute, Pittsburgh, Pa. (1973).

27. N. P. Cernansky and R. F. Sawyer, NO and NO_2 formation in a turbulent hydrocarbon/air diffusion flame, in *Fifteenth Symposium (International) on Combustion*, pp. 1039–1050, The Combustion Institute, Pittsburgh, Pa. (1975).

28. M. J. Oven, W. J. McLean, and F. C. Gouldin NO–NO$_2$ Measurements in a Methane-Fueled Swirl-Stabilized Combustion, Paper presented at Spring Technical Meeting, Central States Section, The Combustion Institute, Cleveland, Ohio (1977).

29. W. H. Wiser, Conversion of coal to liquids—Research opportunities, in *Research in Coal Technology: University's Role*, pp. 73–94, Report CONF-741091, U.S. ERDA, Washington, D.C. (1975).

30. G. L. Tingey and J. R. Morrey, *Coal Structure and Reactivity*, Battelle Energy Program Report, Battelle Northwest Laboratories, Richland, Washington (1973).

31. R. D. Hauck, The genesis and stability of nitrogen in peat and coal, in *Proceedings of 169th National Meeting of American Chemical Society, Division of Fuel Chemistry*, Vol. 20, pp. 85–93, American Chemical Society, Washington, D.C. (1975).

32. W. H. Hill, Recovery of ammonia, cyanogen, pyridine, and other nitrogenous compounds from industrial gases, in *Chemistry of Coal Utilization* (H. H. Lowry, ed.), Vol. 2, pp. 1008–1135, John Wiley and Sons, Inc., New York (1945).

33. J. Klein and H. Jüntgen, Studies on the emission of elemental nitrogen from coals of different rank and its release under geochemical conditions, in *Advances in Organic Geochemistry*, pp. 647–656, Pergamon Press, Oxford (1972).

34. D. W. Blair, J. O. L. Wendt, and W. Bartok, Evolution of nitrogen and other species during controlled pyrolysis of coal, in *Sixteenth Symposium (International) on Combustion*, pp. 475–489, The Combustion Institute, Pittsburgh, Pa. (1977).

35. P. R. Solomon, *The Evolution of Pollutants during the Rapid Devolatilization of Coal*, Report R76-952588-2, United Technologies Research Center, East Hartford, Conn. (1977).

36. D. W. Pershing and J. O. L. Wendt, Relative Contributions of Volatile Nitrogen and Char Nitrogen to NO$_x$ Emissions from Pulverized Coal Flames, Paper presented at 83rd National Meeting of AIChE, Houston, Texas (1977).

37. J. O. L. Wendt and D. W. Pershing, Physical mechanisms governing the oxidation of volatile fuel nitrogen in pulverized coal flames, *Combust. Sci. Technol.* **16**, 111–121 (1977).

38. J. H. Pohl and A. F. Sarofim, Devolatilization and oxidation of coal nitrogen, in *Sixteenth Symposium (International) on Combustion*, pp. 491–501, The Combustion Institute, Pittsburgh, Pa. (1977).

39. J. O. L. Wendt and O. E. Schulze, The effect of Diffusion–Reaction Interactions on Fuel Nitrogen Conversion during Coal Char Combustion, Paper presented at Fall Meeting of the Eastern States Section, The Combustion Institute, Silver Spring, Maryland (1974).

40. A. G. Sharkey, Jr., J. L. Shultz, and R. A. Friedel, Gases from flash and laser irradiation of coal, in *Coal Science, Advances in Chemistry Series*, Vol. 55, pp. 643–649, American Chemical Society, Washington, D.C. (1966).

41. R. L. Bond, W. R. Ladner, and G. I. T. McConnell, Reaction of coals under conditions of high energy input and high temperature, in *Coal Science, Advances in Chemistry Series*, Vol. 55, pp. 650–665, American Chemical Society, Washington, D.C. (1966).

42. A. E. Axworthy, G. R. Schneider, M. D. Shuman, and V. H. Dayan, *Chemistry of Fuel Nitrogen Conversion to Nitrogen Oxides in Combustion*, Report EPA-600/2-76-039, U.S. Environmental Protection Agency, Washington, D.C. (1976).

43. B. S. Haynes, The Formation and Behavior of Nitrogen Species in Fuel Rich Hydrocarbon Flames, Ph.D. Thesis, The University of New South Wales, Sydney (1975).

44. B. S. Haynes, Reactions of ammonia and nitric oxide in the burnt gases of fuel-rich hydrocarbon–air flames, *Comb. Flame* **28**, 81–89 (1977).

45. B. S. Haynes, D. Iverach, and N. Y. Kirov, The behavior of nitrogen species in fuel-rich hydrocarbon flames, in *Fifteenth Symposium (International) on Combustion*, pp. 1103–1112, The Combustion Institute, Pittsburgh, Pa. (1975).

46. W. E. Kaskan and D. E. Hughes, Mechanism of decay of ammonia in flame gases from NH$_3$/O$_2$/N$_2$ flames, *Comb. Flame* **20**, 381–388 (1973).

47. A. F. Sarofim, G. C. Williams, M. Modell, and S. M. Slater, Conversion of Fuel Nitrogen to Nitric Oxide in Premixed and Diffusion Flames, Paper presented at AIChE 66th Annual Meeting, Philadelphia, Pa. (1973).

48. J. N. Mulvihill and L. F. Phillips, Breakdown of cyanogen in fuel-rich $H_2/N_2/O_2$ flames, in *Fifteenth Symposium (International) on Combustion*, pp. 1113–1122, The Combustion Institute, Pittsburgh, Pa. (1975).

49. B. S. Haynes, The oxidation of hydrogen cyanide in fuel-rich flames, *Comb. Flame* **28**, 113–122 (1977).

50. C. P. Fenimore, Reactions of fuel-nitrogen in rich flame gases, *Comb. Flame* **26**, 249–256 (1976).

51. D. I. McLean and H. G. G. Wagner, The structure of the reaction zones of ammonia–oxygen and hydrozine-decomposition flames, in *Eleventh Symposium (International) on Combustion*, pp. 871–878, The Combustion Institute, Pittsburgh, Pa. (1967).

52. W. Bartok, V. S. Engleman, R. Goldstein, and E. G. del Valle, Basic Kinetic Studies and Modeling of Nitrogen Oxide Formation in Combustion Processes, Paper presented at AIChE 70th Annual Meeting, Atlantic City, N.J. (1971).

53. G. D. Ulrich. J. W. Riehl, B. R. French, and R. Desrosiers, The Mechanism of Sub-micron Fly Ash Formation in a Cyclone, Coal-fired Boiler, Paper presented at Engineering Foundation Conference on Ash Deposits and Corrosion Due to Impurities in Combustion Gases, Henniker, N.H. (1977).

54. E. J. Schulz, R. B. Engdahl, and T. T. Frankenberg, Submicron particles from a pulverized coal fired boiler, *Atmos. Environ.* **9**, 111–119 (1975).

55. R. C. Flagan, Ash Particle Formation in Pulverized Coal Combustion, Paper No. 77-4, Spring Meeting of the Western States Section, The Combustion Institute, Seattle, Wash. (1977).

56. R. L. Davison, D. F. S. Natusch, J. R. Wallace, and C. A. Evans, Jr., Trace Elements in fly ash–Dependence of concentration on particle size, *Environ. Sci. Technol.* **8**, 1107–1113 (1974).

57. J. W. Kaakinen, R. M. Jorden, M. H. Lawasani, and R. E. West, Trace element behavior in coal-fired power plants, *Environ. Sci. Technol.* **9**, 862–869 (1975).

58. A. S. Padia, A. F. Sarofim, and J. B. Howard, The Behavior of Ash in Pulverized Coal under Simulated Combustion Conditions, Paper presented at Spring Meeting of the Central States Section, The Combustion Institute, Columbus, Ohio (1976).

59. W. H. Ode, Coal analysis and mineral matter, in *Chemistry of Coal Utilization* (H. H. Lowry, ed.), Supplementary Volume, pp. 150–201, John Wiley and Sons, Inc., New York (1963).

60. R. S. Mitchell and H. J. Gluskoter, Mineralogy of ash of some american coals: Variations with temperature and source, *Fuel* **5**, 90–96 (1976).

61. H. J. Gluskoter, Inorganic sulfur in coal, in *Proceedings of 169th National Meeting of the American Chemical Society, Division of Fuel Chemistry*, Vol. 20, pp. 94–96, American Chemical Society, Washington, D.C. (1975).

62. P. H. Given and J. R. Jones, Experiments on the removal of sulfur from coal and coke, *Fuel* **45**, 151–158 (1966).

63. D. K. Fleming, Purification of intermediate streams in coal gasification, in *Clean Fuels from Coal Symposium II*, pp. 653–680, Institute of Gas Technology, Chicago, Ill. (1976).

64. A. Attar, A. H. Corcoran, and G. S. Gibson, Transformation of sulfur functional groups during pyrolysis of coal, in *Proceedings of 172nd National Meeting of the American Chemical Society, Division of Fuel Chemistry*, Vol. 21, pp. 106–111 American Chemical Society, Washington, D.C. (1976).

65. W. D. Halstead and E. Raask, The behavior of sulfur and chlorine compounds in pulverized-coal-fired boilers, *Inst. Fuel* **42**, 344–349 (1969).

66. W. M. Swift, A. F. Panek, G. W. Smith, G. J. Vogel, and A. A. Jonke, *Decomposition of Calcium Sulfate: A Review of the Literature*, Report ANL-76-122. Argonne National Laboratory, U.S. ERDA, Argonne, Ill. (1976).

67. G. H. Gronhovd, P. D. Tufte, and S. J. Selle, Some studies on stack emissions from lignite-fired powerplants, in *Proceedings of Bureau of Mines—University of North Dakota Symposium: Technology and Use of Lignite*, pp. 83–102, Report No. IC 8650, U.S. Bureau of Mines, Washington, D.C. (1973).

68. Tennessee Valley Authority, *Full-Scale Desulfurization of Stack Gas by Dry Limestone Injection*, Report EPA-650/2-73-019, Environmental Protection Agency, Washington, D.C. (1973).

69. A. R. Ramsden, Application of electron microscopy to the study of pulverized-coal combustion and fly-ash formation, *Inst. Fuel* **41**, 451–454 (1968).

70. A. R. Ramsden, A microscopic investigation into the formation of fly-ash during the combustion of a pulverized bituminous coal, *Fuel* **48**, 121–137 (1969).

71. E. Raask, Cenospheres in pulverized-fuel ash, *Inst. Fuel* **41**, 339–344 (1968).

72. E. Raask, Fusion of silicate particles in coal flames, *Fuel* **48**, 366–374 (1969).

73. P. J. Street, R. P. Weight, and P. Lightman, Further investigations of structural changes occurring in pulverized coal particles during rapid heating, *Fuel* **48**, 343–365 (1969).

74. G. D. Ulrich, Theory of particle formation and growth in oxide synthesis flames, *Combust. Sci. Technol.* **4**, 47–57 (1971).

75. G. D. Ulrich, B. A. Milnes, and N. S. Subramanian, Particle growth in flames. II. Experimental results for silica particle, *Combust. Sci. Technol.* **14**, 243–249 (1976).

76. R. I. Bishop, The formation of alkali-rich deposits by a high-chlorine coal, *J. Inst. Fuel* **41**, 51–65 (1968).

77. C. P. Fenimore, Two modes of interaction of NaOH and SO_2 in gases from fuel-lean H_2-air flames, in *Fourteenth Symposium (International) on Combustion*, pp. 955–963, The Combustion Institute, Pittsburgh, Pa. (1973).

78. R. A. Durie, G. M. Johnson, and M. Y. Smith, Gas phase reactions of sodium species with sulfur species in hydrocarbon flames, in *Fifteenth Symposium (International) on Combustion*, pp. 1123–1133, The Combustion Institute, Pittsburgh, Pa. (1975).

79. H. A. Gollmar, Removal of sulfur compounds from coal gas, in *Chemistry of Coal Utilization* (H. H. Lowry, ed.), Vol. 2, pp. 947–1007, John Wiley and Sons, Inc., New York (1945).

80. C. T. Bowman and L. G. Dodge, Kinetics of the thermal decomposition of hydrogen sulfide behind shock waves, in *Sixteenth Symposium (International) on Combustion*, pp. 971–982, The Combustion Institute, Pittsburgh, Pa. (1977).

81. C. F. Cullis and M. F. R. Mulcahy, The kinetics of combustion of gaseous sulfur compounds, *Combust. Flame* **18**, 225–292 (1972).

82. S. W. Benson, D. M. Golden, R. S. Lawrence, and R. W. Woolfolk, *Estimating the Kinetics of Combustion*, Report EPA-600/2-75-019, U.S. Environmental Protection Agency, Washington, D.C. (1975).

83. A. Levy, E. L. Merryman, and W. T. Reid, Mechanisms of formation of sulfur oxides in combustion, *Environ. Sci. Technol.* **4**, 653–662 (1970).

84. C. P. Fenimore and G. W. Jones, Sulfur in the burnt gas of hydrogen–oxygen flames, *J. Phys. Chem.* **69**, 3593–3597 (1965).

85. R. F. Hampson, Jr., and D. Garvin, *Chemical Kinetic and Photochemical Data for Modeling Atmospheric Chemistry*, Technical Note No. 866, National Bureau of Standards, Washington, D.C. (1975).

86. W. T. Reid, *External Corrosion and Deposits, Boilers and Gas Turbines*, American Elsevier Publishing Co., Inc., New York (1971).

87. E. L. Merryman and A. Levy, Sulfur trioxide flame chemistry—H_2S and COS flames, in *Thirteenth Symposium (International) on Combustion*, pp. 427–436, The Combustion Institute, Pittsburgh, Pa. (1971).

88. R. E. Barrett, J. D. Hummell, and W. T. Reid, Formation of SO_3 in a noncatalytic combustor, *J. Eng. Power* **88**, 165–171 (1966).

89. J. O. L. Wendt, T. L. Corley, and J. T. Morcomb, Interactions between sulfur oxides and nitrogen oxides in combustion process, in *Proceedings of the Second Stationary Source Combustion Symposium*, Report EPA-600/7-77-073d, U.S. Environmental Protection Agency, Research Triangle Park, N.C. (1977).

90. D. F. Becker and B. N. Murthy, *Feasibility of Reducing Fuel Gas Cleanup Needs. Phase I. Survey of the Effect of Gasification Process Conditions on the Entrainment of Impurities in the Fuel Gas*, Contract Report No. FE-1236-15, U.S. ERDA, Washington, D.C. (1976).

91. J. A. Gray, P. J. Donatelli, and P. M. Yavorsky, Hydrogasification kinetics of bituminous coal and char, in *Proceedings of 171st National Meeting of American Chemical Society, Division of Fuel Chemistry*, Vol. 20, pp. 103–154, American Chemical Society, Washington, D.C. (1975).

92. E. M. Magee, *Evaluation of Pollution Control in Fossil Fuel Conversion Processes*, Report EPA-600/2-76-101, U.S. Environmental Protection Agency, Washington, D.C. (1976).

93. A. J. Forney, W. P. Haynes, S. J. Gasior, R. M. Kornosky, C. E. Schmidt, and A. G. Sharkey, *Trace Element and Major Component Balances Around the Synthane PDU Gasifier*, Report PERC/TPR 75/ U.S. ERDA Pittsburgh Energy Research Center, Pittsburgh, Pa. (1975).

94. M. L. Lee and R. L. Coates, personal communication (1977).

95. P. Suresh Babu (ed.), *Trace Elements in Fuel*, Advances in Chemistry Series, Vol. 141, American Chemical Society, Washington, D.C. (1975).

96. J. F. Farnsworth, Clean Environment with K–T Process, Paper presented at EPA Meeting, Environmental Aspects of Fuel Conversion Technology, St. Louis, Missouri (1974).

97. C. W. Zielke, G. P. Curran, E. Gorin, and G. E. Goring, Desulfurization of low temperature char by partial gasification, *Ind. Eng. Chem.* **46**, 53–56 (1954).

98. P. S. Maa, C. R. Lewis, and C. E. Hamrin, Jr., Sulfur transformation and removal for western Kentucky coals, *Fuel* **54**, 62–69 (1975).

99. M. L. Vestal, A. G. Day, III, J. S. Synderman, G. J. Fergusson, F. W. Lampe, R. H. Essenhigh, and W. H. Johnston, *Kinetic Studies on the Pyrolysis, Desulfurization and Gasification of Coals with Emphasis on the Non-isothermal Kinetic Method*, Report No. SRIC 70-14, Scientific Research Instruments Corp., Baltimore, Maryland (1969).

100. A. L. Yergey, F. W. Lampe, M. L. Vestal, A. G. Day, G. J. Fergusson, W. H. Johnston, J. S. Snyderman, R. H. Essenhigh, and J. E. Hudson, Non-isothermal kinetics studies of the hydrodesulfurization of coal, *Ind. Eng. Chem. Process Des. Dev.* **13**, 233–240 (1974).

101. W. J. McMichael, A. J. Forney, W. P. Haynes, J. P. Strakey, S. J. Gasior, and R. M. Koronosky, *Synthane Gasifier Effluent Streams*, Report PERC/RI-77/4, U.S. ERDA, Pittsburgh Energy Research Center, Pittsburgh, Pa. (1977).

102. A. J. Forney, W. P. Haynes, S. J. Gasior, G. E. Johnson, and J. P. Strakey, Jr., *Analysis of Tars, Chars, Gases and Water Found in Effluents from the Synthane Process*, Technical Progress Report No. 76, Bureau of Mines, Washington, D.C. (1974)

12

Modeling Pulverized-Coal Reaction Processes

L. Douglas Smoot and Philip J. Smith

1. General Approach

In this chapter, a generalized model of coal reaction processes, including devolatilization, char oxidation, gas-phase oxidation, and gas–particle interchange, is presented. Coal and gas physical properties are also summarized. The resulting generalized model, which describes the response of a coal particle to its thermal, chemical, and radiative environment, is required for the modeling of propagating coal-dust flames (Chapter 13) and performance of combustors and gasifiers (Chapters 13 and 14).

Development of an analytical treatment of pulverized coal-char behavior in reacting systems is based largely on experimental observations and kinetic parameters deduced from these observations, as summarized in Chapters 7 through 11. Table 1 gives a summary of the most relevent observations. Since there are still unresolved questions regarding the kinetics of coal reaction, an attempt has been made to formulate a general reaction scheme that can accommodate results of future measurements and improved kinetic parameters.

The description that follows applies to pulverized-coal reaction processes, where particles are small (<100 μm) and heating rates are high (10^3–10^5 K s^{-1}). Such a treatment would not necessarily apply to fixed or

L. Douglas Smoot • Professor of Chemical Engineering, Brigham Young University, Provo, Utah. *Philip J. Smith* • Ph.D. Candidate, Chemical Engineering Department, Brigham Young University, Provo, Utah

Table 1.   Observations Regarding Coal-Particle Reactions

1. Reaction of coal particles often occurs in at least two stages; the pyrolysis or devolatilization stage, and the char oxidation stage.[1-6] Combustion may proceed simultaneously in both stages.[4-7] At very high heating rates, it is possible that the pyrolysis stage does not occur.[7]
2. During the devolatilization stage, the coal particles swell and may become porous,[2,6,8,9] and burning occurs on both internal and external surfaces.[8-10] Average particle diameter increase is often on the order of 10%.[8,11]
3. The extent of devolatilization and the composition of the gaseous products are functions of final temperature and possibly the heating rate.[12-14] Percent volatiles determined from proximate analysis is ordinarily less than that which results from pyrolysis at high heating rates.[3,4,12,14,15]
4. Coal devolatilization has been treated with some success as an activated process that takes place by a single overall kinetic step.[12,14,16] However, such results do not apply outside the measured temperature ranges or to coals or sizes other than those measured.[12,14]
5. More recent models of the devolatilization process consider a sequence of activated reactions.[14,15,17] Such models can account for changes in percent devolatilization as a function of heating rate or final temperature but are still limited to specific coals and temperatures analyzed.
6. Electron micrographs of several pulverized coals of varying volatiles content suggest that when coal is devolatilized at high rates, the coal particles soften, become spherical, swell slightly, and are marked with several surface holes, presumably resulting from escaping volatiles. While cenospheres form occasionally, they do not necessarily represent the predominant behavior.[6,11]
7. Composition of volatile matter from coals at moderate heating rates of 10^3 K s^{-1} has been reported[18] but information at higher heating rates is lacking. Recent data[15] suggest that hydrogen and soot are major products at higher heating rates (up to 10^5 K s^{-1}).
8. There is some evidence[19,20] that CO and H_2 are the major products of the rapid oxidation of devolatilized products and that the rate-controlling step is subsequent oxidation of the carbon monoxide.
9. Recent flame-propagation measurements and scanning electron micrographs, together with model predictions for coal devolatilization at high heating rates (10^4 K s^{-1}), confirm that the smaller coal particles in a distribution of particles are particularly important in the devolatilization process.[21]
10. Recent measurements based upon char ultimate analyses of samples obtained from propagating coal–air flames[11,21] show that the hydrogen–carbon ratio of devolatilized products decreases with time during the process, and that significant fractions of coal nitrogen and sulfur are lost during devolatilization.[22]
11. There is some question as to which process is controlling during the relatively fast devolatilization step. Models have been based on diffusion-limited vaporization[5] and heat transfer to the particle,[2] but activated pyrolysis reactions[3,4,12,14] are the most common explanation.
12. There is general agreement that the char combustion for small particles is heterogeneous, and at least partly controlled by chemical reaction steps. The char-combustion stage dominates the total time required for coal-particle burnout.[1-4,23]
13. Oxygen has received the most attention as an oxidizer of chars, coals, carbon, and cokes. However, results from 15 investigations[24,25] for various fuels, temperature ranges, and particle sizes show a wide variation in kinetic parameters, including reaction order
14. Less emphasis has been given to steam, H_2O, or CO_2 reactions with coal. Limited references for mostly carbon and coke oxidation also show a variation of kinetic parameters, and rates much slower than for O_2.[24,25] Significant concentrations of water vapor in oxygen seemed to have little effect on coal combustion at atmospheric pressure.[26]

continued

Table 1 (Continued)

15. For the residual char-combustion stage, the controlling chemical reaction step is not well established and probably varies depending upon system conditions. Reactions that are first- to zero-order in oxygen have been postulated, corresponding to oxygen adsorption and product desorption, respectively. Activation energies also varied greatly. Values are strongly dependent upon coal type.[9,24,26]

16. During the heterogeneous combustion stage, oxygen reaction with the particles takes place on both internal and external surfaces, with the particle diameter remaining nearly constant until the particle fragments.[8]

17. Some data[27,28] suggest that carbon monoxide is the most likely combustion product for small char particles (< 50 μm) and that subsequent oxidation of CO to CO_2 takes place mostly outside the particle boundary layer.

18. Recent work[29] suggests that char particles break into several fragments during the final burnout phases of char combustion.

19. Part of the mineral matter in coals can sometimes be volatilized, depending upon the temperature.[29]

20. There has been considerable emphasis on coal-particle ignition temperature. Early work assumed this parameter to be constant[2,30] and equal to the coal dissociation temperature. Subsequent analysis[1] has shown this ignition temperature to be related to several parameters of the system.

21. For plane coal-dust flames, radiant heating of the coal dust has been assumed to be the dominant factor in flame propagation.[1,23,30] However, more recent work has suggested the importance of diffusional processes.[6,31]

---

fluidized-bed processes, but is intended for application to pulverized-coal furnaces, entrained coal gasifiers, MHD power generators, and to propagation and explosion in pulverized coal–air mixtures. Table 2 summarizes key assumptions made for the general coal reaction model. Assumptions can be altered for application to a specific coal reaction problem.

A general sequence of chemical reactions is shown in Table 3 for a pulverized-coal system, considering an arbitrary number (J) of particle sizes or coal types, an arbitrary number (M) of parallel, activated devolatilization reactions for each of the coal types, and subsequent char oxidation by steam, oxygen, hydrogen, or carbon dioxide (L reactions), and an arbitrary number (M) of gaseous hydrocarbon–oxygen oxidative pyrolysis reactions for each of the coal-particle classifications. The treatment provides for different devolatilization products from each coal-particle type or size, for each devolatilization reaction (V_{jm}). For a particular hydrocarbon product from a single activated devolatilization reaction, there is only one gaseous hydrocarbon oxidative pyrolysis reaction. Only the pyrolyzate hydrocarbon undergoes a global irreversible reaction. Other pyrolyzate gases, such as N_2, O_2, and NO undergo detailed elementary gas-phase reaction steps as specified in Chapter 10. Figure 1 shows a schematic of the reacting coal particle.

Table 2. Coal-Particle Reaction Model Assumptions

1. Particle reaction takes place by two competing processes:
   (a) devolatilization, which is dominated by activated first-order pyrolysis of raw coal and is initiated above a specific threshold temperature.
   (b) particle surface reaction, which is controlled jointly by oxidizer diffusion and surface reaction of order between zero and unity. Reaction can be by oxygen, steam, carbon dioxide, or hydrogen.
2. The particles are a mixture of discrete sets of particle sizes and types, each with specific properties.
3. The volume occupied by the dust is small compared to the gas volume.
4. Interactions or collisions among particles is negligible except for radiation.
5. The temperature of any given particle is uniform throughout at any given time.
6. A particle is assumed to be composed of ash, moisture, raw coal, and char. Ash is inert and any volatile mineral matter is included in the volatiles.
7. The extent of swelling of coal particles is directly proportional to the extent of devolatilization.
8. Devolatilization produces a fuel-rich product whose gases block diffusion and heat transfer by surface transpiration.
9. The devolatilization products are of specified composition, ordinarily richer in hydrogen than the coal as a whole. The carbon–hydrogen ratio is determined from ultimate char analysis where such data are available. The product of surface oxidation is carbon monoxide, which diffuses away from the particle and reacts further in the gas phase.
10. The activated devolatilization process is described by a sequence of pyrolysis reactions, each with specified stoichiometric coefficients and kinetic parameters.
11. The composition of volatiles, which may contain hydrocarbons, nitrogen, oxygen, sulfur, etc., is specified as input for each of the devolatilization reactions, and is based on material balance information where available.
12. The heats of devolatilization and char oxidation are specified parameters, to be obtained from experimental data and/or from coal decomposition computations.

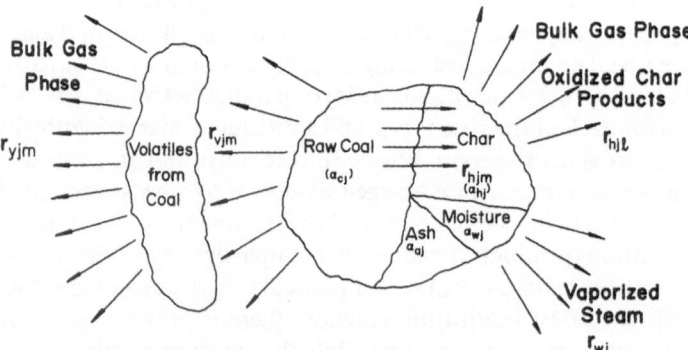

Figure 1. Schematic of jth particle, illustrating constituents and reaction processes.

Table 2 (Continued)

13. Particle velocities are not assumed to be identical to the gas velocity.
14. The change in char-particle diameter during char burnout following devolatilization must be specified. Char remaining after fragmentation is neglected.
15. Char, which may contain hydrogen, oxygen, nitrogen, sulfur, etc., in addition to carbon, is consumed at the rate governed by carbon-oxidizer reaction kinetics, with other elements reacted at a rate which is proportional to their percentage in the char.
16. The heat required to react the coal heterogeneously (endothermic) or that liberated from the heterogeneous coal reaction (exothermic) can be supplied by or delivered to the gas or particle phase.
17. The surface area for surface reaction is taken as a constant parameter times the particle area for an equivalent spherical particle. The particle burns out when the total mass is consumed, which may not be when the particle diameter goes to zero. For example, for a thin hollow sphere, with internal and external burning, the particle diameter would be nearly constant until burnout and the area factor would be 2.0. Only the spherical surface area is used in heat- and mass-transfer rate computations.
18. Local heat transfer between the gas and the particles includes convective and conductive components. Required physical properties for the gas mixtures such as conductivity are evaluated at the local arithmetic mean temperature of the gas and particles.
19. The diffusion coefficient of a given species in the gas mixture is evaluated assuming the other mixture species are stationary, or alternatively, that the diffusing species is very dilute.
20. Moisture loss is controlled by heat transfer to the particle and by vapor diffusion from the coal surface to the bulk gas.
21. The gas-phase reactions are considered as possible rate-limiting steps. A global reaction in the gas phase describes the oxidative pyrolysis of hydrocarbon fractions from the devolatilization process to produce H_2 and CO. A sequence of kinetic reactions expresses the further reaction of H_2 and CO, together with other possible gas-phase species.

Table 3. Postulated Reactions for Generalized Pulverized-Coal Reaction Sequence

(A) Coal pyrolysis (ash, moisture-free basis) ($M \times J$ reactions)

$$C_j(S) \longrightarrow Y_{jm}V_{jm} + (1 - Y_{jm})S_{jm}, \qquad j = 1, J; m = 1, M$$

(B) Char oxidation (carbon) ($L \times J$ reactions)

$$l = 1: \quad C_j(S) + \tfrac{1}{2}O_2 \longrightarrow CO$$

$$l = 2: \quad C_j(S) + H_2O \longrightarrow CO + H_2$$

$$l = 3: \quad C_j(S) + CO_2 \longrightarrow 2CO$$

$$l = L: \quad C_j(S) + O \longrightarrow CO$$

(C) Volatiles oxidation ($M \times J$ reactions)

$$C_{n\,jm}H_{m\,jm} + (n_{jm}/2)O_2 \rightarrow n_{jm}CO + (m_{jm}/2)H_2, \qquad j = 1, J; m = 1, M$$

2. Pulverized-Coal Model Equations

Based upon the assumptions of the preceding section and following the general kinetic scheme of Table 3, a series of coal model equations has been formulated. These equations are divided into several classifications: (1) coal reaction rates, (2) gas–particle rate processes, (3) gas-mixture physical properties, and (4) coal physical properties. A detailed definition of symbols with units is included in the Notation section (Section 3).

2.1. Coal Reaction Rates

Reaction rates are described as the mass of substance reacted per unit volume of gas per unit time.

2.1a. Devolatilization

For the devolatilization reactions of Table 3, the kinetic rates of production of volatiles for the mth reaction are given as[12,14]

$$r_{vjm} = Y_{jm} k_{jm} \rho_{cj}, \qquad j = 1, J \tag{1}$$

where $k_{jm} = A_{jm} \exp(-E_{jm}/RT_j)$ and ρ_{cj} is the mass density of the raw coal for the jth particle per unit volume of gas. For a sequence of M reactions, the total production of volatiles is

$$r_{vj} = \sum_m Y_{jm} k_{jm} \rho_{cj} \tag{2}$$

It is possible that the pyrolysis products can accumulate at the particle surface and in the particle pores, with the rate of volatiles depletion being affected by diffusion or convection from particle blow holes into the gas phase, in competition with condensation or repolymerization. In this case, the diffusion-controlled reaction rate for a given hydrocarbon product may be expressed as

$$r_{vjmd} = k_{cjm} A_j (\rho_{vjm} - \rho_{vgm}) n_j / [1 - (r_j \rho_{vjm} / r_{vjmd} \rho_{cj})] \tag{3}$$

where r_j is the total rate of production of gases from the jth particle and is defined by Eq. (18). The kinetic process is expressed as

$$r_{vjmk} = Y_{jm} k_{jmf} \rho_{cj} - k_{jmr} \rho_{vjm} \tag{4}$$

For steady state, the following relation is valid:

$$r_{vjmd} = r_{vjmk} = r_{vjm} \tag{5}$$

Equations (3) and (4) may be used to evaluate the two unknown quantities r_{vjm} and ρ_{vjm} (the mass density of hydrocarbon volatiles at the particle surface).

For a negligible reverse-reaction rate, the result simplifies to Eq. (1). Diffusional effects only become important when the volatiles can repolymerize or condense prior to diffusion into the gas phase. Diffusion may also be enhanced by convective jets of volatiles through porous blow holes in the coal surface. Because of uncertainties in the reverse-rate coefficients, the nature of the diffusional process, and because the techniques used to correlate existing pyrolysis data most often have utilized Eq. (1),[12,14] Eq. (1) will be emphasized in describing the devolatilization rate.

Char is also produced in competition with volatiles production, and from Table 3, can be expressed as

$$r_{hjm} = r_{vjm}(1 - Y_{mj})/Y_{mj} \tag{6}$$

Further, the volatiles and char products contain a variety of chemical elements that are specified as input for the model. Then, from this specified composition, together with the computed reaction rates, species production rates from the particle to the gas phase by devolatilization can be specified

$$r_{vijm} = \phi_{vijm} r_{vjm} \tag{7}$$

where ϕ_{vijm} is the stoichiometric coefficient for the ith gas species.

2.1b. Char Oxidation

Char is assumed to oxidize heterogeneously by a gaseous oxidizer that diffuses to the particle, adsorbs, reacts with the carbon, and desorbs as carbon monoxide. Two rate-limiting steps are considered for this process: (1) oxidizer diffusion and (2) surface reaction. For the diffusion step

$$r_{hjld} = M_{hj}\phi_l k_{cjl}(C_{og} - C_{oj})n_j A_j/[1 + (M_{hj}\phi_l r_j C_{oj}/M_g r_{hjld} C_g)] \tag{8}$$

where r_j is as defined in Eq. (18). For the reaction step for a general nth-order reaction, the following expression is used where n is usually between 0 and 1

$$r_{hjlk} = \phi_l M_{nj} k_{jl}\zeta A_j C_{oj}^n n_j \tag{9}$$

and where

$$k_{jl} = A_{jl} \exp(-E_{jl}/RT_j) \tag{10}$$

For steady state, $r_{hjld} = r_{hjlk} = r_{hjl}$, the two unknown quantities r_{hjl} and C_{oj} can be evaluated or eliminated. When $n = 1$, this result is

$$r_{hjl} = (A_j n_j)^2 M_{hj} M_g \phi_l k_{cjl} k_{jl}\zeta_j C_{og} C_g/[M_g A_j n_j C_g(\zeta_j k_{jl} + k_{cjl}) + r_j] \tag{11}$$

For a given chemical species entering the gas phase via char oxidation,

$$r_{ihjl} = \phi_{hij} r_{hjl} \tag{12}$$

Further, the net char reaction rate combines production from coal devolatilization and depletion by heterogeneous oxidation:

$$r_{hj} = \sum_l r_{hjl} - \sum_m r_{hjm} \tag{13}$$

2.2. Hydrocarbon Oxidation

Each of the gaseous hydrocarbons produced by the M devolatilization reactions is oxidized in the gas phase according to the general global irreversible oxidation reaction

$$r_{ymj} = A_{mj} p^{n_1} T^{n_2} M_{ym} C_{ymg}^{n_3} C_o^{n_4} \exp\left(-E_{ymg}/RT_g\right) \tag{14}$$

where n_1, n_2, n_3, and n_4 are assigned parameters. According to Siminski et al.,[32] for correlation of oxidation data for heavy long-chain hydrocarbons by oxygen, this reaction, which is discussed in detail in Chapter 10, is

$$r_{ymj} = (\text{const}) \, p^{0.3} T M_{ym} C_{ymg}^{0.5} C_o \exp(-12{,}200/T_g) \tag{15}$$

The corresponding results are shown in Chapter 10 for cyclic hydrocarbons. Alternatively, if a detailed hydrocarbon kinetic mechanism is known (for example, methane), then a generalized, reversible reaction scheme can be so specified in detail, as discussed in Chapter 10.

At the present time, no allowance has been made for formation and subsequent reaction of condensed hydrocarbons other than char, such as tars or soot, which could occur in much different particle sizes and compositions from char. Experimental evidence[21] suggests this may be important in fuel-rich, coal–air flames.

2.3. Moisture Vaporization

Some coals contain high percentages of moisture. Diffusion-limited vaporization of moisture from the coal particle is described by

$$r_{wj} = M_w N u_{im} C_g D_{wm} n_j A_j (X_{wj} - X_{wg})/d_j (1 - X_{wj} r_j/r_{wj}) \tag{16}$$

where, for particle temperatures well below the boiling point of water, $X_{wj} = p_{wej}/p$. As the particle continues to heat up, heat transfer to the particle by radiation and convection may limit the rate of moisture vaporization such that

$$(r_{wj})_{\max} = (Q_j + Q_{fj})/\Delta h_w \tag{17}$$

Combining the rate expressions of Eqs. (1), (11), and (16), the total rate of production of gases from the jth particle is then the sum

$$r_j = \sum_m r_{vjm} + \sum_l r_{hjl} + r_{wj} \tag{18}$$

2.4. Gas–Particle Rates

The sequence of kinetic and other rate expressions in the preceding section introduced parameters describing gas–particle interchange that are summarized in this section. The terms, Q_j, k_{cjm}, Nu_m, and Nu_{im} deal with heat and mass transfer between the particles and the gases as outlined in Chapter 6. Radiation between the gases and the particles has been neglected. Radiation among particles is considered in Chapter 5.

2.4a. Convective–Conductive Heat Transfer

Assuming spherical particles, heat-transfer rates per unit volume of gas have been based on the heat-transfer coefficient

$$Q_j = Nu_{jm} k_{gb} A_j (T_j - T_g) n_j / d_j \tag{19}$$

where Nu_{jm}, the mean Nusselt number for heat transfer, is influenced by particle blowing. From film theory, this is calculated as

$$(Nu_{jm}/Nu_{j0m}) = B_j / [\exp(B_j) - 1] \tag{20}$$

and from reference 9, or Eq. (34) of Chapter 6,

$$Nu_{j0m} = (h_{jm} d_j / k_{gb}) = 2.0 + 0.65 (Re_{jb})^{0.5} (Pr_b)^{1/3} \tag{21}$$

where $Re_{jb} = (d_j |v_j - v_g| \rho_g / \mu_g)_b$ and $Pr_b = (C_{pg} \mu_j / k_g)_b$. The blowing parameter, B_j, is related to the total production of gases from the jth particle, r_j, by the expression

$$B_j = r_j C_{pgb} / (2\pi d_j k_{gb} n_j) \tag{22}$$

Equations (20) and (22) form the classical approximate correction for transpiration cooling effects as derived from film theory.

Film theory assumes that there is a stagnant film on the surface of the particle. At a large distance from the surface the temperature and concentrations are those of the bulk fluid. Steady state is assumed and the only motion through the film is taken to be that created by the diffusing species. Although film theory is instructive it leads to large errors in many cases. It is generally adequate when there is no forced convection (i.e., $Re_{jb} = 0$). Hoffman and Ross[33] have chosen a reasonable stream function and conducted a theoretical investigation on the interaction of the radial flow rising from mass transfer and the forced-convection flow field around an evaporating droplet in order to elucidate the effect of mass transfer on heat transfer. Their model indicates that for Schmidt numbers between 0.7 and 10 and for Reynolds numbers up to 400, good results should be obtained from

$$(Nu_{jm}/Nu_{j0m}) = (1 + B')^{-0.6} \tag{23}$$

where B' is the conventional Spalding number. In terms of the preceding nomenclature,

$$(Nu_{jm}/Nu_{j0m}) = \exp(-0.6B_j) \tag{24}$$

Both this theory and film theory reduce the correction factor to 1 when $B_j = 0$.

2.4b. Convective–Diffusive Mass Transfer

Assuming spherical particles, the mean Nusselt number for mass transfer between the gas and particles can be calculated from film theory:

$$(Nu)_{im}/(Nu)_{im0} = B_{jm}/[\exp(B_{jm}) - 1] \tag{25}$$

and

$$Nu_{im0} = (k_{co}d_j/D_{im}) = (k_{xo}d_j/C_gD_{im}) = 2.0 + 0.65(Re_{jb})^{0.5}(Sc_b)^{1/3} \tag{26}$$

where $Sc_b = (\mu_g/\rho_g D_{im})_b$. The blowing parameter for mass transfer, B_{jm}, is related to the total production of gases from the jth particle by

$$B_{jm} = r_j(2\pi D_{im}\rho_g d_j n_j) \tag{27}$$

As discussed in the preceding section, the correction factor as calculated from film theory should be useful for $Re_{jb} = 0$. For Reynolds numbers up to 400 the following equation is suggested:

$$(Nu)_{im}/(Nu)_{im0} = \exp(-0.6B_{jm}) \tag{28}$$

Where the Reynolds assumption is applicable ($Sc = Pr = 1$), then

$$B_j = B_{jm} \tag{29}$$

2.5. Gas-Mixture Properties

Coal reaction rate expressions require evaluation of several gas-phase physical properties, including density, heat capacity, thermal conductivity, and diffusivity. These properties are all functions of temperature and composition. When the property is for the bulk gas phase, $T = T_g$, while for particle–gas interchange, $T_b = (T_g + T_j)/2$ is recommended. For an ideal gas,

$$p/\rho_g = RT_g/M_g \tag{30}$$

and

$$C_g = p/RT_g \tag{31}$$

where

$$M_g = \left(\sum_i \omega_i/M_{ig}\right)^{-1} \tag{32}$$

Heat capacity is conventionally expressed as a polynomial expansion in temperature

$$C_{pi} = \sum_{k} a_{ki} T^k \tag{33}$$

and for the gas mixture

$$C_{pg} = \sum_{i} \omega_i C_{pi} \tag{34}$$

Transport coefficients are evaluated using conventional statistical theory results[34,35] as outlined in Chapter 2 and summarized below:

Gas-species viscosity:

$$\mu_i = 2.67 \times 10^{-4} (M_i T_{gb})^{1/2} / (\sigma_i^2 \Omega_\mu) \tag{35}$$

Gas-mixture viscosity:

$$\mu_g = \sum_{i} \left(X_i \mu_i \middle/ \sum_{k} X_k \phi_{ik} \right) \tag{36}$$

Interaction parameter:

$$\phi_{ik} = (1/8)^{1/2} [1 + (M_i/M_k)]^{-1/2} [1 + (\mu_i/\mu_k)^{1/2} (M_k/M_i)^{1/4}]^2 \tag{37}$$

Gas-species conductivity:

$$k_i = [C_{pi} + (5M_i/4)] \mu_i \tag{38}$$

Gas-mixture conductivity:

$$k_g = \sum_{i} \left(X_i k_i \middle/ \sum_{k} X_k \phi_{ik} \right) \tag{39}$$

Species diffusivity:

$$D_{ik} = 1.84 \times 10^{-12} T_b^{3/2} [(1/M_k) + (1/M_i)]^{1/2} / (p \sigma_{jk}^2 \Omega_d) \tag{40}$$

Mixture diffusivity:

$$D_{im} = (1 - X_i) \middle/ \sum_{k \neq i} (X_k/D_{ik}) \tag{41}$$

These coefficients are functions of the local values of temperature and composition (diffusivity is also a function of pressure), all of which can vary with time and position. Repeated evaluation can lead to long computation times. Where approximations are appropriate, all diffusivities can be taken to be equal and evaluated from the conductivity, assuming unity Lewis number. In some computational schemes, especially where the steady-state solution is sought, these transport coefficients need only be reevaluated every several iterations. Further, Bartlett *et al.*[36] have applied a bifurcation concept for evaluating transport coefficients which can greatly reduce computational time for complex applications.

2.6. Coal Physical Properties and Parameters

Additional required coal-properties relationships are summarized in this section. Particle diameter for the spherical, swelling particles is assumed to increase at a rate proportional to the extent of devolatilization:

$$d_j = d_{j0}[1 + \gamma(\alpha_{cj0} - \alpha_{cj})/\alpha_{cj0}] \tag{42}$$

It is anticipated that γ will be a function of the percent volatiles (Y_m) and would be near zero for a nonvolatile coal such as anthracite. The effective surface area of the particle for reaction is taken to be a multiple of the spherical area:

$$A_j = \pi d_j^2, \qquad A_j \zeta_j = \pi \zeta_j d_j^2 \tag{43}$$

Particle component density and mass and number relationships are

Particle density:

$$\rho_j = \rho_{cj} + \rho_{hj} + \rho_{aj} + \rho_{wj} \tag{44}$$

Coal density:

$$\rho_{cj} = \alpha_{cj} n_j \tag{45}$$

Moisture density:

$$\rho_{wj} = \alpha_{wj} n_j \tag{46}$$

Char density:

$$\rho_{hj} = \alpha_{hj} n_j \tag{47}$$

Ash density:

$$\rho_{aj} = \alpha_{aj} n_j \tag{48}$$

Particle mass:

$$\alpha_j = \alpha_{hj} + \alpha_{cj} + \alpha_{aj} + \alpha_{wj} \tag{49}$$

By material balance, particle mass can be related to reaction rates by

$$d\alpha_{cj}/dt = -r_{cj}/n_j \tag{50}$$

$$d\alpha_{hj}/dt = -r_{hj}/n_j \tag{51}$$

$$d\alpha_{wj}/dt = -r_{wj}/n_j \tag{52}$$

where

$$r_{cj} = \sum_m (r_{vjm}/Y_{jm}) \tag{53}$$

Since ash is taken as inert, α_{aj} is a constant. Further, particle mass fractions can be related to particle component masses by

$$\omega_{aj} = \alpha_{aj}/\alpha_j \tag{54}$$

$$\omega_{hj} = \alpha_{hj}/\alpha_j \tag{55}$$

$$\omega_{wj} = \alpha_{wj}/\alpha_j \tag{56}$$

Particle heat capacity and enthalpy are taken as weighted averages over each of the particle components:

Particle heat capacity:

$$C_{pj} = \omega_{hj}C_{phj} + \omega_{cj}C_{pcj} + \omega_{wj}C_{pwj} + \omega_{aj}C_{paj} \tag{57}$$

Particle enthalpy:

$$h_j = (\alpha_{cj}h_{cj} + \alpha_{hj}h_{hj} + \alpha_{aj}h_{aj} + \alpha_{wj}h_w)/(\alpha_{cj} + \alpha_{hj} + \alpha_{aj} + \alpha_{wj}) \tag{58}$$

where component heat capacities and enthalpies are given by

Particle heat capacity:

$$C_{pj\alpha} = \sum_k b_{\alpha k}T_j^k \tag{59}$$

Coal enthalpy:

$$h_{cj} = \int C_{pcj}\,dT_j + h_{cj}^0 \tag{60}$$

Char enthalpy:

$$h_{hj} = \int C_{phj}\,dT_j + h_{hj}^0 \tag{61}$$

Moisture enthalpy:

$$h_{wj} = \int C_{pwj}\,dT_j + h_{wj}^0 \tag{62}$$

Ash enthalpy:

$$h_{aj} = \int C_{paj}\,dT_j + h_{aj}^0 \tag{63}$$

The enthalpy of the coal reaction products entering the gas phase is also taken as a reaction-rate weighted average:

Particle products enthalpy:

$$h_{jg} = \left(\sum_m r_{jvm}h_{jvmg} + \sum_l r_{jhl}h_{jhlg} + r_{jw}h_{jwg}\right) \Big/ \left(\sum_m r_{jvm} + \sum_l r_{jhl} + r_{jw}\right) \tag{64}$$

where volatiles product enthalpy in gas phase is

$$h_{jvgm} = h_{jvm} + \Delta h_{vm} \tag{65}$$

Char product enthalpy in gas phase is

$$h_{jhlg} = h_{jhl} + \Delta h_{hl} \tag{66}$$

and moisture enthalpy in gas phase is

$$h_{jwg} = h_{jw} + \Delta h_{w} \tag{67}$$

Depending upon the specific assumptions and the model applications, the above sequence of equations will be altered. However, this outlines a general scheme that can serve as a starting point for treatment of coal reactions in complex combustion processes.

3. Notation

A_j	Particle area (m^2)	r	Particle reaction rate (kg m^{-3} s^{-1})
A	Preexponential factor	Re	Reynolds number
a	Heat-capacity coefficient [(Eq. (33)] ($J\ kg^{-1}\ K^{-(k+1)}$)	Sc	Schmidt number
		T	Temperature (K)
b	Heat-capacity coefficient [(Eq. (59)] ($J\ kg^{-1}\ K^{-(k+1)}$)	X	Mole fraction
		Y	Pyrolysis coefficient
B	Blowing parameter	α	Mass of coal component (kg)
C	Molar concentration (kmol m^{-3})	γ	Particle swelling parameter
C_p	Heat capacity ($J\ kg^{-1}\ K^{-1}$)	ϕ	Interaction parameter
D	Binary diffusivity ($m^2\ s^{-1}$)	ϕ	Stoichiometric coefficient
d	Particle diameter (m)	ζ	Particle internal area factor
E	Activation energy ($J\ kmol^{-1}$)	σ	Collision diameter (Á)
h	Heat-transfer coefficient ($J\ s^{-1}\ m^{-2}\ K^{-1}$)	μ	Viscosity (kg $m^{-1}\ s^{-1}$)
h	Enthalpy ($J\ kg^{-1}$)	ρ	Density (kg m^{-3})
Δh	Heat of reaction ($J\ kg^{-1}$)	ω	Mass fraction
k	Reaction-rate coefficient (s^{-1}, ms^{-1})		
k_c	Mass-transfer coefficient (ms^{-1})		**Subscripts**
k_g	Gas conductivity ($J\ s^{-1}\ m^{-1}\ K^{-1}$)		
k_x	Mass-transfer coefficient (kmole $m^{-2}\ s^{-1}$)	a	ash
M	Molecular weight (kg $kmol^{-1}$)	b	boundary
Nu	Nusselt number	c	coal
n	Particle number density (m^{-3})	d	diffusion
Pr	Prandtl number	e	equilibrium
p	Pressure (Pa)	f	flame; forward
Q	Particle–gas heat transfer rate ($J\ m^{-3}\ s^{-1}$)	g	gas
		h	char
R	Gas constant ($J\ kmol^{-1}\ K^{-1}$)	i	gas species; summation index
		j	particle; surface of particle

k	kinetic; summation index	r	reverse
l	char oxidation reaction	v	volatiles
m	mean; mixture; devolatilization reaction; mass transfer	y	hydrocarbon
		α	coal component
n	reaction order	μ	viscosity
o	oxidizer	0	initial value; without blowing

4. References

1. S. Bandyopadhyay and D. Bhaduri, Prediction of ignition temperature of a single coal particle, *Comb. and Flame* **18**, 411 (1972).
2. M. M. Baum and P. J. Street, Predicting the combustion behavior of coal particles, *Comb. Sci. and Tech.* **3**, 231–243 (1971).
3. R. Essenhigh, Dominant Mechanisms in the Combustion of Coal, ASME Paper No. 70-WA/Fu-2 (1970).
4. D. Gray, G. Cogoli, and R. H. Essenhigh, Problems in pulverized coal and char combustion, Paper No. 6, in *Advances in Chemistry Series No. 131*, pp. 72–91, American Chemical Society, Washington, D.C. (1976).
5. M. Hertzberg, The Expansion and Application of Laminar Flame Theory to the heterogeneous Combustion of Coal Dust, unpublished note, Bureau of Mines, Pittsburgh, Pa. (1974).
6. L. D. Smoot, M. D. Horton, and G. A. Williams, Propagation of laminar pulverized coal-air flames, in *Sixteenth Symposium (International) on Combustion*, pp. 375–387, The Combustion Institute, Pittsburgh, Pa. (1976).
7. J. B. Howard and R. H. Essenhigh, Mechanism of solid-particle combustion with simultaneous gas-phase volatiles combustion, in *Eleventh Symposium (International) on Combustion*, p. 399, The Combustion Institute, Pittsburgh, Pa. (1967).
8. D. Anson, F. D. Moles and P. J. Street, Structure and surface area of pulverized coal during combustion, *Comb. and Flame* **16**, 265 (1971).
9. I. W. Smith, The kinetics of combustion of pulverized semi-anthracite in the temperature range 1900–2200 K, *Comb. and Flame* **17**, 421 (1971).
10. M. A. Field, Measurements of the effect of rank on combustion rates of pulverized coal, *Comb. and Flame* **14**, 237–248 (1970).
11. L. D. Smoot and M. D. Horton, *Exploratory Studies of Flame and Explosion Quenching*, 3rd Interim Annual Report, Bureau of Mines Contract 10122052, Brigham Young University, Provo, Utah (Sept. 30, 1975).
12. D. B. Anthony, J. B. Howard, H. C. Hottel, and H. P. Meissner, Rapid devolatilization of pulverized coal, in *Fifteenth Symposium (International) on Combustion*, pp. 1303–1317, The Combustion Institute, Pittsburgh, Pa. (1974).
13. H. Juntgen and K. H. VanHeek, Gas release from coal as a function of rate of heating, *Fuel* **47**, 103 (1967).
14. H. Kobayashi, J. B. Howard, and A. F. Sarofim, Coal devolatilization at high temperatures, in *Sixteenth Symposium (International) on Combustion*, pp. 411–425, The Combustion Institute, Pittsburgh, Pa. (1976).
15. S. K. Ubhayakar, D. B. Stickler, C. W. von Rosenberg, Jr., and R. E. Gannon, Rapid devolatilization of pulverized coal in hot combustion gases, in *Sixteenth Symposium (International) on Combustion*, pp. 427–436, The Combustion Institute, Pittsburgh, Pa. (1976).
16. S. Badzioch and P. G. W. Hawksley, Kinetics of thermal decomposition of pulverized coal particles, *Ind. Eng. Chem. Process Design Develop.* **9**, 521 (1970).

17. H. Reidelbach and M. Summerfield, Kinetic Model for Coal Pyrolysis Optimization, paper given at the Eastern States Section Meeting, The Combustion Institute, Pittsburgh, Pa. (1974).
18. R. Loison and R. Chauvin, Pyrolyse rapide de charbon, *Chim. Ind.* (*Paris*) **91**, 269–275 (1964).
19. H. C. Hottel, G. C. Williams, and M. L. Baker, Combustion studies in a stirred reactor, in *Sixth Symposium* (*International*) *on Combustion*, pp. 398–411 The Combustion Institute, Pittsburgh, Pa. (1957).
20. J. P. Longwell and M. A. Weiss, High temperature reaction rates in hydrocarbon combustion, *Ind. Eng. Chem.* **47**, 1634–1637 (1955).
21. L. D. Smoot and M. D. Horton, *Exploratory Studies of Flame and Explosion Quenching*, 4th Interim Annual Report, Bureau of Mines Contract H0122052, Brigham Young University, Provo, Utah (Dec. 1976).
22. J. H. Pohl and A. F. Sarofim, Devolatilization and oxidation of coal nitrogen, in *Sixteenth Symposium* (*International*) *on Combustion*, pp. 491–501, The Combustion Institute, Pittsburgh, Pa. (1977).
23. D. Bhaduri and S. Bandyopadhyay, Combustion in coal dust flames, *Comb. and Flame* **17**, 15–24 (1971).
24. C. Crompton, Summary of Kinetic Parameters for Devolatilization and Heterogeneous Oxidation of Carbonaceous Solids, unpublished report, Brigham Young University, Chemical Engineering Dept., Provo, Utah (1975).
25. H. H. Lowry, editor, *Chemistry of Coal Utilization*, Supplementary Volume, John Wiley and Sons, New York (1963).
26. M. A. Field, Rate of combustion of size-graded fractions of char from a low-rank coal between 1200 K and 2000K, *Comb. and Flame* **13**, 237–252 (1969).
27. J. R. Arthur, Reactions between carbon and oxygen, *Trans. Faraday Soc.* **47**, 164–178 (1951).
28. M. A. Field, D. W. Gill, B. B. Morgan, and P. G. S. Hawksley, *Combustion of Pulverized Coal*, BCURA, 2nd edition, Institute of Fuel, London, England (1974).
29. A. S. Padia, A. F. Sarofim, and J. B. Howard, The Behavior of Ash in Pulverized Coal under Simulated Combustion Conditions, paper given at the Spring Meeting, Central States Section, The Combustion Institute, Pittsburgh, Pa. (April 5, 1976).
30. R. H. Essenhigh and J. Casaba, The thermal radiation theory for plane flame propagation in coal-dust clouds, in *Ninth Symposium* (*International*) *on Combustion*, pp. 111–125, The Combustion Institute, Pittsburgh, Pa. (1963).
31. R. H. Essenhigh, Combustion and flame propagation in coal systems—Review, in *Sixteenth Symposium* (*International*) *on Combustion*, pp. 353–374, The Combustion Institute, Pittsburgh, Pa. (1976).
32. V. J. Siminski, F. J. Wright, R. Edelman, D. Economos, and O. Fortune, *Research on Methods of Improving the Combustion Characteristics of Liquid Hydrocarbon Fuels*, AFAPL TR 72-24, Vols. I and II, Air Force Aeropropulsion Laboratory, Wright–Patterson Air Force Base, Ohio (Feb. 1972).
33. T. W. Hoffman and L. L. Ross, A theoretical investigation of the effect of mass transfer of heat transfer to an evaporating droplet, *Int. J. Heat Mass Transfer* **15**, 599–617 (1972).
34. R. B. Bird, W. E. Stewart, and E. N. Lightfoot, *Transport Phenomena*, John Wiley and Sons, New York (1960).
35. R. C. Reid and T. K. Sherwood, *The Properties of Gases and Liquids*, 2nd edition, McGraw–Hill, New York (1966).
36. E. P. Bartlett, R. M. Kendall, and R. A. Rindal, *A Unified Approximation for Mixture Transport Properties for Multicomponent Boundary Layer Applications*, NASA CR-1063, Itek Corp., Palo Alto, Calif. (1968).

Part IV
Mathematical Modeling of
Coal Conversion Processes

13

Modeling One-Dimensional Systems

L. Douglas Smoot and Philip J. Smith

1. Propagating Pulverized Coal–Air Flames

1.1. Background

In this section, a model for predicting properties of premixed propagating methane–air and coal–air flames is described. This model is formulated to treat the general case of propagation in multiphase systems containing finely divided particles or droplets; however, the application emphasized herein is for pulverized coal–air systems. The model can also be used to describe gaseous systems, and has been applied to methane–oxidizer systems. The development that follows emphasizes laminar flows, but treatment of turbulent, propagating flames is also briefly discussed. The model is also two-dimensional, with one dimension being time and the other being distance along the flame; however, since the emphasis here is on the steady-state aspects, it is considered in this chapter on one-dimensional systems.

A major reason for interest in premixed coal–air flames relates to coal-mine fires and explosions. Such explosions result from ignition and combustion of methane–air mixtures. As the flame propagates in the mine tunnels, pulverized-coal dust is entrained, thus reinforcing the combustion and pro-

L. Douglas Smoot • Professor of Chemical Engineering, Brigham Young University, Provo, Utah. *Philip J. Smith* • Ph.D. Candidate, Chemical Engineering Department, Brigham Young University, Provo, Utah

moting explosive propagation in the mine. Figure 1 schematically illustrates this process. In such explosions, both methane and coal-dust flames are of interest. Grumer[1] and Richmond and Liebman[2] have recently reviewed and described in detail the nature of these explosions. Propagating flames are also of interest in the analysis of ignition and stability of gaseous and pulverized-coal mixtures in furnaces, gasifiers, and other conversion processes.

Much work has been done to determine the properties of laminar methane–air flames and to understand the mechanisms by which they propagate. Temperature and concentration profiles have been determined experimentally by several investigators for specific methane–air flames,[3-7] usually at low pressures. Andrews and Bradley[8,9] and Bradley and Hundy[10] have measured flame velocities, and have also summarized literature values of flame velocity and thickness for flames of varying pressures, methane concentrations, and unburned gas temperatures.

Williams provides a review of the models of propagating gaseous flames up to 1965.[11] More recent modeling work has also been reported for hydrocarbon–air flames. Browne et al.[12] used general conservation equations to formulate a one-dimensional, steady-state, propagating-flame model which predicted the temperature and concentration profiles of acetylene–oxygen flames. Cooke and Williams[13] studied ignition of methane–oxygen–argon in a shock tube and compared predicted concentration and temperature as a function of time (but not distance) with measurements. Cordeiro et al.[14] used the conservation equations to formulate a one-dimensional, steady-state model, which uses flame velocity as input to predict the temperature and concentration profiles of Peeters'[7] low-pressure methane–oxygen flame. The model outlined in this section is a transient or unsteady-state formulation which predicts profiles, flame thickness, and propagation velocity of complex flames.

Previous theoretical models for propagating coal-dust flames have been based upon the assumption that radiative heat transfer from the hot-flame region to the incoming coal–air mixture is the dominant step in the propagation rate,[15,16] while gas-phase diffusional processes have been neglected. However, experimental data for chemical suppression of laminar, coal–air flames show characteristics strikingly similar to gaseous flames,[17] where radiation is not an important factor.[18] Predictions and measurements[19] also suggest that other rate-limiting processes may also be important in laminar pulverized coal–air flame propagation. Optimum control, enhancement, and suppression of coal–air ignition and reaction processes require identification of these dominant processes. This section deals with these rate-limiting processes, and includes the development of a laminar propagation model for two-phase mixtures with application to coal–air systems. Comparison to new experimental data is also included.

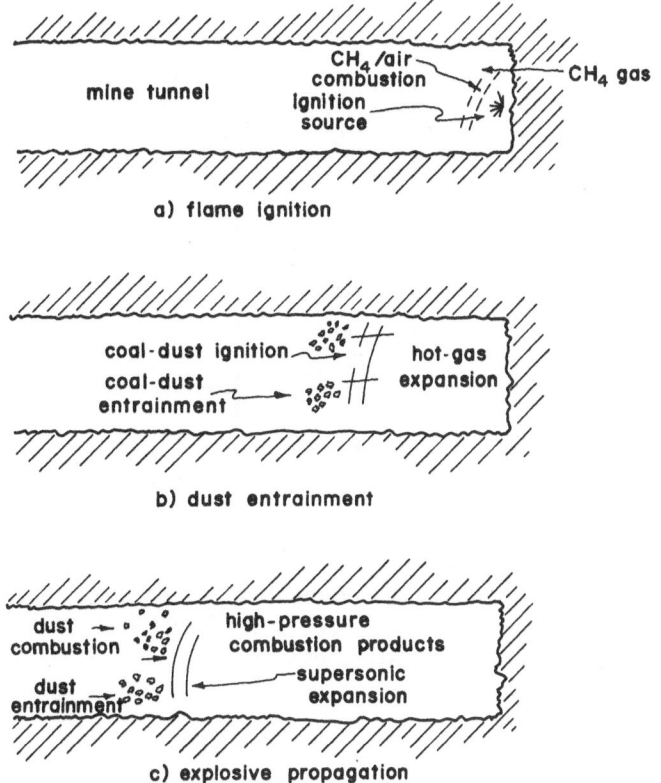

Figure 1. Propagating flames in coal-mine explosions.

1.2. Model Assumptions and Equations

Spalding and co-workers developed a unique computational method for prediction of gas-phase flames,[20] which has served as a guide in this study. More recently, Essenhigh and Csaba[15] and Bhaduri and Bandyopadhyay[16] have based propagating coal–air theories on thermal radiation. In this study, basic unsteady-state conservation equations for laminar, multicomponent, compressible gas–particle mixtures, which were developed in Chapter 2, were used to formulate a one-dimensional, propagating-flame model. A schematic diagram of this flame is shown in Figure 2. The particles were taken to be spherical and sufficiently small and numerous to be treated as a continuous medium, while particle volume, particle diffusion, and collisions among particles were neglected. Pressure was assumed to be uniform while the particle velocities were taken to be equal to the local gas velocity in the flame. Effects of gravity, viscous dissipation, forced diffusion,

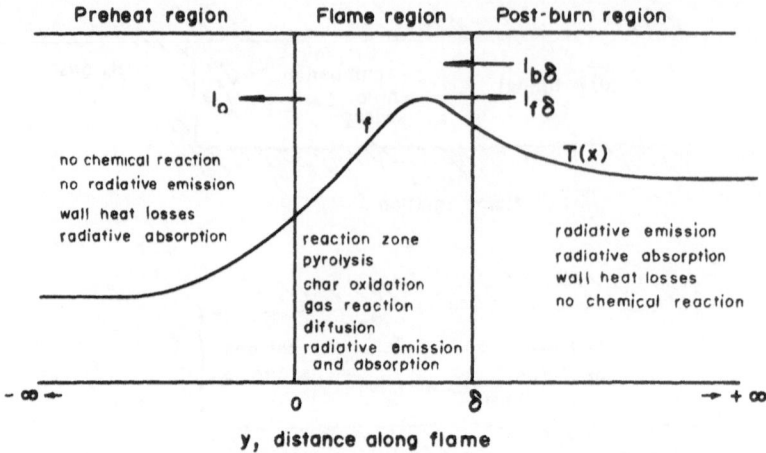

Figure 2. Schematic diagram of laminar, particle-laden flame model.

thermal diffusion, and temperature gradients within particles were neglected. Resulting differential equations, mostly from Table 1 of Chapter 2, and transformed into streamline coordinates, are

Gas-species conservation:

$$\rho_g(\partial\omega_{ig}/\partial t)=\rho_t(\partial/\partial\psi)(\rho_g\rho_t D_{ig}\partial\omega_{ig}/\partial\psi)+r_{ig}+\sum_j (r_{pi})_j-\left(\sum_j r_{pj}\right)\omega_{ig} \quad (1)$$

Gas-phase thermal energy:

$$\rho_g(\partial h_g/\partial t)=\rho_t(\partial/\partial\psi)[k_g\rho_t/C_{pg})(\partial h_g/\partial\psi)]+\sum_j (Q_{pj})-Q_{sg}+\sum_j [r_{pj}(h_{pjg}-h_g)]$$
$$+\rho_t(\partial/\partial\psi)\left\{\rho_t\sum_i [\rho_g D_{ig}-(k_g/C_{pg})]h_{ig}(\partial\omega_{ig}/\partial\psi)\right\}-(\rho/\rho_g)\left(\sum_j r_{pj}/\rho_{pjg}\right)$$

$$(2)$$

Particle thermal energy:

$$\rho_{pj}(\partial h_{pj}/\partial t)=Q_{pf}-Q_{pj}-Q_{ps}-r_{pj}[(h_{pjh}-h_{pj})-(p/\rho_{pgj})] \quad (3)$$

Particle mass:

$$\partial m_{pj}/\partial t=-r_{pj}/n_{pj} \quad (4)$$

Particle-species conservation:

$$\rho_p(\partial\omega_{pj}/\partial t)=\left(\omega_{pj}\sum_j r_{pj}\right)-r_{pj} \quad (5)$$

Particle-number conservation:

$$\rho_g(\partial n_{pj}/\partial t) = n_{pj}\left[(\partial \rho_g/\partial t) - \sum_j r_{pj}\right] \tag{6}$$

where $\partial \psi/\partial t = -\rho_t v$, $\partial \psi/\partial y = \rho_t$, and $\rho_t = \rho_g + \Sigma_j \rho_{pj}$ for the gas–particle mixture. Equations (1)–(6) are a set of $(I + 4J + 1)$ equations, describing the dependent variables ω_i, h_g, h_{pj}, m_{pj}, ω_{pj} and n_{pj} as functions of time (t) and position $\psi(t, y)$, where I is the number of gas-phase chemical species and J is the number of particle phases (discrete sizes or types). These equations reduce to those of Smoot et al.[18] and Spalding et al.[20] for gas-phase systems.

1.3. Auxiliary Equations

Equations (1)–(6) contain a number of variables for which algebraic expressions must be postulated for the problem to be completely formulated. These variables are often a function of additional parameters which must be specified. For example, r_{pj}, the reaction rate of the jth particle class, includes both coal-devolatilization and char-oxidation components, and contains the mass-transfer coefficient k_c as a parameter, which must be further specified. Essentially all of these required expressions, referred to as auxiliary equations, have been outlined in Chapters 6 and 12. Table 1 provides a detailed reference to the appropriate auxiliary expressions used in this model.

Products of devolatilization were specified as an arbitrary hydrocarbon, $C_n H_m$, together with oxygen, nitrogen, etc. The carbon–hydrogen ratio in the hydrocarbon can be obtained from material balance considerations. Measurements of char and gas composition reported in Chapters 8 and 10, together with data for pyrolysis products,[21,22] have been useful in specifying the volatiles composition. The irreversible rate of oxidation of the hydrocarbon to carbon monoxide and hydrogen was given by Eq. (5) of Chapter 10.

Some limitations in the above method for predicting propagating flames are identified. Velocities of particles greater than about 20 μm will lag behind the gases in regions of the flame where gases are accelerating. The motion of the particles can be influenced also by buoyancy and gravity effects as well as by drag effects due to the accelerating gases. Depending upon the configuration of the flame, these effects may tend to counterbalance. Radiation interchange among the particles, neglecting gaseous radiation, assumed graybody emission from each of the increments of the flame, with absorption of the exponentially decaying radiative flux being related to the local size and concentration of particles. This approximate treatment is outlined in reference 21.

Table 1. Summary of Auxiliary Equations for Premixed Laminar Flame-Propagation Model

Symbol	Variable	Source	Comments
	Reaction Rates		
r_{ig}	production of ith species in gas	Eq. (7), Ref. 19	general gas mechanism
r_{pj}	reaction of jth particle	Eqs. (10) and (11), Ref. 19	devolatilization, oxidation, and moisture loss
r_{pij}	production of ith gas species from jth particle reaction	Ref. 19	from stoichiometry
r_{vjm}	devolatilization reaction rate	Chap. 12, Eq. (1)	activated process
r_{hjl}	char oxidation rate	Chap. 12, Eq. (11)	coupled reaction and diffusion
r_{wj}	moisture loss rate	Chap. 12, Eq. (16)	controlled by heat and mass transfer
	Heat-Transfer Rates		
Q_j	gas–particle conduction	Chap. 12, Eq. (19)	neglects convection; no drag
Q_{pf}	particle–particle radiation	Ref. 19	neglects gas radiation
Q_{sg}	heat loss from gas	Eq. (12), T7[a]	convective gas loss to surroundings
Q_{ps}	heat loss from particles	Eq. (13), T7[a]	radiative particle loss to surroundings
	Gas Physical Properties		
C_{pg}	gas heat capacity	Ch. 12, Eq. (34)	temperature dependent
D_{im}	ith gas-species diffusivity	Ch. 12, Eq. (41)	concentration and temperature dependent
h_{ig}	partial enthalpy of ith species	$\int C_{pi}\, dT + h_i^0$	
k_g	gas-mixture conductivity	Ch. 12, Eq. (39)	concentration and temperature dependent
p	pressure	input	constant
ρ_t	total density	$\rho_g + \Sigma_j \rho_j$	
ρ_g	gas density	pM/RT	ideal gas
	Particle Physical Properties		
h_{jg}	jth particle product enthalpy	Ch. 12, Eq. (64)	
ρ_{pjg}	jth-particle product density		ideal gas
d_j	particle diameter	Ch. 12, Eq. (42)	can increase with swelling

[a]Modeled as shown in equation referenced, where T=Table in this chapter.

1.4. Solution Technique

For computer solution, Eqs. (1)–(6) were further transformed into S coordinates, following the approach of Spalding et al.,[20] where

$$S(\psi, t) = [\psi - \psi_u(t)]/[\psi_b(t) - \psi_u(t)] \qquad (7)$$

This transform reduces the time-dependent flame thickness to a fixed grid between $S=0$ and $S=1$ but introduces the mass-flow rate at the boundaries into the differential equations. In S coordinates, all of Eqs. (1)–(6) take the general form

$$\partial\zeta/\partial t = -(a+bS)(\partial\zeta/\partial S) + c(\partial/\partial S)(d\partial\zeta/\partial S) + e \qquad (8)$$

The equations were solved simultaneously using a numerical finite-difference scheme to iterate to the steady-state solution. The gas-composition equations [Eq. (1)] were solved using a mixed numerical scheme,[23] where diffusion terms were treated explicitly and linearized kinetic terms were treated implicitly.[24] The gas-enthalpy equation [Eq. (2)] was solved using a Crank–Nicholson implicit technique. Finally, the hyperbolic particle equations [Eqs. (3)–(6)] were solved by explicit techniques. These solution techniques are discussed in detail by Williams.[25]

Boundary conditions at the leading or unburned edge of the flame specify the conditions (pressure, temperature, velocity, composition, particle size) of the gas–particle mixture. At the burned edge of the flame, boundary conditions require that the change with distance in the dependent variables be zero (i.e., $\partial\zeta/\partial S=0$).

To start the solution, a flame thickness and temperature profile are assumed, together with arbitrary concentration distributions for major chemical species across the flame. Equilibrium solutions or quasi-equilibrium solutions assuming inert char are useful in estimating initial conditions at the burned end of the flame. Radical concentrations are initially taken to be zero.

The local, time-dependent mass-flow rate in the flame, M_t, is evaluated following each iteration in time by solving Eq. (1) for a particular major gas species in y-coordinates, assuming quasi-steady state. In practice, the mass-flow rates were evaluated for different species at different locations in the flame. Equality in mass-flow values was one criterion of final convergence to the steady-state solution. The flame velocity is then

$$v_f = M_t/\rho_{tu} \qquad (9)$$

The solution proceeds in time until the flame thickness, flame velocity, mass-flow rates, temperature profile, and species concentration profiles are acceptably stable. Hecker[26] discusses the criteria for convergence in more detail. Converged solutions were shown to be independent of variation in assumed initial conditions and independent of species chosen for evaluating mass-flow rate. However, convergence was difficult to obtain when the assumed initial flame thickness varied greatly from the final value. Predictions were in good agreement with those of Spalding et al.[20] for a single, three-reaction hydrazine decomposition flame. Fifty grid segments spanning the flame were generally found to be adequate.

1.5. Predictions and Comparisons with Measurements

1.5a. Gas-Phase Systems

Radiative effects, which were included in the model formulation, were neglected in all of the computations for gaseous systems. Smoot et al.[18] provide detailed comparisons of model predictions vs. measurements for methane–air and methane–oxygen systems. The methane predictions were made using the recommended gas-phase kinetic mechanism in Table 2 of Chapter 10. Comparison of the predicted and measured effects of equivalence ratio (ϕ) on flame velocity is shown in Figure 3. Predicted results compare well except at the higher ϕ values. Measured effects of pressure and initial temperature were also well predicted by the model.[18] Figure 4 shows comparisons of measured and predicted species profiles for a low-pressure CH_4–O_2 flame. Agreement is remarkably good for all but the CH_3 and HO_2 radicals. These and other comparisons[18] provide a suitable evaluation of model accuracy for gaseous systems of interest.

1.5b. Coal–Air Systems

References 19 and 27 discuss measurements and comparison with model predictions in some detail and only a summary is presented here. Radiative effects have been neglected in several of the computations for pulverized coal–air systems. Most of the coal–air predictions were made for Pittsburgh, high-volatile B bituminous coal. Devolatilization has been treated with a single reaction step while char oxidation was assumed to be first order. Input parameters for this coal included: (1) coal heat of formation, -7.12×10^5 J kg^{-1}; (2) $E_v = 7.41 \times 10^7$ J kmol^{-1} K and $A_v = 5 \times 10^5$ s^{-1} [28];

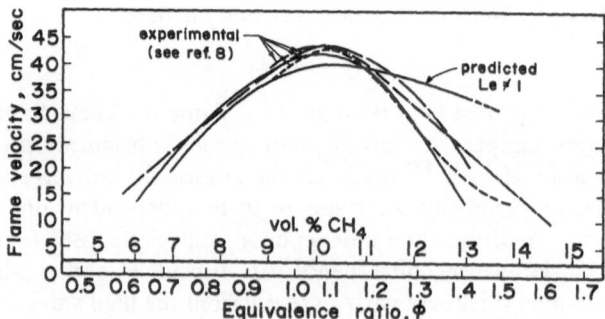

Figure 3. Comparison of predicted and experimental methane–air flame velocities with varying equivalence ratio ($P = 101$ kPa, $T_u = 298$ K). (Data shown with permission from Andrews and Bradley.[8])

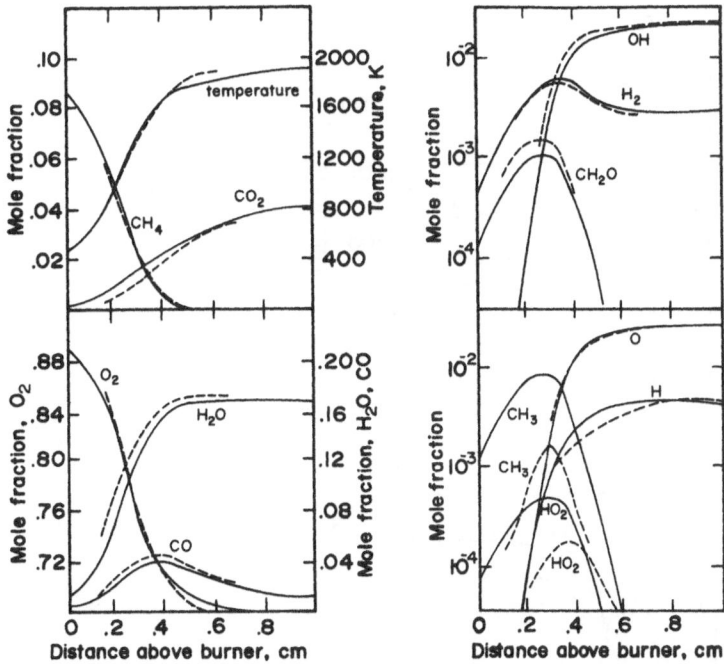

Figure 4. Comparison of predicted and experimental temperature; methane and carbon dioxide profiles for Peeters' low-pressure methane–oxygen flame ($P = 5.35$ kPa, 45S grid increments; 9.5% CH_4 with oxygen; Le = 1.0.) Experimental data (– – –), model predictions (—). (Data from Peeters and Mahnen.[7])

(3) $E_c = 1.24 \times 10^8$ J kmol^{-1} and $A_c = 3.3 \times 10^4$ ms^{-1}; (4) 50% volatiles with hydrocarbon molecular weight of 67.5; (5) coal density, 1.3×10^9 kg m^{-3}; (6) pressure, 101 kPa; (7) initial temperature, 298 K; (8) heat of devolatilization, -1.73 MJ kg^{-1} [29]; (9) char heat of combustion (carbon), 3.94–4.35 MJ kg^{-1}; (10) coal-swelling factor, 10%; (11) char surface-area factor, $\phi = 1.0$; (12) coal heat capacity, 1.3 kJ kg^{-1}; (13) volatiles products, C_nH_m, CO, H_2, N_2, OH; (14) gas-phase species, C_nH_m, CO, H_2, H_2O, OH, H, HO_2, CO_2, O, O_2; (15) gas-phase reaction mechanism and rate constants of Table 2 in Chapter 10.

Calculations were made for coal concentrations from 100 to 1200 g m^{-3} of air and for particle sizes of 10 and 33 μm. For 300 g m^{-3}, predicted profiles are shown in Figure 5. These flames are approximately 1 cm in thickness, which agrees well with the thickness determined from measured temperature profiles. The particle has been assumed to swell only slightly during devolatilization, as is suggested by experimental evidence (see Figure 5 in Chapter 8).

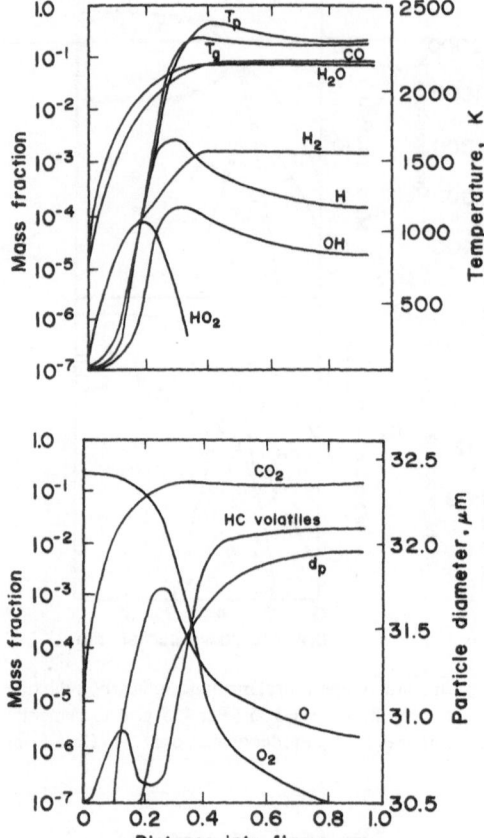

Figure 5. Predicted temperature (top) and species (top and bottom) profiles for high-volatile B bituminous coal (Pittsburgh) (300 g m^{-3}, 30 μm).

This prediction has neglected radiative-transfer effects and heat losses, so the predicted temperature is well above measured values.

According to the predictions of Figure 5, the particle does not swell or devolatilize in the early region of the flame (0–0.2 cm), while in this same region, several gas species derived from the coal are vigorously reacting. Thus, coal-derived gas-phase products exist in the flame where the coal has not yet reacted. According to these model predictions, particles penetrate through the reacting flame and pyrolyze downstream in an oxygen-lean environment. The pyrolysis products then back-diffuse while reacting with oxygen to produce CO, H_2, and other species. These secondary products also diffuse toward the incoming oxygen where they react to produce a high gas temperature, which in turn heats the incoming coal particles by conduction.

Flame-velocity predictions are compared with measured values in Figure 6 for several coal concentrations and two particle sizes. All of these predictions were made using a single coal-particle size and assuming a

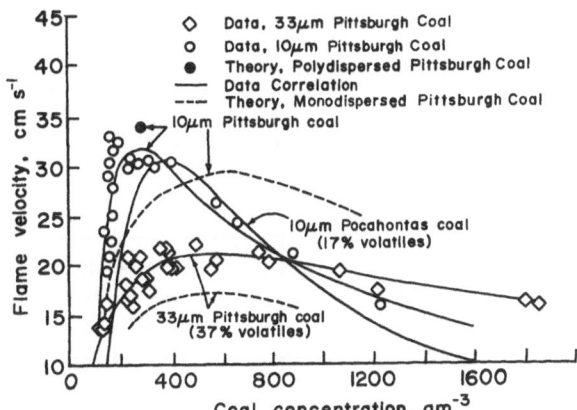

Figure 6. Measured and predicted effects of coal-particle size and coal concentration.

Lewis number of unity. The actual particle-size distribution is better approximated by considering polydispersed pulverized coal, as shown by the single computation of Figure 6. Considering nonunity Lewis number for the 33 μm, 300 g m^{-3} case increased the predicted flame velocity by 20% to 0.19 m s^{-1}.

Figure 7 compares predicted and measured temperature and species profiles for a coal–air flame.[30] This prediction included heat losses and radiative effects. The close agreement for these thin laminar flames provides strong evidence in support of the model assumptions.

Earlier treatments of pulverized-coal flame propagation have been based principally on radiative transfer as the governing propagation mechanism,[15,16] and have neglected diffusion effects. The present predictions suggest that diffusion is of primary importance in propagating laminar coal–air flames. However, recent computations performed with this model show that radiation increases the propagation velocity of a coal-dust–air mixture somewhat.[30] The principal influence of radiation is the upstream radiation from the hot-particle cloud to the incoming coal–air mixture, which preheats the mixture, thus raising the initial temperature.

1.6. Propagation Mechanism

A series of 11 predictions were made for coal–air flames in order to identify mechanisms which, according to the model, control the rates of propagation. A summary of these calculations is shown in Table 2. The reference case is very similar to the case illustrated in Figure 5, being 300 g m^{-3} of 30-μm Pittsburgh coal, with a flame velocity of 0.141 m s^{-1}.

In the other 10 cases, one parameter was adjusted at a time. From these computations, and for this set of conditions, the rate of hydrocarbon oxidation in the gas phase and the rate of heterogeneous reaction of the carbon (char) were not rate-limiting. Computations also indicate that uncertainties

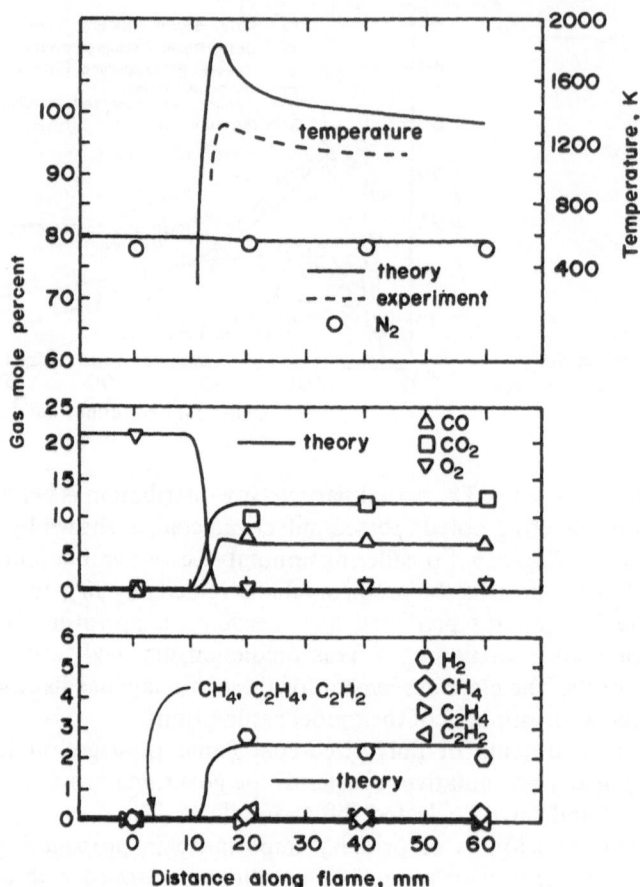

Figure 7. Comparisons of measured temperature and species concentration profiles with predicted values for atmospheric coal–air flame. Measured: 0.38 kg m^{-3}, 21-μm particle diameter. Theoretical: 0.30 kg m^{-3}, 30-μm particle diameter.

in the value of the heat of pyrolysis may not be important in the propagation process. Gas-phase reaction (other than hydrocarbon oxidation) and coal pyrolysis were significant, while conduction and molecular diffusion in the gas phase and conduction between the particle and gas phases were of major importance. The importance of diffusion is further illustrated in Table 2, column 7, where the Lewis number was taken as nonunity. This change increases the diffusivity of species with low molecular weights without changing thermal conductivity. The resultant flame velocity was higher by 21% According to recent data,[31] the activated rate of devolatilization for this coal type and particle size may be much lower than considered in these

Table 2. Summary of Coal–Air Mechanism Computations

Property variable	1	2	3	4	5	6	7	8	9	10	11
	Base	Percent volatiles	Volatiles mol. wt.	Molecular transport	HC oxidation	Gas reaction	Lewis number	Coal pyrolysis	Char reaction	Heat of pyrolysis	Pressure
Variation	Reference	50→67%	4x	0.25x	0.25x	All 0.25x	Le≠1	0.25x	0.25x	Endo→exo	101→10.1 kPa
Effect		Significant	Nil	Major	Nil	Significant	Significant	Significant	Nil	Small	Major
Time[a] (s)	0.023	0.011	0.023	0.23	0.023	0.023	0.023	0.028	0.023	0.033	0.014
Temperature (K)											
final-gas	2078	1899	2078	2079	2078	2076	2080	2121	2078	2045	1572
final-particle	2080	1901	1932	2084	2080	2078	2083	2119	2080	2046	1567
peak-gas	2104	2004	2105	2079	2103	2076	2080	2159	2104	2045	1793
peak-particle	2164	2153	2159	2181	2162	2102	2102	2186	2163	2046	1808
Mass flux (g cm^{-2} s^{-1})											
mean-reactant	0.0217	0.0251	0.0216	0.00838	0.0218	0.0183	0.0251	0.0148	0.0217	0.0210	0.0059
mean-product	0.0202	0.0264	0.0200	0.00856	0.0204	0.0171	0.0254	0.0159	0.0202	0.0216	0.0052
velocity[b] (cm s^{-1})	14.1	17.4	14.04	5.71	14.2	11.9	17.0	10.3	14.4	14.4	35.8
thickness[b] (cm)	0.99	0.79	0.99	0.92	1.0	0.94	0.98	1.20	1.0	0.87	4.1
% char-reacted	0.3	0.1	0.3	0.3	0.3	0.3	0.3	0.2	0.3	0.1	0.0
% volatiles reacted	99.4	99.3	99.4	99.9	99.4	99.9	99.2	99.6	98.6	99.9	99.9
Species mole fraction[c]											
HC	0.0029/84	0.037/100	0.000975/84	—	0.0039/100	0.0037/100	0.0040/76	0.0019/100	0.0039/81	0.0053/79	0.024/94
H_2O	0.120/100	0.057/100	0.121/100	0.120/100	0.120/100	0.120/100	0.118/100	0.120/100	0.120/100	0.120/100	0.096/100
CO_2	0.108/31	0.169/31	0.108/31	0.107/28	0.108/31	0.099/37	0.111/33	0.116/31	0.108/31	0.105/30	0.083/28
CO	0.077/100	0.092/46	0.077/100	0.077/100	0.077/100	0.084/58	0.078/100	0.070/100	0.077/100	0.078/75	0.142/69
O	0.0028/28	0.0019/28	0.0028/28	0.0025/24	0.0027/28	0.0041/29	0.0024/27	0.0018/30	0.0018/28	0.0027/30	0.0099/25
OH	0.0054/31	0.0028/28	0.0054/31	0.0046/28	0.0053/31	0.0060/37	0.0045/27	0.0097/33	0.0053/31	0.0053/30	0.006/28
H	0.0037/31	0.0002/28	0.0036/31	0.0038/28	0.0037/31	0.0062/37	0.0031/30	0.0021/37	0.0037/31	0.0043/34	0.030/32
H_2	0.021/100	0.0002/31	0.021/100	0.021/100	0.021/100	0.024/38	0.020/100	0.017/100	0.021/100	0.022/71	0.074/100
Onset of coal reaction[e]	25	25	25	21	25	26	24	27	24	18	

[a] Kinetic time

[b] Flame.

[c] (Peak mole fraction)/(peak location) in flame, percentage of distance to burned edge.

[d] 300 g m^{-3} of Pittsburgh coal, 30 μm, 50 increments in flame.

[e] Percentage of distance through flame.

computations, which would make the devolatilization process more important as a rate-controlling step than shown by Table 2.

1.7. Turbulent Flames

The propagation model described in Section 1.6 was for laminar methane–air and coal–air flames. The study of laminar flames has several advantages, including simplicity of experimental equipment and tractability of analytical models; however, the relationship of laminar flame velocities to propagation velocities in turbulent flames, as occur in coal-mine explosions, has not been resolved. This question has increased in importance with observations that suppressants such as $KHCO_3$, which are chemically active and very efficient under some conditions in laminar laboratory flames, have been relatively inefficient in suppressing coal-dust mine explosions.[1,2] In this section, characteristics of propagation in turbulent flames and in coal-mine explosions are outlined and their relationships to laminar flame propagation are indicated.

In a recent review, Andrews et al.[32] noted that there is not yet a comprehensive theory of turbulent burning. However, predictive equations from some 14 investigators all express the turbulent flame velocity as a function of the laminar flame velocity. This result in itself suggests a close relationship between laminar and turbulent flame propagation. Andrews et al.[32] sug-

Table 3. Mixture Compositions for Figure 8

Mixture number	Mixture	v $(mm^2 s^{-1})$	u_1 $(m s^{-1})$	Temperature[b] (K)	Reference[c]
1	$CH_4 + 2O_2 + 4.93N_2$	15.90	0.7	2497	150
2	$CH_4 + 6O_2 + 9.07$ He	34.00	0.7	1999	150
3	$2H_2 + O_2 + 17Ar$	15.63	0.29	1408	159
4	$C_3H_8 + 5O_2 + 28.5Ar$	13.78	0.45	2288	159
5	$CH_4 + 2O_2 + 4.9Ar$	14.91	0.7	2694	159
6	$2H_2 + O_2 + 17He$	95.35	0.7	1337	159
7	$C_3H_8 + 5O_2 + 28.5He$	54.31	0.74	2308	159
8	$C_3H_8 + 5O_2 + 18Ar$	13.63	0.8	2578	159
9	$C_3H_8 + 5O_2 + 18He$	42.77	1.4	2581	159
10	$CH_4 + 2O_2 + 4.50N_2$	15.91	0.89	2543	187
11	$2C_2H_2 + 5O_2 + 27.15N_2$	15.22	0.90	2247	187
12	$2C_2H_2 + 5O_2 + 18.48N_2$	15.09	1.55	2550	187
13	$CH_4 + 2O_2 + 2.79N_2$	16.00	1.55	2747	187
14	$CH_4 + 2O + 1.96N_2$	16.07	1.93	2818	187
15	$2C_2H_2 + 5O_2 + 14.88N_2$	15.01	1.98	2681	187

[a]Table used with permission from Andrews et al.[32]
[b]Adiabatic flame temperature.
[c]References are as originally cited in Andrews et al.[32]

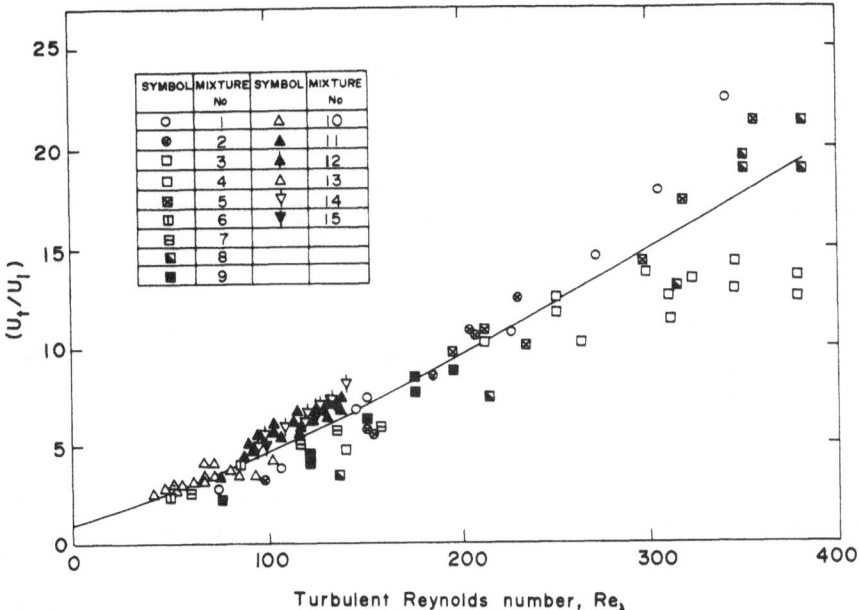

Figure 8. Influence of turbulent Reynolds number on turbulent burning velocities (see Table 3 for mixture compositions). (Figure used with permission from Andrews *et al.*[32])

gested that the ratio of the turbulent to laminar burning velocities, u_t/u_l, be correlated as a function of the turbulent Reynolds number,

$$\text{Re}_\lambda = u'\lambda/\nu \tag{10}$$

where $u' = (\overline{u^2})^{1/2}$, the velocity scale or turbulent intensity; λ is the microscale of turbulence, or the mean size of small eddies; and $\nu = \mu/\rho$, the kinematic viscosity of the gas.

Results of this correlation for 15 chemical systems summarized in Table 3 for $40 < \text{Re}_\lambda < 400$ are shown in Figure 8.

Some observations concerning Figure 8 are in order. First, while there is scatter in the data, especially at higher Re_λ values, the correlation extends over a wide range of turbulence levels, covering both the small-eddy region ($\lambda < \lambda_f$), and the wrinkled-flame region ($\lambda > \lambda_f$).

Secondly, the Re_λ can be related directly to the eddy viscosity

$$\text{Re}_\lambda = (u'\lambda/\nu) - (lu'\lambda)(l\nu) = \varepsilon(\lambda/l)/\nu \tag{11}$$

where $\varepsilon = u'l$, the eddy viscosity; and l is the macroscale of turbulence or mean size of large eddies. From Dryden,[33] for isotropic turbulence,

$$\lambda/l = 48.64(\nu/\lambda u') = 48.64/\text{Re}_\lambda \tag{12}$$

Then, combining Eqs. (11) and (12),

$$Re_\lambda = 6.97(\varepsilon/v)^{1/2} \tag{13}$$

This result implies that

$$u_t/u_l \sim (\varepsilon/v)^{1/2} \tag{14}$$

which corresponds to the original Damkohler postulation for fine-scale turbulence,[34] but which appears to hold over a wider range of turbulence levels as well.

From simple laminar flame-propagation theory,[11]

$$u_l \sim k^{1/2}r^{1/2} \sim (k/\rho C_p)^{1/2}r^{1/2} \tag{15}$$

Since $v \sim k/\rho C_p$, combining Eqs. (14) and (15) implies that

$$u_t \sim \varepsilon^{1/2}(r_t)^{1/2} \tag{16}$$

where $(r_t)^{1/2}$ is the overall reaction rate for turbulent flow. Borghi,[35] as discussed in Chapter 4, has shown for a nonpremixed, reacting, turbulent flow that

$$r_t \sim X_i\bar{r}_t \sim X_i r_l \tag{17}$$

where X_i has been called the "contact index" and is related to the fluctuating properties of the flow. Then, for turbulent flow,

$$u_t \sim \varepsilon^{1/2}X_i r_l \tag{18}$$

$$u_t/u_l = (\varepsilon/v)^{1/2}X_i \tag{19}$$

This later result suggests a revised correlation for u_t/u_l, where the turbulent chemical effects (X_i) as well as the fluid mechanical effects (ε/v) are accounted for. Since X_i is difficult to compute and may not be generally applicable, this revised correlation has not been attempted.

Even without the turbulent kinetic correction term X_i, the correlation of Figure 8 and Eq. (14) suggests that the laminar propagation model can be used to predict turbulent flame propagation by replacing k with $\varepsilon\rho C_p$. In other words, the molecular coefficients are replaced by the turbulent counterparts. Since these turbulent properties have been measured for some pertinent systems, this proposition has been tested by using the laminar flame-propagation model of the previous section, modified with turbulent transport coefficients. The properties of a methane–oxygen system at high Re_λ values, where data are available in the literature,[36] have been predicted. The system treated was mixture number 14 of Table 3. For laminar flow, the predicted flame velocity was 1.67 m s^{-1}, which was 14% below the measured value. For turbulent flow, using the measured Re_λ value of 141 to deduce the value of ε/v for use in the flame propagation model, the predicted turbulent propagation velocity was 27.8 m s^{-1}, compared to the experimental value of 15.8 m s^{-1}. Additional details of these two computations are summarized in Table 4. These computations show that converged solutions can be obtained with the propagation model, modified to consider turbulent flow.

2. Pulverized-Coal Combustion and Gasification

2.1. Introduction

This model is one-dimensional in nature and requires input specification of the rates of jet-mixing and recirculation. The potential uses and advantages of this model include: (1) low computer costs for investigating effects of operating and configurational variables on coal reaction processes; (2) a method for evaluating various descriptions of coal kinetics together with model parameters; (3) a basis for evaluation of multidimensional solutions of coal reaction processes; (4) computational results to assist in evaluation of experimental data; (5) a single model for describing both combustion and gasification processes for pulverized-coal systems; (6) potential use as a component in systems codes that describe processes with a combustion or gasification step. This one-dimensional model is not well suited for comparison with local measurements in two or more dimensions inside combustors or gasifiers. Further, it would not be applicable to the detailed design and/or optimization of such systems.

Table 4. Summary of Measurements and Predictions for the $CH_4-O_2-N_2$ System

	Laminar		Turbulent	
	Theory	Experimental[32]	Theory	Experimental[32]
Initial composition (mol %)				
CH_4	20.4	20.4	20.4	20.4
O_2	40.2	40.2	40.2	40.2
N_2	39.4	39.4	39.4	39.4
Temperature (K)	2364	—	2732	—
Turbulent Re_λ	0	0	141	141
Flame velocity (cm/s)	167	193	2780	1578
Flame thickness (cm)	0.101	—	2.48	—
Ratio of turbulent to laminar flame velocities (u_t/u_l)				
1. Experimental	8.2			
2. Theoretical (this study)	16.4			
3. Andrews *et al.*[32]				
correlation	6.5			

The general approach used in development of this model is to provide for detailed description of the physical and chemical processes that take place in the reacting regions within combustors and gasifiers. Experimental observations and model parameters obtained from data correlations from independent investigators are being used wherever possible. For example, coal reaction parameters from devolatilization studies and char oxidation studies, as summarized in Chapters 8 and 9, are used as input for model predictions.

2.2. Model Basis

This model, referred to as 1-DICOG (*1-DI*mensional Combustion Or Gasification), uses the integrated or macroscopic form of the general conservation equations (see Chapter 2) for a volume element inside the gasifier or furnace, as illustrated in Figure 9. The following aspects of pulverized-coal combustion and gasification have been included in the model: (1) mixing of primary and secondary streams (specified as input); (2) recirculation of reacted products (specified as input); (3) pyrolysis and swelling of coal; (4) oxidation of the char by oxygen, steam, or carbon dioxide; (5) heat transfer between the coal-char particles, gases, and walls by convection, conduction, and radiation; (6) variation in composition of inlet gases and solids; (7) variation in coal-char particle size; (8) reaction of the pyrolysis products in the gas phase. A summary of key model assumptions and conditions, which are based largely upon observations summarized in Table 1 of Chapter 11, are given in Table 5.

The following are considered to be the major limitations of the one-dimensional model (1-DICOG):

(1) The model does not predict local mean or fluctuating properties within the pulverized-coal reactor as a function of both radial and axial position.

(2) The model does not predict rates of jet-mixing or recirculation; rather, these values are required input.

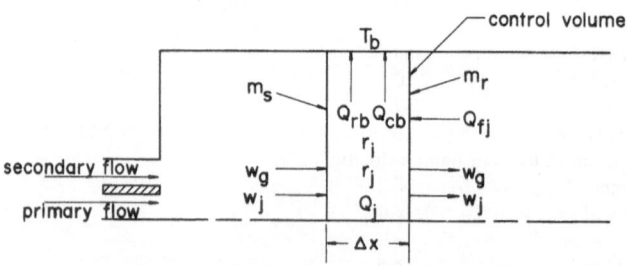

Figure 9. Schematic diagram of control volume in a pulverized-coal combustor or gasifier.

Table 5. *Summary of Key Assumptions and Conditions of Macroscopic Coal Reaction Model (1-DICOG)*

1. Steady-state, compressible gas, with uniform pressure and specified reactor-area variation.
2. Particles and gases in dynamic equilibrium.
3. Secondary gases and recirculated products input along reactor with instantaneous mixing at each interval of specified reactor area.
4. Multiple particle sizes or types.
5. Each particle size and type and gas are all distinct phases.
6. Negligible gas conduction, molecular diffusion, thermal diffusion, gravity effects, particle interactions, wall friction, viscous dissipation, work on surroundings, gas-phase radiation, particle-phase convective losses, kinetic energy, particle volume.
7. Coal reactions take place by competing processes of devolatilization and char oxidation.
8. Rate-limiting steps include upstream radiative transport, rate of gross oxidizer–fuel mixing, rate of product recirculation, rate of coal-particle pyrolysis, and rate of char oxidation (with O_2, CO_2, or H_2O).
9. Gas phase in local equilibrium.
10. Particles are composed of raw coal, char, ash, and moisture.
11. Ash is inert and remains with the particle. Any volatile mineral matter is considered part of volatiles content.
12. Coal pyrolysis is by an arbitrary sequence of irreversible, parallel, activated processes with specified activation energies.
13. Coal-particle swelling is proportional to extent of pyrolysis.
14. Spherical particles are of uniform local particle temperature with arbitrary change in char-particle diameter during reaction and with internal and external surface reactions according to specified burn area per particle.
15. Char contains specified proportions of other elements which enter the gas phase at a rate proportional to the carbon reaction.
16. Coal moisture loss is controlled by vapor diffusion and heat transfer.
17. Devolatilization produces a fuel-rich product whose gases block diffusion and conduction by surface transpiration.

(3) The detailed behavior of coal reactions is not yet well understood, which leads to uncertainty in the kinetic description and parameters for the pulverized-coal systems.

(4) Some details of the pulverized-coal gasification and combustion processes have been neglected to reduce model complexity and computation time. These include particle-velocity lag effects, micromixing processes, gas-phase, rate-limiting reactions, and gas-phase radiation.

2.3. Model Equations

Differential mass, energy, and momentum balances developed for particle and gas phases are based upon the macroscopic equations of Chapter 2 and are summarized in Table 6. This set of first-order, nonlinear equations also requires a large number of auxiliary algebraic equations which describe the following aspects of the coal reaction process: (1) enthalpy–temperature

Table 6.   Summary of Model Differential Equations

Equation number	Type	Equation	Number of equations
1	Gas element continuity (kth)	$d(w_g \omega_k)/dx = A\Sigma_j\, r_{jk} + m_{sgk} + m_{\rho gk}$	K
2	jth-Particle phase continuity	$dw_j/dx = -Ar_j + m_{sj} + m_{\rho j}$	J
3	Gas energy	$d(w_g h_g)/dx = h_{sg}m_{sg} + h_{\rho g}m_{\rho g}$ $\qquad + A(\Sigma_j Q_j - Q_{cb} + \Sigma_j r_j h_{jg})$	1
4	Particle energy	$d(w_j h_j)/dx = m_{sj}h_{sj} + m_{\rho j}h_{\rho j}$ $\qquad + A(Q_{fj} - Q_{rbj} - Q_j - r_j h_j)$	J
5	Particle number	$d(vn_j A)/dx = (m_{\rho j}/\alpha_j) + (m_{sj}/\alpha_j)$	J
6	Total gas continuity^a	$d(w_g)/dx = A\Sigma_j\, r_j + m_{sg} + m_{\rho g}$	1
7	Coal mass	$d(\alpha_{cj})/dx = (r_{cj}/n_j v)$	J
8	Char mass	$d(\alpha_{hj})/dx = (r_{hj}/n_j v)$	J
9	Moisture mass	$d(\alpha_{wj})/dx = -(r_{wj}/n_j v)$	J

^{a}Sum of Eqs. (1) over the K elements.

relationships; (2) physical properties including heat capacity, thermal conductivity, diffusivity, and viscosity; (3) radiative interchange inside the gasifier; (4) equations of state and mass-flow continuity; (5) convective and conductive heat interchange among the gases, particles, and walls; (6) rates of pyrolysis and oxidation of coal and char. A summary of these auxiliary equations, which have largely been developed in Chapter 11, is given in Table 7. A detailed notation is provided in Section 3.

2.4. Solution Technique

A general scheme for computer solution of the equations has been developed, which incorporates an existing generalized thermochemical program for calculating equilibria in complex gases.[37] The solution technique requires an initial assumption for the axial temperature profile along the reactor. From this assumption, radiative heat transfer within the reactor can be computed, and then the set of differential equations can be integrated along the reactor. The revised temperature profile to be used in the next iteration is part of the output from the previous iteration. This process is repeated until the solution converges. The integration technique used in this code is an Adams–Moulton predictor–corrector system. The corrector portion is repeated until the particle-energy computation converges.

2.5. Predictions and Comparisons with Measurements

1-DICOG has been successfully applied to predicting characteristics of laboratory pulverized-coal combustors and entrained-flow gasifiers. In

Table 7. Auxiliary Equations for Coal Reaction Model (*1-DICOG*)

Equation number	Type[a]	Equation
10	Gas velocity	$v = w_g/\rho_g A$
11	Particle–gas heat transfer [see Eqs. (19), (20), (21)]	$Q_j = 2\pi\{B_j/[\exp(B_j) - 1]\}k_g(T_j - T_g)\,d_j n_j$
12	Convective heat exchange with surroundings	$Q_{cb} = 4h_m(T_g - T_b)/D_s$
13	Radiative heat exchange with surroundings	$Q_{ibj} = (4\sigma\varepsilon_b/D_s)(T_j^4 - T_b^4)$ $\times \{1 - \exp[-E_\lambda(\pi/4)(D_s/2)d_j^2 n_j]\}$
14	Transpiration parameter for heat transfer [see Eq. (22)]	$B_j = r_j C_{pg}/2\pi d_j k_g n_j$
15	Radiative heat transfer in reactor	$Q_{fj} = (\pi/4)I_{qt}n_j d_j^2\varepsilon_j - I_{q0j}/\Delta x_p$
16	Local radiative intensity	$I_{qt} = \Sigma_p\, I_{pq}$
17	Radiative emission	$I_{p0j} = (a_j T_j^4)_p\sigma\Delta x_p$
18	Radiative intensity component	$I_{pq} = (\Sigma_j I_{p0j})\exp\left[-\Sigma_{p \neq q}^{q}(\Sigma_j a_j)_p\,\Delta x_p\right]$
19	Radiative adsorption coefficient	$a_j = (\pi/4)\varepsilon_j n_j d_j^2$
20	Particle density [see Eq. (44)]	$\rho_j = \rho_{cj} + \rho_{hj} + \rho_{aj} + \rho_{wj}$
21	Coal density [see Eq. (45)]	$\rho_{cj} = \alpha_{cj}n_j$
22	Moisture density [see Eq. (46)]	$\rho_{wj} = \alpha_{wj}n_j$
23	Char density [see Eq. (47)]	$\rho_{hj} = \alpha_{hj}n_j$
24	Ash density [see Eq. (48)]	$\rho_{aj} = \alpha_{aj}n_j$
25	Particle mass [see Eq. (49)]	$\alpha_j = \alpha_{hj} + \alpha_{cj} + \alpha_{aj} + \alpha_{wj}$
26	Particle reaction rate to gas phase [see Eq. (18)]	$r_j = \Sigma_l\, r_{hjl} + \Sigma_m\, r_{vjm} + r_{wj}$
27	Net char reaction rate [see Eq. (13)]	$r_{hj} = \Sigma_m\, r_{hjm} - \Sigma_l\, r_{hjl}$
28	Oxidizer–char reaction rate [see Eq. (11)]	$r_{hjl} = (M_h\phi_l A_j n_j)^2 k_{cjl}k_{jl}\zeta C_{olg}C_g/$ $\times[M_h\phi_l A_j n_j C_g(\zeta k_{jl} + k_{cjl}) - r_j]^{-1}$
29	Kinetic char rate [see Eq. (10)]	$k_{jl} = A_{jl}T_j^m\exp(-E_{jl}/RT_j)$
30	Transpiration parameter for mass transfer [see Eq. (27)]	$B_{mj} = r_j/(2\pi D_{om}\rho_g d_j n_j)$
31	Char mass-transfer coefficient [see Eqs. (25), (26)]	$k_{cjl} = 2D_{om}B_{mj}/\{d_j[\exp(B_{mj}) - 1]\}$
32	Total coal reaction rate [see Eq. (53)]	$r_{cj} = \Sigma_m\, r_{cjm} = -[\Sigma_m(r_{hjm} + r_{vjm})]$
33	Volatiles reaction rate [see Eq. (1)]	$r_{vjm} = k_{mi}Y_{mi}\rho_{cl}$
34	Char production rate [see Eq. (6)]	$r_{hjm} = r_{vjm}(1 - Y_{mj})/Y_{mj}$
35	Gas element production	$r_{jk} = \phi_{wjk}r_{wj} + \Sigma_l\,\phi_{hkl}r_{hjl} + \Sigma_m\,\phi_{vjkm}r_{vjm}$
36	Coal kinetic rate	$k_{mj} = A_{mj}\exp(-E_{mj}/RT_j)$
37	Gas-species conductivity [see Eq. (38)]	$k_i = [C_{pi} + (5R/4M_i)]\mu_i$
38	Gas-mixture conductivity [see Eq. (39)]	$k_g = \Sigma_i\,(X_i k_i/\Sigma_k\, X_k\phi_{ik})$

continued overleaf

Table 7 (Continued)

Equation number	Type	Equation
39	Gas-species viscosity [see Eq. (35)]	$\mu_i = 2.67 \times 10^{-4}(M_i T_g)^{1/2}/\sigma_i^2 \Omega_\mu$
50	Interaction parameter [see Eq. (37)]	$\phi_{ik} = (1/8)^{1/2}[1+(M_i/M_k)]^{-1/2}[1+(\mu_i/\mu_k)^{1/2} \times [1+(\mu_i/\mu_k)^{1/2}(M_k/M_i)^{1/4}]^2$
41	Gas-mixture viscosity [see Eq. (36)]	$\mu_g = \Sigma_i (X_i \mu_i / \Sigma_k X_k \phi_{ik})$
42	Heat-transfer coefficient wall losses	$h_m D_s/k_g = 0.023(\text{Re}_g)^{0.8}(\text{Pr}_g)^{1/3}$
44	Reynolds number	$\text{Re}_g = D_s v \rho_g / \mu_g$
44	Prandtl number	$\text{Pr}_g = C_{pg}\mu_g/k_g$
45	Particle diameter [see Eq. (42)]	$d_j = d_{j0}[1+\gamma(\alpha_{cj0}-\alpha_{cj})/\alpha_{cj0}]$
46	Species diffusivity [see Eq. (40)]	$D_{ik} = 1.84 \times 10^{-12}(T_g)^{3/2}[(1/M_k)+(1/M_i)]^{1/2} \times (p\sigma_{ik}^2 \Omega_d)^{-1}$
47	Mixture diffusivity [see Eq. (41)]	$D_{im} = (1-X_i)/\Sigma_{i \neq k} (X_k/D_{ik})$
48	Particle temperature	$\Delta T_j = \Delta h_j / C_{pj}$
49	Particle heat capacity [see Eq. (57)]	$C_{pj} = \omega_{hj}C_{phj} + \omega_{cj}C_{pcj} + \omega_{wj}C_{pwj} + \omega_{aj}C_{paj}$
50	Volatiles product enthalpy in gas phase [see Eq. (65)]	$h_{vjm} = \int C_{pvjm}\, dT_j + h_{fvjm}^0 + \Delta h_{vjm}$
51	Particle product (gas) enthalpy [see Eq. (64)]	$h_{jg} = (\Sigma_m r_{vjm}h_{vjm} + h_{hj}\Sigma_i r_{hjl} + r_{wj}h_{wj})/r_j$
52	Particle area [see Eq. (43)]	$A_j = \pi d_j^2$
53	Moisture vaporization rate [see Eq. (16)]	$r_{jw} = \{2\pi B_{mj}/[\exp(B_{mj})-1]\}C_q M_w \times (X_{wj}-X_{wg})d_j n_j D_{wg} + x_{wj}r_j$
54	Char enthalpy [see Eq. (60)]	$h_{hj} = \int C_{phj}\, dT_j + h_{hj}^0$
55	Moisture enthalpy [see Eq. (62)]	$h_{wj} = \int C_{pwj}\, dT_j + h_{wj}^0$
56	Molar gas concentration [see Eq. (30)]	$C_g = p/RT_g$
57	Moisture equilibrium mole fraction	$X_{wj} = p_{ej}/p$

[a]All equation numbers referenced in this table are from Chapter 12.

particular, the model has been applied to predicting performance in the Brigham Young University (BYU) rate-resolution combustor and gasifier.[38] Selected results are presented here.

The BYU rate-resolution combustor is a laboratory-scale, pulverized-coal combustor developed to: (1) identify the basic rate-controlling processes in combustion of pulverized coal and the influence of key experimental variables on these rate processes; and (2) provide basic local experimental measurements of concentration and temperature inside the pulverized-coal furnace. 1-DICOG has been used to predict the behavior of this combustor.

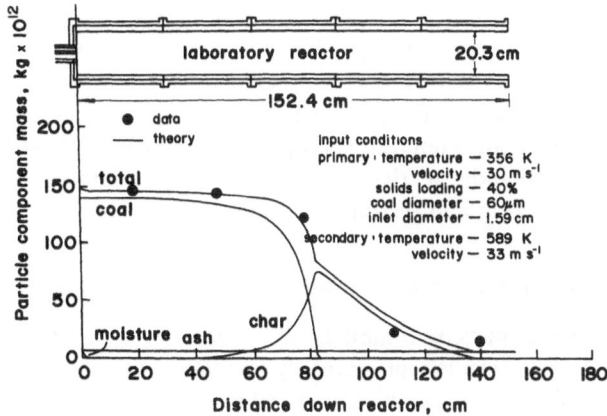

Figure 10. Predictions and measurements of coal-particle burnout.

Figure 10 shows a comparison between model predictions and combustor measurements of individual coal-particle burnout as a function of reactor length, for the specified conditions, as well as total particle burnout. The model predictions for the mass of each component of the coal are given. Model predictions and observed burnout are in good agreement. Similarly, Figure 11 shows the points of experimentally measured gas composition plotted on the profiles predicted by the model. The input conditions are those summarized in Figure 10.

1-DICOG has also been compared with data from the BYU rate-resolution gasifier. Though experimental results are in the process of being obtained, the predictions of the model provide useful insight. The results of

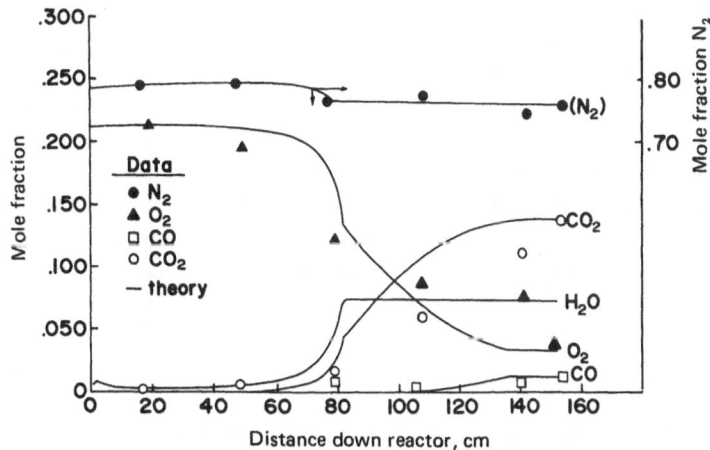

Figure 11. Predictions and measurements of gas composition.

these calculations indicate that pyrolysis and subsequent oxidation of the pyrolyzate is very rapid in the gasifier. Oxygen is depleted in this devolatilization phase, and the oxidation of the char with steam and carbon dioxide follows slowly. The calculations indicate that with a single coal-particle size of 60 μm, char burnout is not quite completed in the gasifier length of 1.12 m. These results also indicate that gasification in the laboratory reactor is affected by the mixing and recirculation processes.

2.6. Other Pulverized-Coal Combustion and Gasification Models

Several one-dimensional models have been developed for combustion of pulverized-coal systems, but relatively few have been applied to gasification. The complex fluid mechanics involved in such systems must be coupled with the local heterogeneous and homogeneous gas- and particle-phase reactions. Overcoming these types of complexities has forced most modeling efforts to look at a more macroscopic level of approximation where larger areas of the reactor may be characterized by simplifying assumptions rather than defining the detailed mechanisms throughout the reactor.

Models for pulverized-coal combustors include those of Vulis,[39] Hedley and Jackson,[40] Horn et al.,[41] and Field et al.[21] These models may be considered to be plug-flow in nature where some have been modified to include the effect of recirculation. Beer and Lee[42] have a model which incorporates a combination of perfectly stirred reactors and plug-flow reactors. Lewellen et al.[43] have recently developed a model for a swirling-coal combustor wherein they identified the dominant physical processes in four regions of interest.

Entrained-flow gasification models are only recently being developed. Batchelder et al.[44] presented some of the earliest work in this area. Mehta[45] has developed a model combining perfectly stirred reactors and plug-flow reactors to describe a three-staged entrained gasifier. Gasification models presently being developed include those of George et al.[46] They are developing two models, one of which is a combination of perfectly stirred reactors and plug-flow reactors, and the other is based on a characteristic time approach. Lewis et al.[47] are preparing a gasification model consisting of two parts, one part containing the single-particle chemistry, etc., the other part containing the fluid mechanics and homogeneous chemistry. Blake[48] has initiated development of a generalized, multidimensional, gasification model, which is more closely related to that outlined in the next chapter.

3. Notation

A	Effective duct area (m^2)	a_j	Particle adsorption coefficient (m^{-1})
A_j	jth particle area (m^2)		

A_{mJ} Surface reaction preexponential factor $(s^{-1}, m\ s^{-1})$

B_j Transpiration parameter for jth particle type for heat transfer

B_{mj} Transpiration parameter for jth particle type for mass transfer

C_p Heat capacity $(J\ kg^{-1}\ K^{-1})$

C Total molar concentration $(kmol\ m^{-3})$

d Particle diameter (m)

D_s Reactor diameter (m)

D_{om} Oxidizer diffusivity in mixture $(m^2\ s^{-1})$

D_{ik} Binary diffusivity $(m^2\ s^{-1})$

D_{im} Diffusivity of ith species in mixture $(m^2\ s^{-1})$

E Adsorption efficiency factor

$E(E_{mj})$ Activation energy for (m_j^{th}) reaction $(J\ kmol^{-1})$

h Static enthalpy $(J\ kg^{-1})$

h Heat of reaction $(J\ kg^{-1})$

h_i Partial molar enthalpy $(J\ kg^{-1})$

h_m Mean heat-transfer coefficient $(J\ m^2\ s\ K^{-1})$

I Radiation intensity $(J\ m^{-2}\ s^{-1})$

k_g Gas thermal conductivity $(J\ m^{-1}\ s^{-1}\ K^{-1})$

k_i Gas species i thermal conductivity $(J\ m^{-1}\ s^{-1}\ K^{-1})$

k_c, k_0 Mass-transfer coefficient $(m\ s^{-1})$

k_f Gas reaction-rate coefficient (s^{-1})

k_q Gas thermal conductivity $(J\ s^{-1}\ m^{-1}\ K^{-1})$

k Surface kinetic-rate coefficient $(m\ s^{-1})$

k_{mj} Pyrolysis reaction coefficient (s^{-1})

K_p Equilibrium constant

K_v Volatiles diffusion parameter $(m^2\ s^{-1})$

Le   Lewis number, $[k_g/(\rho_g C_{pg} D_{im})]$

m Rate of mass addition $(kg\ m^{-1}\ s^{-1})$

m Particle mass flux $(kg\ m^{-2}\ s^{-1})$

m Mass (kg)

m Temperature dependence, exponent

M Molecular weight $(kg\ kmol^{-1})$

M_i Mass flux $(kg\ m^{-2}\ s^{-1})$

n Number density (m^{-3})

n Reaction order

Nu   Nusselt number

p Static pressure (Pa)

Pr_g Prandtl number, $(C_{pg}\mu_g/k_g)$

Q Volumetric heat-transfer rate $(J\ m^{-3}\ s^{-1})$

R Universal gas constant $(J\ kmol^{-1}\ K^{-1})$

r Volumetric reaction rate $(kg\ m^{-3}\ s^{-1})$

Re_g Reynolds number, $(d_p|u_g - u_p|\rho_g/\mu_g)$

S Dimensionless distance along flame

T Temperature (K)

t Time (s)

u Flame velocity $(m\ s^{-1})$

v Velocity in x-direction $(m\ s^{-1})$

w Mass-flow rate $(kg\ s^{-1})$

x Distance along burner (m)

X Mole fraction-gas phase

X_i Contact index

y Distance along flame (m)

γ Swelling coefficient

Δ Incremental change

ε Emissivity

μ Viscosity $(kg\ m^{-1}\ s^{-1})$

ρ Density $(kg\ m^{-3})$

v_{ik} Kinematic viscosity $(m^2\ s^{-1})$

σ Collision diameter (Å)

σ Stefan–Boltzmann constant $(J\ m^{-2}\ s^{-1}\ K^{-4})$

ψ Streamline coordinate $(kg\ m^{-1})$

ϕ Stoichiometric coefficient

ϕ_{ik} Interaction parameter

$\int$ Particle surface-area factor

ω_i Mass fraction

Ω Collision integral

ζ General dependent variable

Subscripts

a ash

b boundary; wall surroundings; or burned

c convection or coal

d diffusion

e equilibrium or vapor

f flame (hot region)

g gas

h char; carbon

i ith species

j jth particle type or phase

k kinetic; kth element; kth volatile species; kth gas phase reaction

l	lth reaction	u	unburned, initial
m	mth reaction; mixture or product	v	volatiles
o	oxidizer	w	moisture
p	pth increment in x-direction; particle	y	hydrocarbon
q	qth increment in x-direction	μ	viscosity
r	radiation	ρ	recirculation
s	secondary; solid; surroundings	0	initial
t	total		

4. References

1. J. Grumer, Recent research concerning extinguishment of coal dust explosions, in *Fifteenth Symposium (International) on Combustion*, pp. 103–114, The Combustion Institute, Pittsburgh, Pa. (1975).
2. J. K. Richmond and I. Liebman, A physical description of coal mine explosions, in *Fifteenth Symposium (International) on Combustion*, pp. 115–126, The Combustion Institute, Pittsburgh, Pa. (1975).
3. J. C. Biordi, C. P. Lazzara, and J. F. Papp, Flame structure studies of CH_3Br-inhibited methane flames. II. Kinetics and mechanisms, in *Fifteenth Symposium (International) on Combustion*, pp. 917–932, The Combustion Institute, Pittsburgh, Pa. (1975).
4. J. W. Hastie, Mass spectrometric analysis of 1 atm flames: Apparatus and the CH_4–O_2 system, *Comb. Flame* **21**, 187–194 (1973).
5. C. P. Lazzara, J. C. Biordi, and J. F. Papp, Concentration profiles for radical species in a methane–oxygen–argon flame, *Comb. Flame* **21**, 371–382 (1973).
6. R. M. Fristrom, C. Grunfelder, and S. Favin, Methane–oxygen flame structure. I. Characteristic profiles in a low-pressure, laminar, lean, premixed methane–oxygen flame, *J. Phys. Chem.* **64**, 1386–1392 (1960).
7. J. Peeters and G. Mahnen, Reaction mechanisms and rate constants of elementary steps in methane–oxygen flame, in *Fourteenth Symposium (International) on Combustion*, pp. 133–146, The Combustion Institute, Pittsburgh, Pa. (1973).
8. G. E. Andrews and D. Bradley, The burning velocity of methane–air mixtures, *Comb. Flame* **19**, 275–288 (1973).
9. G. E. Andrews and D. Bradley, Limits of flammability and natural convection for methane–air mixtures, in *Fourteenth Symposium (International) on Combustion*, pp. 1119–1128, The Combustion Institute, Pittsburgh, Pa. (1973).
10. D. Bradley and G. F. Hundy, Burning velocities of methane–air mixtures using hot-wire anemometers in closed-vessel explosions, in *Thirteenth Symposium (International) on Combustion*, pp. 575–583, in The Combustion Institute, Pittsburgh, Pa. (1971).
11. F. A. Williams, *Combustion Theory*, Addison–Wesley Publishing Co., Reading, Mass. (1965).
12. W. G. Browne, R. P. Porter, J. D. Verlin, and A. H. Clark, A study of acetylene–oxygen flames, in *Twelfth Symposium (International) on Combustion*, pp. 1035–1047, The Combustion Institute, Pittsburgh, Pa. (1969).
13. D. F. Cooke and A. Williams, Shock tube studies of the ignition and combustion of ethane and slightly rich methane mixtures with oxygen, in *Thirteenth Symposium (International) on Combustion*, pp. 757–775 The Combustion Institute, Pittsburgh, Pa. (1971).
14. A. A. Cordeiro, P. M. Becker, and R. J. Heinsohn, Computer Simulations of Two Lean Premixed Methane–Oxygen Flames, Unpublished paper, Pennsylvania State University (1974).
15. R. H. Essenhigh and J. Csaba, The thermal radiation theory for plane flame propagation in

coal dust clouds, in *Ninth Symposium (International) on Combustion*, pp. 111–125, The Combustion Institute, Pittsburgh, Pa. (1963).

16. D. Bhaduri and S. Bandyopadhyay, Combustion in coal dust flames, *Comb. Flame 17*, 15–25 (1971).
17. L. D. Smoot, M. D. Horton, F. P. Goodson, G. A. Williams, and W. C. Hecker, Measurement and Prediction of Laminar Flame Propagation in Methane/Coal/Air Suppressant Systems, Paper 74-1112, AIAA 10th Propulsion Conference, San Diego, California (Oct. 21, 1974).
18. L. D. Smoot, W. C. Hecker, and G. A. Williams, Predicting propagating methane–air flames, *Comb. Flame* **26**, 323–342 (1976).
19. L. D. Smoot, M. D. Horton, and G. A. Williams, Propagation of laminar coal–air flames, in *Sixteenth Symposium (International) on Combustion*, pp. 375–387, The Combustion Institute, Pittsburgh, Pa. (1976).
20. D. B. Spalding, P. L. Stephenson, and R. G. Taylor, A calculation procedure for the prediction of laminar flame speeds, *Comb. Flame* **17**, 55–64 (1971).
21. M. A. Field, D. W. Gill, B. B. Morgan, and P. G. W. Hawksley, *Combustion of Pulverized Coal*, The British Coal Utilization Research Association, Leatherhead, Surrey, England (1967).
22. E. M. Suuberg, W. A. Peters, and J. B. Howard, Product composition and kinetics of lignite pyrolysis, *ACS Div. Fuel Chem. Preprints*, **22**(1), 112–136 (1977).
23. R. Edelman and O. Fortune, Some Recent Developments on the Analysis of Exhaust Plume Afterburning, paper given at the Aerospace Science Plume Symposium, San Bernardino, California (1966).
24. H. B. Moretti, A new technique for the numerical analysis of nonequilibrium flows, *AIAA J.* **3**, 223–229 (1965).
25. G. A. Williams, The Numerical Solution of a Particle-Laden Propagating Flame Model, M.Sc. Thesis, Brigham Young University, Provo, Utah (1975).
26. W. C. Hecker, A Theoretical Study of the Kinetics, Propagation, and Suppression of Methane–Air Flames, M.Sc. Thesis, Brigham Young University, Provo, Utah (1975).
27. M. D. Horton, F. P. Goodson, and L. D. Smoot, Characteristics of flat, laminar, coal-dust flames, *Comb. Flame* **28**, 187–195 (1977).
28. S. Badzioch and P. G. W. Hawksley, Kinetics of thermal decomposition of pulverized coal particles, *Ind. Eng. Chem. Process Design Develop.* **9**, 521–530 (1970).
29. D. B. Stickler, R. E. Gannon, and H. Kobayashi, Modeling of Coal, paper given at the Eastern States Section Meeting, The Combustion Institute Pittsburgh, Pa. (1975).
30. L. D. Smoot and M. D. Horton, Propagation of laminar pulverized coal–air flames, *Prog. Energy Comb. Sci.* **3**, 235–258 (1977).
31. H. Kobayashi, J. B. Howard, and A. F. Sarofim, Coal devolatilization at high temperatures, in *Sixteenth Symposium (International) on Combustion*, The Combustion Institute, Pittsburgh, Pa. (1976).
32. G. E. Andrews, D. Bradley, and S. B. Lwakabamba, Turbulence and turbulent flame propagation—A critical appraisal, *Comb. Flame* **24**, 285–304 (1975).
33. H. L. Dryden, A review of the statistical theory of turbulence *Q. Appl. Math.* **1**, 7–42 (1943).
34. G. Z. Damkohler, *Z. Elecktrochem.* **46**, 601 (1940).
35. R. Borghi, *Chemical Reaction Calculations in Turbulent Flows: Application to a C-Containing Turbojet Plume*, AGARD Publication, ONERA, Paris, France (1974).
36. J. Vinckier and A. Van Tiggelen, Structure and burning velocity of turbulent premixed flames, *Comb. Flame* **12** 561–568 (1968).
37. C. Selph, *Generalized Thermochemical Equilibrium Program for Complex Mixtures*, Rocket Propulsion Laboratory, Edwards AFB, California (1965).
38. L. D. Smoot, J. R. Thurgood, and P. O. Hedman, The BYU rate-resolution coal combustor, *Comb. Sci. and Tech.* (1978), submitted for publication.

39. L. A. Vulis, *Thermal Regimes of Combustion*, McGraw–Hill, New York (1961).
40. A. B. Hedley and W. E. Jackson, Simplified mathematical model of a pulverized coal flame showing effects of recirculating on combustion rate, *J. Inst. Fuel* **39**, 208 (1966).
41. G. Horn, J. Csaba, and D. J. Street, Combustion experiments using a pulverized coal-fired superheater, *J. Inst. Fuel* **39**, 521 (1966).
42. J. M. Beer and K. B. Lee, The effect of residence time distribution on the performance and efficiency of combustion, in *Tenth Symposium (International) on Combustion*, pp. 1187–1202, The Combustion Institute, Pittsburgh, Pa. (1965).
43. W. S. Lewellen, H. Segur, and A. K. Varma, *Modeling Two Phase Flow in a Swirl Combustor*, Final Report, ERDA Contract No. EY-76-C-02-4062, Aeronautical Research Associates of Princeton, Inc., Princeton, New Jersey (1977).
44. H. R. Batchelder, R. M. Busche, and W. P. Armstrong, Kinetics of coal gasification, *I. and E.C.* **45**, 1856–1878 (1953).
45. A. K. Mehta, *Mathematical Modeling of Chemical Processes for Low BTU/Gasification of Coal for Electric Power Generation*, Final Report, ERDA Contract No. E(49-18)-1545, Combustion Engineering, Inc., Windsor, Connecticut.
46. P. E. George, R. C. Lenzer, J. F. Thomas, J. S. Barnhart, and N. M. Laurendeau, *Gasification in Pulverized Coal Flames*, Second Annual Progress Report, ERDA-FE-2029-6, The Combustion Laboratory, School of Mechanical Engineering, Purdue University, West Lafayette, Indiana (Aug. 1977).
47. P. Lewis, G. Simons, M. Finson, and E. Tomaszewski, A Model for Entrained Flow Gasifiers, paper presented at Fall Technical Meeting 1977, Eastern Section, The Combustion Institute, Pittsburgh, Pa. (1977).
48. T. R. Blake, *Computer Modeling of Coal Gasification Reactors*, QPR-FE-1770-23, ERDA Contract EX-76-C-01-1770, System, Science and Software, La Jolla, California (Feb. 1977).

14

Modeling Multidimensional Systems

John J. Wormeck

1. Generalized Pulverized-Coal Reactor Model

1.1. Introduction

The material presented in earlier chapters has focused on theoretical conservation equations, phenomenological laws, and development of various models. In this chapter, many of these components will be systematically organized into what will be called the generalized model. The numerical algorithms of each component will be outlined and the overall strategy required to obtain a solution will be discussed.

Historically, processes that occur inside pulverized-coal furnaces and gasifiers have been too complex to model in detail. Therefore, the techniques reported in the literature for such analyses have not included formal treatment of these processes and thus do not provide for detailed design and optimization of these furnaces and gasifiers. Bueters et al.[1] and Lowe et al.[2] recently reviewed the status of techniques for predicting performance in tangential burners. These investigators have emphasized radiative heat transfer and employed simple empirical correlations or approximate analyses to describe the flow field and combustion processes.

The one-dimensional model developed in Chapter 13 does not predict multidimensional variation of local properties inside a pulverized-coal

John J. Wormeck • Senior Engineer, Department of Mechanical Engineering, University of Utah, Salt Lake City, Utah

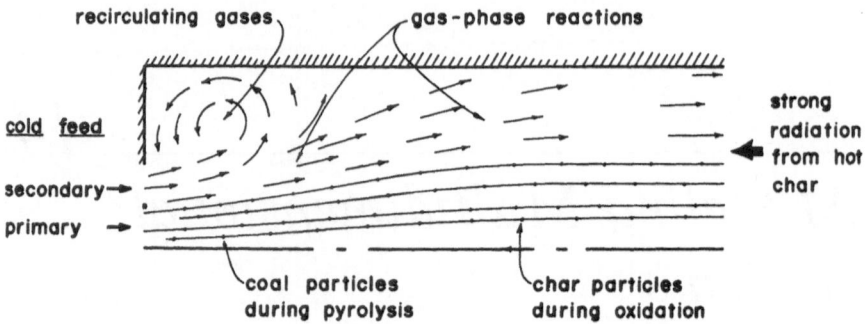

Figure 1. A schematic of the two-dimensional furnace.

reactor that can be directly compared with experimental measurements, nor does it have the potential capability for use as a detailed design technique for coal gasifiers and furnaces.

Recent advancements in computing recirculating flows, gas-phase kinetics, and the dynamics of multiphase and coal combustion have suggested new approaches for more detailed description for pulverized-coal reactors. The generalized model attempts to bring together these new, advanced technologies and ultimately predict the characteristics of pulverized-coal furnaces and gasifiers, as illustrated in Figure 1.

The generalized model has been developed as a group of modules organized in a specific manner. This modular approach provides (1) a clear overall view to the reader, (2) improved program organization, (3) better suitability for computer overlays (techniques to reduce computer core requirements), and (4) maintains the identity of the various codes developed for this effort. Each module corresponds to a component of the model and each is reviewed briefly below.

1.1a. Turbulent Mixing Model

An excellent summary of the computation techniques available up to 1972 for solving fluid mechanics problems is given by Roache.[3] The computational technique chosen to solve the general Navier–Stokes equations in this model was the method developed by Patankar and Spalding[4] and colleagues at Imperial College in London. The choice of this hydrodynamic code was based on the author's familiarity with both their stream function-vorticity formulation,[5] and with their primitive-variable, steady-state version, TEACH.[6] The Imperial College hydrodynamic codes have been applied to a wide range of problems. These codes are written to conserve mass, and at certain times, to employ upwind differencing, thus guaranteeing stability at all Reynolds numbers. Also, it is a straightforward task to implement various other additions to the basic hydrodynamic program, such as turbulence (see Chapter 3) and radiation (see Chapter 5).[7−10]

1.1b. Gas-Phase Chemical Kinetics Component

Pratt and Wormeck[11] have developed a generalized technique for extending hydrodynamic computer codes to include generalized gas-phase kinetics. This code, consisting of five subroutines, is called CREK and utilizes a Newton–Raphson iteration technique to solve the thermochemical equations. Details are given in Chapter 1 and 4.

1.1c. Two-Phase Flow Component

Crowe *et al.*[8] have developed a numerical technique for extending the TEACH hydrodynamic code to include two-phase flow with mass, momentum, and energy-coupling between the phases. This component has been extended to include dispersion of particles due to random motion of the gas, which has been shown experimentally to be an important effect in such two-phase flows. See Chapter 6 for details.

1.1d. Radiation Component

Chu and Churchill[12] have formulated a flux method for radiation which includes nonisotropic, multiple-scattering from particles. From various radiation theories, this method can be extended to include coal, char, and ash particulate radiation, which are important to heat-transfer rates in pulverized-coal furnaces and gasifiers. The solution of these flux equations is accomplished by the general hydrodynamic technique as developed by Gosman and Lockwood.[7] Chapter 5 contains relevant background material.

1.1e. Coal Reaction Component

Techniques for treating coal pyrolysis and heterogeneous oxidation in coal flames are developed in Chapter 12. The approach has included use of correlations of coal-char pyrolysis and combustion data which is documented in Chapters 7 through 10. Effects of swelling, internal burning in porous particles, variations in particle type and size, diffusion, activated pyrolysis, and oxidation are included in the model. Much of this information is employed in the one-dimensional model of Chapter 13, since the description of coal-particle physics and chemistry is common between the one-dimensional model and the generalized model.

1.1f. Gas-Mixture Properties

Molecular transport coefficients, including viscosity, thermal conductivity, and diffusivity, are evaluated using traditional statistical theory as functions of the local values of temperature, pressure, and composition as outlined in Chapter 2. Heat capacity, enthalpy, and entropy are described with standard fifth-order polynomial curve-fits, as given in Chapter 1.

1.2. Overall Organization

Although the generalized model requires solving many equations to describe the performance of combustors and gasifiers, these equations fall basically into two categories: gas phase and solid phase. Furthermore, because of physical and numerical considerations, all of the equations dealing with the gas phase will be solved in a Eulerian frame of reference, whereas the Lagrangian (following the motion) approach is employed for the description of everything involving the solid phase. Chapter 2 discusses these frames of reference in more detail.

The overall organization of the generalized model is best understood by studying the program flowchart in Figure 2, which illustrates the linkage between the various program modules or elements. The MAIN routine controls the frequency and order of calling each of the program modules which

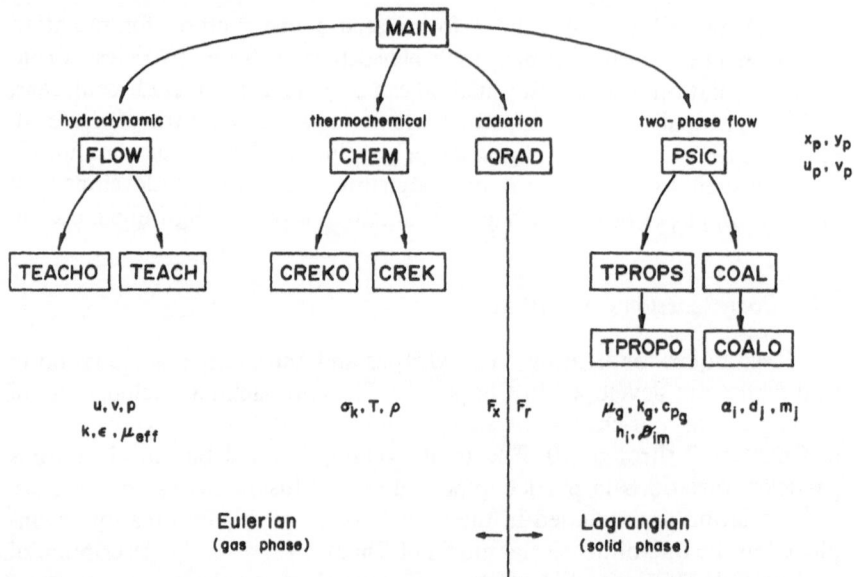

Figure 2. The modular flow chart of the generalized model.

comprise the generalized model. The Eulerian component consists of modules FLOW, CHEM, and QRAD, which solve the hydrodynamics, thermo-chemical, and radiation equations, respectively. These modules will be discussed in detail in subsequent sections. Similarly, the Lagrangian description consists of modules PSIC, which computes particle trajectories, and COAL, in which the physical and chemical processes of pulverized coal are calculated. The remaining submodules pertain to initialization and are explained below in connection with their parent modules.

The generalized model should be viewed as a general computational framework to which various submodels can be attached to describe the performance of pulverized-coal combustors and gasifiers. Thus the predictions from the code depend on the quality of the submodels utilized. An attempt has been made to devise a sufficiently general framework of modules such that, as particular new experimental or theoretical models become available, only the relevant module needs to be updated. For example, QRAD can be altered to employ other radiation models, while SLAG can be added as a new module to describe slagging on the reactor walls.

It is important for computational efficiency that each submodel contain a similar level of complexity for quantitative description of the physics. For example, it would be inefficient to perform complex, finite-rate chemistry if the application was such that infinite-rate or equilibrium assumptions gave similar results. For this reason, the assumptions of the generalized model belong properly to the individual module, which will be described in subsequent sections.

1.3. Eulerian Equations (Gas-Phase)

The Eulerian equations described in Chapter 2, that must be solved, can be cast into the standard form

$$\text{div} \, (\rho \mathbf{v} \phi) - \text{div} \, (\Gamma \, \text{grad} \, \phi) = S^{\phi} \qquad (1)$$

where ϕ is the particular dependent variable of immediate interest, $\mathbf{v}$ is the velocity vector, ρ is the fluid mass density, Γ is the exchange coefficient in the transport law, and S^{ϕ} is the generalized source term for ϕ which depends on the geometry, transport coefficients, and the other dependent variables. In terms of cylindrical–polar coordinates appropriate to the cylindrical geometry of many laboratory furnaces and gasifiers, including the Brigham Young University rate resolution reactors, Eq. (1) may be written as

$$[(\partial/\partial x)(r\rho u\phi) + (\partial/\partial r)(r\rho v\phi)] \quad [(\partial/\partial x)(r\Gamma\partial\phi/\partial x) + (\partial/\partial r)(r\Gamma\phi/\partial r)] = rS^{\phi} \qquad (2)$$

where u and v are the axial (x) and radial (r) gas-phase velocity components.

The terms in the first set of brackets in Eq. (2) are the convection terms, describing the transport of ϕ by the mass flow. In the next set of brackets are the diffusion terms, which neither create nor destroy ϕ, but rather just redistribute it about the flow domain due to gradient-driven diffusion. The last term is the source term, describing the production and destruction of ϕ. These equations are coupled, nonlinear, elliptic, partial differential equations for each of the dependent variables considered. All of the equations represented by Eq. (2) are coupled to the continuity equation [Eq. (2) with $\phi = 1$] by the products in the convection terms, and additional coupling and non-linearities are present, since the coefficients in all the terms may be complicated functions of the dependent variables.

In addition to the system of Eqs. (2), all parameters and transport exchange coefficients must be calculated or specified. Boundary conditions must also be specified to have the mathematical problem well posed. Elliptic equations require conditions on each dependent variable at all points of a closed boundary surrounding the field, and must be one of the following types:

(1) Specification of the value of ϕ along the boundary (Dirichlet type);
(2) Specification of the normal derivative of ϕ ($\partial\phi/\partial n$) along the boundary (Neumann type);
(3) Any algebraic relationship between the variable ϕ, and its normal derivative $\partial\phi/\partial n$, along the boundary of the field (Robbin's type).

These three types of boundary conditions can be used in any combination, and each equation in the system may be constrained by any of the three types.

The Eulerian equations are solved in the generalized model by three different modules. The hydrodynamic equations, consisting of overall mass continuity, two momentum equations, and the two-equation model of turbulence, are solved by module FLOW. The energy equation and conservation of individual chemical species equations comprise the thermochemical set and are solved by module CHEM. Similarly, QRAD solves the equations of the radiation model.

Table 1 identifies the Eulerian or gas-phase variables to be solved, together with the corresponding expressions in the general Eq. (2) for exchange coefficients Γ and the source term S^ϕ.

The solution technique utilized in the generalized model to solve the set of Eulerian equations employs two different computational schemes, each appropriate to the nature of the particular conservation equation being solved, in order to solve the complete set of conservation and transport equations with maximum efficiency.[11]

Appendix D contains the derivation of the general Eulerian finite-difference equation in the standard form

$$\sum_d A_d \phi_P = \sum_d A_d \phi_d + S^\phi \tag{3}$$

where the subscript d (for "direction") refers to the adjacent nodes E, W,

Table 1. Eulerian Conservation Equations: Identification of Terms in Equation (*1*)

Conservation equation	ϕ	Γ	S^ϕ
Hydrodynamic			
Mass (continuity)	1	0	0
Axial momentum	u	μ_{eff}	$-\partial P/\partial x + \{(\partial/\partial x)[r\mu_{\text{eff}}(\partial u/\partial x)] + (\partial/\partial r)[r\mu \ (\partial v/\partial x)]\}/r + {}_pS^u$
Radial momentum	v	μ_{eff}	$-\partial P/\partial r + \{(\partial/\partial x)[r\mu_{\text{eff}}(\partial u/\partial r)] + (\partial/\partial r)[r\mu_{\text{eff}}(\partial v/\partial r)]\}/r - 2\mu_{\text{eff}}\,v/r^2 + {}_pS^v$
Turbulent kinetic energy	k	$\mu_{\text{eff}}/\sigma_k$	$G - \rho\varepsilon + {}_pS^k$, where $G = 2\mu_{\text{eff}}\{(\partial u/\partial x)^2 + (\partial v/\partial r)^2 + (v/r)^2 + [\partial v/\partial x + (\partial u/\partial r)^2/2]\}$ and $\varepsilon = C_\mu \rho k^2/\mu_{\text{eff}}$
Dissipation rate of turbulent kinetic energy	ε	$\mu_{\text{eff}}/\sigma_1$	$(\varepsilon/k)(C_1 G - C_2\rho\varepsilon) + {}_pS^1$
Thermochemical			
Energy	h	$\mu_{\text{eff}}/\sigma_h$	div $(\mathbf{q}_r) + {}_pS^h$, where div $(\mathbf{q}_r) = 2k_a(F_x + F_r - 2E_b)$
Concentration of species k	σ_k	$\mu_{\text{eff}}/\sigma_s$	$-\sum_{j=1}^{JJ}(\alpha_{kj}^i - \alpha_{kj}'')(R_j - R_{-j}) + {}_pS^{\sigma_k}$

N, and S of the computational cell shown in Figure 1 of Appendix D, and the A's are the convection and diffusion coefficients.

The source term S^ϕ in Eq. (3) is frequently linearized,

$$S^\phi = S_U^\phi + S_P^\phi \phi_P \tag{4}$$

and the resulting equations for any variable ϕ are solved either point-by-point by Gauss–Seidel iteration (GS), with Eq. (3) in the form

$$\phi_P = \left(\sum_d A_d \phi_d + S_U\right)\Big/\left(\sum_d A_d - S_P\right) \tag{5}$$

or line-by-line, by means of the tridiagonal matrix algorithm (TDMA), with Eq. (3) in the form

$$\phi_P = (A_E\phi_E + A_w\phi_w + S_U)\Big/\left(\sum_d A_d - S_P\right) + A_N\phi_N\Big/\left(\sum_d A_d - S_P\right)$$

$$+ A_S\phi_S\Big/\left(\sum_d A_d - S_P\right) \tag{6}$$

either row-by-row or column-by-column.

These two practices have been successful when applied to solution of the hydrodynamic conservation equations, in which the source terms are linear or mildly nonlinear. TDMA is generally faster than GS iteration, presumably because there is simultaneous solution of *many* values of ϕ, while GS updates only *one* value of ϕ at a time. In physical terms, the linearization of Eq. (4) and solution by GS or TDMA gives satisfactory rates of convergence when local variations in ϕ_P depend more strongly on the values of the *same* variable ϕ_d at neighboring nodes than on the values of *other* variables at the same point.

However, for conservation of chemical species in cases where chemical reaction, rather than turbulent mixing, dominates local variations in mole numbers (such as in laminar flames or slow reactions with fast mixing) the homogeneous, chemical–kinetic source term for mole numbers of species i is obtained from Chapter 4:

$$S_{\sigma_i} = -\sum_{j=1}^{JJ} (\alpha'_{ij} - \alpha''_{ij})(R_j - R_j) \tag{7}$$

where R_j is the rate of forward reaction j, given by the modified Arrhenius form:

$$R_j = 10^{B_j} T^{N_j} \exp\left(-T_j/T\right)(\rho\sigma_m)^{\bar{\alpha}_j} \sum_{k=1}^{NS} (\rho\sigma_k)^{\alpha_{kj}} \tag{8}$$

It may be supposed from inspection of Eqs. (7) and (8) that $\sigma_{i,P}$ may possibly depend more strongly on *all* the mole numbers and temperature at point P than on the values of $\sigma_{i,d}$ at adjacent nodes. If this is true, then two principles for solving the σ_i equations may be postulated:

(1) Solutions of the equations for $(\sigma_i, i = 1, NS)$ should be *simultaneous, point-by-point*, and coupled with T, through some form of the energy equation.

(2) Due to the nonlinearities of the rate equation [Eq. (8)], the use of derivative information may be valuable to achieve fast convergence when large local gradients in $(\sigma_i, i = 1, NS)$ or T occur.

Summarizing, all of the Eulerian differential equations are first expressed in the standard finite-difference form given by Eqs. (2) and (3), respectively. The hydrodynamic and radiation sets are then solved line-by-line by TDMA iteration, while the thermochemical set is solved simultaneously, point-by-point, by Newton–Raphson iteration. The following subsections will outline the numerical techniques employed and the various assumptions utilized in the solution of these sets of equations. The complete solution strategy and starting estimates will be presented after the development of the solution of the Lagrangian equations.

1.3a. Hydrodynamic Equations

The hydrodynamic equations are solved by module FLOW by modifying the TEACH computer program developed by Gosman and Pun,[6] which utilizes the SIMPLE algorithm of Patankar and Spalding.[7] A complete, rigorous derivation of this numerical procedure is available, including the staggered grid definition, formulation and solution of the four different types of equations, underrelaxation control, and examples of various boundary conditions.[9] Appendix D illustrates a portion of the procedure by developing the finite-difference equation for the general ϕ equation.

In the language of computational fluid dynamics, TEACH can be described as a finite-difference, steady-state formulation of the primitive, compressible flow variables in a multidimensional Eulerian description. TEACH solves nonuniform orthogonal geometries, utilizing appropriate differencing schemes for the convection or advection terms depending on the local cell Reynolds number: central-differencing if $R_c < 2$, and upwind-differencing if $R_c > 2$. The diffusion terms are central-differenced, and used only when the magnitude of the diffusion terms are comparable to that of the convection terms. A "weak" pressure equation is solved simultaneously with all of the momentum equations, with mass being conserved for each iteration. The weak or incomplete pressure equation (the basis of the SIMPLE algorithm) requires an indirect iterative solution with underrelaxation to converge. The source terms are linearized and central-differenced. The resulting finite-difference equations are solved by a line-by-line technique, where a whole line of node points are solved simultaneously by means of the tridiagonal matrix algorithm (TDMA) using the latest values determined. For those cells where $R_c < 2$, the method is second-order accurate; otherwise, it approaches first-order accuracy as the convection magnitude increases relative to the diffusion magnitude. The procedure is conservative for all variables (mass, momentum, and energy) and transportive only for cells with large local Reynolds numbers. TEACH is stable for all flow Reynolds numbers, allows specification of all types of boundary conditions, and is valid for both subsonic and supersonic flow (with the latter requiring special density interpolation). The computer code is available in two dimensions only, but can be extended to three-dimensional geometries.

This numerical scheme has proved to be simple, straightforward, and usually convergent, as long as the equations considered are linear or only mildly nonlinear. However, as mentioned earlier, the very highly nonlinear source terms in the equations of conservation of chemical species cause divergence, and these equations should be solved by the procedure described in Chapter 4.

It should be pointed out that the generalized model does not require the use of the SIMPLE algorithm for the solution of the Navier–Stokes equations since any technique yielding the distribution of the two velocities and pressure would suffice.

A grid spacing much larger than the microscale of turbulence was employed so that it was necessary to model the effects of turbulence. Chapter 3 discusses the various turbulence models available and the recommended approach for the generalized model. Following Gosman and Pun[6] and Launder and Spalding,[13] the turbulent phenomena were accounted for by utilizing an "effective turbulent viscosity" (μ_{eff}) or "eddy viscosity" in the laminar flow finite-difference formulation. This effective viscosity is computed locally by means of a two-equation, "k–ε" model,[13] together with Prandtl–Kolmogorov relation (see Chapter 3)

$$\mu_{\text{eff}} = c\rho k^2/\varepsilon \qquad (9)$$

where c is a constant of order 0.1, and k and ε are the turbulent kinetic energy and its rate of dissipation, respectively, and are obtained from solution of two additional transport equations. The equations for k and ε are derived by Reynolds-decomposing the momentum and continuity equations and modeling the resulting correlations,[14] as described in Chapter 3. These two transport equations are reduced to the standard equation form and are then solved by the TEACH procedure (i.e., line-by-line TDMA).

TEACH (see Figure 2) automatically sets up the orthogonal nonuniform grid (3 options) and performs all the initial operations required for the hydrodynamic equations. In general, a submodule with a zero as the last character of its name implies that it is an initialization routine and only required at the beginning of the calculation, thus conveniently allowing a computer overlay to eliminate its core storage requirements.

Submodule TEACH performs the calculations by solving simultaneously the five hydrodynamic equations and their boundary conditions, and determines the effective viscosity of the fluid mixture. Each one of these functions is a separate routine comprising TEACH.

Anytime module FLOW is called in the generalized model, the local distributions of u, v, p, k, ε, and μ_{eff} are updated, using as input the latest values of all variables.

1.3b. Thermochemical Equations (CHEM)

A multicomponent flow consists of a distinct number of separate chemical species: molecules, atoms, ions, and free radicals. The concentration of these species change with position and time, due to convection, molecular and turbulent diffusion, and to chemical reaction. The various levels of description of multicomponent mixtures occurring in a flow situation are as follows, in increasing order of complexity: (1) chemically inert or frozen composition; (2) equilibrium or infinite-rate chemical kinetics; and (3) finite-rate chemical kinetics. The first situation is basically a mixing problem requiring only a fluid flow technique capable of calculating multicomponent mixtures, while the equilibrium calculation is usually performed by minimizing the Gibbs free energy (Chapter 1), and thus only thermodynamic data of the species is required. With detailed finite-rate chemistry, a reaction mechanism with rate data is needed for the general case (Chapter 4) and are only available for the combustion of hydrogen and a few simple hydrocarbon systems as discussed in Chapter 10. A usual simplification is to employ a global reaction for more complex chemical systems (see Chapter 10). In this section, all of these possibilities are considered, with the model utilized in the calculations being selected merely by setting flags (logical variables).

The thermochemical equations consist of the conservation of energy and chemical species and are solved by module CHEM. For a chemically reacting flow consisting of NS distinct chemical species, a continuity equation may be written for each species (see Chapter 2 for development):

$$\text{div}\,(\rho\mathbf{v}\sigma_k) - \text{div}\,(\Gamma_{k,\text{eff}}\,\text{grad}\,\sigma_k) = -r_k + {}_p S^{\sigma k} \tag{10}$$

where r_k (kmol m^{-3} s^{-1}) is the net molar destruction term due to chemical reactions per unit volume per unit time, and ${}_p S^{\sigma k}$ is the source term for species k due to everything except chemical reaction; for example, this includes products from coal devolatilization and char oxidation entering the gas phase from the solid-phase coal-char reactions. Note that Fick's law is employed in Eq. (10) to express the diffusion flux as driven by a concentration gradient and that the exchange coefficient $\Gamma_{k,\text{eff}}$ (kg m^{-1} s^{-1}) is in general different for each species; however, the laminar contribution (related to the gas diffusivity, which varies greatly between species) is totally dominated by the turbulence portion, and therefore can be expressed simply as

$$\Gamma_{i,\text{eff}} = \mu_{\text{eff}}/\sigma_s \tag{11}$$

where σ_s is a Schmidt number of order unity. This assumption is reasonable since the dominant mechanism causing this diffusion is due to the turbulence, the same transport mechanism which diffuses momentum. If better information should become available, the assumption of Eq. (11) would not be required for the solution of the chemical equations.

These equations are elliptic partial differential equations which are nonlinearly coupled due to the Arrhenius destruction term r_k, as well as being coupled to the hydrodynamic (momentum and continuity) equations through the velocity appearing in the convection term, and further coupled to the thermal energy equation through the mass density and temperature. Therefore, all of the conservation equations must be solved simultaneously in the general case. If the reacting flow is turbulent, the turbulent exchange coefficients must also be solved simultaneously with the above equations. Also, a weak coupling between the coal devolatilization and char burnout is present as well.

The total energy equation for the gas in the gas–particle mixture is developed by taking $\theta = 1$ in Eq. (86), Chapter 2, to give

$$\text{div}\,[\rho_g(h_g + v_g^2/2)\mathbf{v}_g] = -\text{div}\,(\mathbf{q}) + q_{P,c} + r_P(\bar{h}_s + v_P^2/2) + \Phi - (\mathbf{v}_P \cdot \mathbf{f}_P) + \rho_g(\mathbf{g} \cdot \mathbf{v}_g) \tag{12}$$

where the unsteady term has been dropped, the dissipation function Φ is substituted for div $(\mathbf{\tau} \cdot \mathbf{v}_g)$, gas-phase radiation is neglected $(q_{r,P}=0)$, $v_P \gg w'$ is assumed, and the regression rate of the particle surface has been neglected (see Chapter 2 for notation). The gas-phase heat transfer includes only two terms,

$$\mathbf{q} = -k \text{ grad } (T_g) + \sum_i \mathbf{j}_i h_i \tag{13}$$

since the diffusional transport of kinetic energy will be neglected compared to diffusional transport of enthalpy. Expressing the Fourier rate in terms of gas-phase enthalpy h_g and introducing the Prandtl number σ_h, the conduction heat transfer becomes

$$\mathbf{q} = -(\mu/\sigma_h) \text{ grad } h_g \tag{14}$$

Again, the turbulence mechanism dominates molecular transfer rates and the usual practice is to employ effective coefficients which, when substituted in Eq. (12), give

$$\text{div } (\rho_g \mathbf{v}_g h_g) - \text{div } [(\mu_{\text{eff}}/\sigma_h) \text{ grad } h_g] = q_{P,c} + r_P(\bar{h}_s + v_P{}^2/2) + \Phi - (\mathbf{v}_P \cdot \mathbf{f}_P)$$
$$+ \rho_g(\mathbf{g} \cdot \mathbf{v}_g) - \text{div } (\rho_g \mathbf{v}_g v_g^2/2) \tag{15}$$

where σ_h is of order unity and the mechanical energy portion is transferred to the right-hand side of the equation (last term) so that the equation is in standard form for the dependent variable h_g, with the source term as

$$S^h = q_{P,c} + r_P(\bar{h}_s + v_P^2/2) + \Phi - (\mathbf{v}_P \cdot \mathbf{f}_P) + \rho_g(\mathbf{g} \cdot \mathbf{v}_P) - \text{div } (\rho_g \mathbf{v}_g v_g^2/2) \tag{16}$$

It is convenient to separate this source term into a gas-phase and solid-phase portion

$$S^h = {}_p S^h + \Phi - \text{div } (\rho_g \mathbf{v}_g v_g^2/2) \tag{17}$$

where

$${}_p S^h = q_{P,c} + r_P(\bar{h}_s + v_P^2/2) + \mathbf{v}_P \cdot (\rho_g \mathbf{g} - \mathbf{f}_P) \tag{18}$$

is the particle energy source term which is calculated and stored in PSIC; further details are given in Section 1.4.

The set of NS species equations given by Eq. (10) and the energy equation [Eq. (15)] are in standard form; their finite-difference formulation are given by Eq. (3) and, in contrast to the hydrodynamic equations, are solved simultaneously point-by-point employing a Newton–Raphson iteration as described in Chapter 4.

Referring to Figure 2, CREK0 is an initialization routine which reads all the thermodynamic curve-fit data, and if finite-rate chemistry is to be considered, reads and processes the reaction mechanism and its associated rate data.

The solution of the thermochemical equations presented above completely ignores micromixing; there is no interaction between the turbulence and the thermochemical field. Such an assumption is reasonable to describe very high-intensity combustion and the above technique has been utilized with success.[9,10] Since the flame in the present application is "slow and lingering," this assumption may lead to considerable error. An alternative approach recommended in Chapter 4 is to utilize the probability density function (PDF) approach by solving two additional transport equations for mean fuel-inlet mixture fraction $\bar{f}$ and its variance g, which allows, after assuming some shape for the PDF, the variation of mean properties over the flow domain. Then the time-average concentrations and temperature follow as

$$\bar{\sigma}_i = \int_0^1 P(f)\sigma_i^*(f) \tag{19}$$

$$\bar{T} = \int_0^1 P(f)T^*(f)\,df \tag{20}$$

Similarly, the various correlations may be computed as

$$\bar{\sigma}_i'\bar{\sigma}_j' = \int_0^1 P(f)\sigma_i^*(f)\sigma_j^*(f)\,df - \bar{\sigma}_i\bar{\sigma}_j \tag{21}$$

$$\bar{\sigma}_i'\bar{T}' = \int_0^1 P(f)T^*(f)\sigma_i^*(f)\,df - \bar{\sigma}_i\bar{T} \tag{22}$$

which allows the contact index to be estimated. Chapter 4 describes this approach in more detail.

For the generalized model, both approaches will be utilized with the user choosing the method by setting the appropriate flag. A separate study is underway to compute the thermochemical fields by both methods for comparison with the experiments described elsewhere in this book.

Each time module CHEM is called, the following variables are updated: T_g, ρ_g, (σ_k, $k=1$, NS), and σ_m. Summarizing, the generalized model has a number of different techniques coded to solve the thermochemical equations. As of this writing, the best method has not been established, but will be reported in the future.[15] The reader should note how this modular approach allows, in a straightforward manner, the testing of models containing different physics.

1.4. Lagrangian Equations (Solid-Phase)

An analysis of combustion and gasification of pulverized coal not only requires modeling the difficult problem of gas-phase combustion, but additional complexity is introduced because of the presence of solid particles. Thus the multicomponent analysis of the preceding sections must be extended to include two-phase flow with mass, momentum, and energy coupling between the phases.

1.4a. Governing Equations

Chapter 2 describes the governing equations for both the gaseous and condensed phases, the various terms which should be important in the present model application, and the recommended approach for predicting the velocity and temperature fields of both the gas and coal-char particles in combustors and gasifiers. This recommended PSI-CELL technique[8] (module PSIC) has been applied with success in a variety of two-dimensional gas–particle flows with all three modes of gas–particle coupling (mass, momenttum, and energy).

The gas-phase conservation equations were summarized in Table 1, with $_pS^\phi$ representing the source terms for contribution of property ϕ from the particle to the gas phase. These gas-phase conservation equations were considered in Eulerian framework; that is, field values of the gas-phase variables are stored and calculated at grid nodes throughout the field by modules FLOW and CHEM. In contrast, conservation equations for the solid phase are considered in a Lagrangian framework; that is, by following the motion of individual particles.

A correct formulation of the two-phase flow equations requires the following two conditions[16] (1) If some property appears as a source in the equations of one phase, it must appear as a sink in the other set. (2) In the two limits of vanishing mass in each phase, the complete set of two-phase flow equations must approach the correct form of the single-phase flow equation set.

The solid-phase conservation equations are different in form from the general equation [Eq. (1)], and are summarized below.

Particle Mass Conservation. From Eq. (19) in Chapter 2, the particle mass-conservation equation is

$$dm_p/dt = -r_p \tag{23}$$

where m_p is the mass of the particle and r_p is the rate of mass loss from the solid phase. This equation simply states that the change in mass of the particles is equal to the mass transfer from the solid to the surrounding gas.

Particle Momentum Conservation. The equation of motion of a particle is, from Eq. (21) in Chapter 2,

$$m_p(d\mathbf{v}_p/dt) = \tfrac{1}{2}\rho C_D(\mathbf{v}_g - \mathbf{v}_p)|\mathbf{v}_g - \mathbf{v}_p|A_p + m_p\mathbf{g} \tag{24}$$

where C_D is the particle aerodynamic drag coefficient, ρ the gas-phase mass density, $\mathbf{v}_g$ the gas velocity, $\mathbf{v}_p$ the particle velocity, A_p the particle cross-sectional area (consistent with C_D), and $\mathbf{g}$ the gravitation acceleration vector. Only gravity and aerodynamic drag forces are considered to change the particle velocity in the present application. Solution of this equation yields the particle trajectories.

Particle Energy Equation. The conservation of particle energy leads to Eq. (92) in Chapter 2,

$$m_p(di_p/dt) = Q_P + (h_s - h_p)(dm_p/dt) \tag{25}$$

where h_s is the enthalpy of the combustion product gases evolving from the particle (specifically, the enthalpy of the gas at the surface of the particle, including heat of formation and sensible enthalpy). Q_P is the radiative, conductive, and convective heat transfer to the particle. The particle internal energy, i_p, is assumed uniform throughout the particle, and

$$di_p/dt = C_{Pp}(dT_P/dt) \tag{26}$$

The particle heat capacity is taken as weighted averages over each of the particle components:

$$C_{PP} = \omega_a C_{P_a} + \omega_c C_{P_c} + \omega_h C_{P_h} + \omega_w C_{P_w} \tag{27}$$

where the ω's are particle component mass fractions (see Chapter 12) and the component heat capacities are specified as quadratic curve-fits.

Heat transfer to the particle is due to conduction, convection, and radiation,

$$Q_P = -[\mathrm{Nu} k_g A_j(T_P - T_g)]/d_p + k_a(F_x + F_r) - \varepsilon_p \sigma T_P^4 \tag{28}$$

where the first term represents the conduction–convection heat transfer, the second term is the radiation absorbed by the particle due to the presence of radiation fluxes, and the last term is the particle radiative emission (see Chapter 5 for details).

Substituting Eqs. (26)–(28) into Eq. (25) yields the particle energy equation

$$dT_P/dt = aT_P^4 + bT_P + c \tag{29}$$

where $a = -\pi d_P \varepsilon_P \sigma/(m_P C_{Pp})$, $b = -\pi d_P \mathrm{Nu}\, k_g/(m_P C_{Pp})$, $c = -aT_g + [\pi d_P^2 k_a \times (F_x + F_r) + L(dm_P/dt)]/(m_P C_{Pp})$, and $L = h_s - h_p$, the heat of vaporization. Since an analytical solution of this nonlinear, first-order differential equation has not been found, a numerical or approximate solution must be utilized.

The three conservation equations describing the solid phase require the composition of the solid mass and the rates at which it is converted to the gaseous state. These rates are obtainable from the physical and chemical models of coal developed in Chapter 12. For reasons which will become apparent later, the physical and chemical processes of coal will be calculated simultaneously in the Lagrangian sense with the three conservation equations given by Eqs. (23), (24), and (29).

Module PSIC (see Figure 2) solves the solid-phase conservation equations, while module COAL deals directly with the physical and chemical processes occurring in the coal particles, and the rates at which mass and energy change from solid to gas phase. Each of these modules will now be discussed in turn.

1.4b. Two-Phase Flow (PSIC)

The one-dimensional models of Chapter 13 consider two-phase flow; however, dynamic equilibrium is assumed (no velocity differences between the phases), which greatly simplifies the analysis. As shown in Eq. (24), the generalized model considers a velocity lag between the phases resulting in an aerodynamic force proportional to the square of this velocity difference. Therefore, the recommended approach (see Chapter 2) to solve Eqs. (23), (24), and (29) is different in the generalized model compared to the one-dimensional models. The PSI-CELL method was taken from other applications[8,17] and incorporated into PSIC to fit the general modular approach of this effort. Some additional features were added to fit this specialized application of coal combustion and gasification.

The basis of PSIC is to obtain the particle trajectories by integrating the equations of motion [Eq. (24)] for the particles in the gas-flow field. At various points along these trajectories, module COAL computes the particle–gas mass and energy-transfer rates. By storing the differences in mass, momentum, and energy of the particles as they enter and leave each computational cell, the particle source terms required in the gas-flow equations of Table 1 may be computed. Therefore, the particles are considered as sources of mass, momentum, and energy to the conveying gaseous phase. With these source terms being incorporated in the solution of the gas-phase equations, the effects of particles on the gas velocity, species, and temperature fields are accounted for.

For theoretical considerations with solid-phase averaging,[16] and for numerical reasons, the particles are represented as entering by a finite number of entry ports. Furthermore, a continuous distribution of particle sizes must be represented by a discretum of a few different particle diameters. Using Crowe's notation,[8] the mass of a particle of diameter d_{p_i} which enters per unit time at port j is

$$m_{P_j}(d_{P_i}) = m_P X_j Y_i \qquad (30)$$

where m_P is the total particle mass inflow rate, X_j the fraction of particle mass which enters at port j, and Y_i is the fraction of coal-particle mass with initial diameter d_{P_i}. As now programmed, NSL($= 5$) is the number of starting locations in the reactor inlet, and at each starting location, there are NPS($= 3$) number of particle sizes. Thus a total of NSL $\times$ NPS($= 15$) trajectories must be computed to adequately describe the particles. Obviously, more starting locations and particle sizes can be considered, but at the expense of greater computational cost. Further, each separate coal type (anthracite, bituminous, lignite, recirculated char, ash, etc.) will require separate trajectories because of the different physical and chemical processes which determine their path.

If the particles are assumed to be spherical with mass density ρ_P, then

$$n_j(d_{P_i}) = 6m_P X_j Y_i / \pi \rho_P d_{P_i}^3 \qquad (31)$$

is the number flow rate of particles of initial diameter d_{P_i} along a given trajectory and will be constant along that trajectory, provided no fragmentation takes place. To account for fragmentation, a relationship giving the new size categories and particle number in each category must be provided, as well as requiring additional trajectories for each fragmented class—a complexity not included in the present formulation.

For each starting location and particle size class, $n_j(d_{P_i})$ particles follow a trajectory obtained by integrating Eq. (24) with the following assumptions (the reader is referred to the original work by Crowe *et al.*[8] for complete details): The drag coefficient from Eq. (10) of Chapter 6 is

$$C_D = (24/\mathrm{Re})[1 + 0.15\mathrm{Re}^{0.687}] \qquad (32)$$

where $\mathrm{Re} = \rho_g |\mathbf{v}_g - \mathbf{v}_p| d_p / \mu_g$ is the Reynolds number based on the gas–particle relative velocity, ρ_g is the gas mixture mass density, and μ_g is the molecular viscosity of the gas mixture. Note that the relative velocity differences between phases will be small. In addition, the largest particle diameters of coal in this application are of the order of 100 μm, which is small compared to the smallest length scale of turbulence (~ 0.1 mm), so that molecular transport rates govern the interaction of coal particles with the gas. Coal pyrolysis can reduce the drag coefficient due to mass fluxes from the surface of the coal,[8] as shown by Eq. (13) of Chapter 6.

Expressing A_P and m_P in terms of particle diameter and density, Eq. (24) may be rewritten as

$$d\mathbf{v}_P/dt = 18\mu f\,(\mathbf{v}_g - \mathbf{v}_P)/\rho_P d_P^2 + \mathbf{g} \qquad (33)$$

where $f = C_D(\mathrm{Re}/24)$. By assuming the gas-velocity constant over the short time interval of integration, this equation can be integrated once formally to give

$$\mathbf{v}_P = \mathbf{v}_g - (\mathbf{v}_g - \mathbf{v}_{P_0})\exp(-\Delta t/\tau) + \mathbf{g}\tau[1 - \exp(-\Delta t/\tau)] \qquad (34)$$

where $\mathbf{v}_{P_0}$ is the initial particle velocity at the beginning of the integration time interval Δt, which is constrained to require a few integrations per computational cell, and τ is the characteristic time defined by

$$\tau = \rho_P d_P^2 / 18 \mu_g f \tag{35}$$

This equation gives the particle velocity at Δt time along the trajectory, from which its position is obtained as

$$\mathbf{x}_P = \mathbf{x}_{P_0} + (\mathbf{v}_P + \mathbf{v}_{P_0})(\Delta t/2) \tag{36}$$

where $\mathbf{x}_{P_0}$ is the particle position at the beginning of the time increment. By repeated application of this integration, the complete trajectory for this initial starting location and particle size distribution is determined. A more accurate formulation, considering a linear variation in the gas velocity before integrating Eq. (33) and/or formally integrating this equation twice by solving a second-order equation, has been completed, but has not been implemented in the code.

During this integration of the particle trajectory equation, the physical and chemical processes of the coal must be accounted for. This is accomplished by calls to module COAL between integration steps. The details are presented in the development of COAL in Section 1.4.3. PSIC will call COAL with the following parameters passed through Fortran COMMON blocks: number flow rate of particles $n_j(d_{P_i})$; particle diameter d_{P_i}; particle mass m_{P_i}; particle composition α_i; time interval Δt; gas–particle relative velocity $\mathbf{v}_g - \mathbf{v}_P$; position in reactor, i.e., the computational cell I, J, which allows COAL to access the gas-phase conditions surrounding these particles, including temperature T_g, pressure P_g, chemical species σ_k, and the radiation fluxes to which these particles are subjected. On return from COAL, the following parameters are updated and passed to PSIC: new particle mass, diameter and composition; the rate of mass transferred from the solid to the gas, r_p; the composition of this mass transferred, $\sigma_{k,s}$; the heat transfer rate between the phases, Q_P, including contributions from conduction, convection, and radiation; and the energy transferred between phases, $r_p h_s$, resulting from mass transfer.

Changes in coal properties are calculated by integrating along particle pathlines (Lagrangian framework). The resulting fluxes of mass, momentum, and energy and the kth species mole numbers are calculated in each computational cell intersected by the particle trajectories, and are stored in the (Eulerian framework) gas-phase source terms $_pS^\phi$, to be considered at a later iteration as source terms in the gas-phase equations. These source terms for particle–gas coupling (see Table 1) are now presented in detail.

Mass:

$$_pS^m = \Sigma \, n_i (\Delta m_p)_i \tag{37}$$

where n_i is the number frequency of passage of particles along trajectory i, $(\Delta m_p)_i = \pi \rho_p(d_{i,out}^3 - d_{i,in}^3)/6$ is the increment in mass of those particles crossing the computational cell along trajectory i. The summation is over all trajectories which pass through the cell under consideration.

u-Momentum:

$$_pS^u = \sum_i n_i \Delta(u_p m_p)_i \tag{38}$$

where u_p is the x-direction (axial) particle velocity along the trajectory and other symbols are as in $_pS^m$.

v-Momentum:

$$_pS^v = \sum_i n_i \Delta(v_p m_p)_i \tag{39}$$

Total energy:

$$_pS^h = q_{p,c} + r_p(\bar{h}_s + v_p^2/2) + \mathbf{v}_p \cdot (p_g \mathbf{g} - \mathbf{f}_p) \tag{40}$$

kth Species:

$$_pS^{\sigma k} = \sum_i n_i \Delta(m_p \sigma_{k,s})_i \tag{41}$$

where $\sigma_{k,s}$ is the amount of kth species transferred from the particles to the gas in the cell of interest.

Turbulent kinetic energy and dissipation rate. $_pS^k$ and $_pS^d$ will initially be taken as zero, since no adequate physical models are available to assume otherwise, and this assumption appears to be a reasonable one.

1.4c. Pulverized Coal (COAL)

Module COAL contains all the physical and chemical processes of pulverized coal that are in the generalized model which was discussed in Chapter 12; these processes are also included in the one-dimensional models of Chapter 13. The reader is referred to those chapters for assumptions of the model, the treatment of coal-char chemical reactions including rate expressions for coal pyrolysis, char oxidation, and particle–gas heat and mass transfer. In what follows, only the differences and extensions required for multidimensional flow are outlined.

The organization of the generalized model requires the gaseous hydrocarbon combustion to be treated separately in module CHEM, and the gas-mixture transport properties are grouped into module TPROPS (which is called by COAL) and thus discussed in Section 1.4d.

A major difference in the one-dimensional and generalized models relates to gas-phase kinetics. The one-dimensional coal combustion and gasification model (1-DICOG) considers only equilibrium processes occurring in the gas phase, which require only knowledge of the elemental composition of the mass transfer between phases. In contrast to this, the generalized model requires the distribution of chemical species because of the capability of calculating finite-rate chemistry in the gas phase. The lack of data requires that heavy hydrocarbons (larger than CH_4) be pyrolyzed to H_2 and CO at specified rate. The generalized model is coded to consider a general number of coal pyrolysis steps, a general number of char oxidation reactions with any and all oxidizers (see Table 3, Chapter 12), and general, power-series curve-fits for coal-particle component enthalpies and specific heat capacities.

Another difference in the models is that the generalized model considers a relative velocity difference between the phases; the effective Nusselt number for both heat and mass transfer must be utilized. The full expression for the Nusselt number for heat transfer (without pore blowing) is

$$Nu_0 = 2 + 0.60 \ (Re)^{0.5} \ (Pr)^{1/3} \tag{42}$$

where Pr is the Prandtl number. For mass transfer,

$$Nu_0 = 2 + 0.60 \ (Re)^{0.5} \ (Sc)^{1/3} \tag{43}$$

where Sc is the Schmidt number of the species being transferred.

This aerodynamic drag is in addition to the blowing of gases from the internal pores of the coal particle undergoing pyrolysis. The blowing from the internal pores has been accounted for as shown in Chapters 6 and 12, for low Reynolds numbers:

$$Nu/Nu_0 = B/(e^B - 1) \tag{44}$$

and for Reynolds numbers up to 400:

$$Nu/Nu_0 = \exp \ (-0.6B) \tag{45}$$

where B is either the heat- or mass-transfer blowing parameter.

Module COAL is called by PSIC at various times and places along the particle trajectories and computes the physical and chemical processes occurring in that coal particle type during time Δt since the last call from PSIC COAL calls module TPROPS for an update in local gas-mixture transport coefficients, returns the new state of the coal to PSIC, and saves various particle parameters needed later in QRAD for the radiation treatment.

The computation time required for PSIC to integrate the particle motion equation is small; however, a call to COAL (and thus TPROPS) is very time-consuming and should be called only if the physical and chemical process rates occurring in the coal are changing rapidly. A coal process time constant may be defined to determine the local frequency of calling module COAL. A minimum of one call per cell is required, on entrance and exit, to determine the gas–particle source terms. Both PSIC and COAL have initialization routines, PSICO and COALO, which read input data and perform preliminary calculations.

1.4d. Transport Properties (TPROPS)

Molecular gas-phase properties are required in the analytical treatment of pulverized coal-char behavior. Module TPROPS performs these calculations whenever it is called. The calling routine must specify temperature, pressure, and gas-phase species concentrations; TPROPS returns the gas-mixture viscosity, thermal conductivity, specific heat capacity at constant pressure and, for each species, the enthalpy and mass diffusivity of that species in the mixture.

There are four different models available in TPROPS, depending on the technique utilized to compute these transport coefficients: (1) constant properties, usually values of air; (2) approximate expressions summarized below, where properties vary as a function of temperature only; (3) a "bifurcation" model[18] which requires only one summation over the number of species; and (4) a statistical-mechanics formulation in which properties depend on the concentration of species in the mixture as well as temperature and pressure. This model, in contrast with model 3, requires a double summation over the species. A still more general model exists[19,20] which could be incorporated as model 5; however, model 4 gives results to within 5% accuracy,[21] which is more than adequate compared to the accuracy of the overall treatment of coal in this generalized model.

Model 4 calculates the mixture properties from the Chapman–Enskog kinetic theory as shown in Chapters 2 and 12:

Gas-mixture viscosity:

$$\mu_g = \sum_i \left(x_i \mu_i \middle/ \sum_k x_k \phi_{ik} \right) \tag{46}$$

Gas-mixture conductivity:

$$k_g = \sum_i \left(x_i k_i \middle/ \sum_k x_k \phi_{ik} \right) \tag{47}$$

Mixture diffusivity:

$$D_{im} = (1 - x_i) \middle/ \sum_{k \neq i} (x_k/D_{ik}) \tag{48}$$

where the interaction parameter is given by

$$\phi_{ik} = (1/8)^{1/2} [1 + (M_i/M_k)]^{-1/2} [1 + (\mu_i/\mu_k)^{1/2} (M_k/M_i)^{1/4}]^2 \tag{49}$$

the species diffusivity is

$$D_{ik} = 0.0188 T^{3/2} [(1/M_i) + (1/M_k)]^{1/2} / (p\sigma_{ik}^2 \Omega_d) \qquad (50)$$

and the gas-species viscosity and conductivity are given by

$$\mu_i = (2.67 \times 10^{-6})(M_i T)^{1/2} / (\sigma_i^2 \Omega_\mu) \qquad (51)$$

and

$$k_i = [C_{p_i} + 1.25(R/M_i)]\mu_i \qquad (52)$$

respectively, with all units in SI. The thermodynamic data are obtained from standard curve-fits[11,22] for each species where usually the NASA thermo-chemical data[20,22] are utilized (see Chapter 1) and the functional form fitted as nondimensional properties as a function of temperature:

$$C_p/R = z_1 + z_2 T + z_3 T^2 + z_4 T^3 + z_5 T^4 \qquad (53)$$

$$h/RT = z_1 + z_2 T/2 + z_3 T^2/3 + z_4 T^3/4 + z_5 T^4/5 + z_6/T \qquad (54)$$

$$s^0/R = z_1 \log T + z_2 T + z_3 T^2/2 + z_4 T^3/3 + z_5 T^4/4 + z_7 \qquad (55)$$

The individual species enthalpy of Eq. (53) includes the standard-state enthalpy of formation and the sensible enthalpy:

$$h = h_{f_{298}}^0 + \int_{298\,K}^T C_p \, dT \qquad (56)$$

The individual species one-atmosphere entropy s^0 defined by Eq. (55) includes the low-temperature absolute entropy

$$s^0 = s_{f_0}^0 + \int_0^T (C_p/T) \, dT \qquad (57)$$

and Eq. (55) thus provides the temperature-dependent contribution to the total ideal-gas entropy at arbitrary pressures,

$$s = \left[s_{f_0}^0 + \int_0^T (C_p/T) \, dT \right] - RT \log(P/P_0) \qquad (58)$$

where P_0 is standard atmospheric pressure, 101.3 kPa.

In evaluating the gas-species viscosity and diffusivity, the collision diameters σ_i and σ_{ik} and collision functions Ω_μ and Ω_d are required in Eqs. (50) and (51). The collision diameter values are fixed for each species and are summarized in Appendix B[23]; however, the nondimensional collision functions require the specification of an intermolecular potential model in order to compute, and furthermore are slowly varying functions of the temperature. The Lennard-Jones "6-12" potential was utilized to curve-fit the collision functions of viscosity, conductivity, and diffusivity for nonpolar molecules, while for polar molecules the viscosity and conductivity were curve-fitted using the Stockmayer potential.[23] All were curve-fit as quadratic log–log functions over the three temperature intervals of the dimensionless temperature kT/ε tabulated by Bird *et al.*[21] and Reid and Sherwood.[23] See Appendix B for data.

The Lennard-Jones parameters for binary mixtures are estimated from

$$\sigma_{ik}=(\sigma_i+\sigma_k)/2 \tag{59}$$

$$\varepsilon_{ik}=(\varepsilon_i\varepsilon_k)^{1/2} \tag{60}$$

Transport coefficients determined by model 4 are accurate, but repeated evaluation can lead to long computation times. As a result, TPROPS has been coded to be as efficient as possible; model 2 may be utilized to compute coefficients much faster but less accurately. The following equations comprise model 2:

$$\mu_g=\mu_0 T^{0.6} \tag{61}$$

$$C_{pg}=C_1+C_2T+C_3T^2 \tag{62}$$

$$k_g=(C_{pg}+C_k)\mu_g \tag{63}$$

$$D_{im}=D_{fi}T^{1.6}/p \tag{64}$$

$$h_i=C_{pg_i}T+h_{fi} \tag{65}$$

where μ_0, C_1, C_2, C_3, C_k, D_{fi}, and h_{fi} are all constants determined by least-squares curve-fits performed on data generated by model 4. Program element TPROP0 performs these fits automatically, as well as other bookkeeping and initialization functions of module TPROPS.

It should be borne in mind that the generalized model is an elliptic formulation, and thus, these answers are only tentative and are updated at every iteration. Therefore, model 2 can be utilized until the last few final iterations, at which time the time-consuming model 4 can be used for most accurate results. Furthermore, these transport properties need not be reevaluated at every iteration.

1.4e. Thermal Radiation (QRAD)

Radiative heat transfer enters the equation set of the generalized model as a source/sink term, div $(\mathbf{q}_r)$, appearing in both the gas- and solid-phase energy equations. Here, $\mathbf{q}_r$ is the vector describing the net radiative heat flux in space, and in two-dimensional cylindrical coordinates is given by

$$\text{div }(\mathbf{q}_r)=\partial Q_x/\partial x+(1/r)[\partial(rQ_r)/\partial r] \tag{66}$$

The purpose of module QRAD is to compute this term. Following the recommended approach in Chapter 5, a flux method is utilized. Chu and Churchill[12] have reduced the general transport equation to a six-flux model comprising a set of differential equations which describe the steady-state scattering the absorption of electromagnetic radiation by dense dispersions of particles. Their flux formulation does include nonisotropic multiple-scattering.

In the present application in two dimensions, a four-flux model, as derived in Appendix C, can be obtained as a degenerate case of the six-flux representation:

$$dE_1/dx = -C_1E_1 + C_2E_2 + C_3(E_3 + E_4) + C_4 \tag{67}$$

$$dE_2/dx = C_1E_2 - C_2E_1 - C_3(E_2 + E_4) - C_4 \tag{68}$$

$$(1/r)(drE_3/dr) = -C_1E_3 + C_2E_4 + C_3(E_1 + E_2) + C_4 \tag{69}$$

$$(1/r)(drE_4/dr) = C_1E_4 - C_2E_3 - C_3(E_1 + E_2) - C_4 \tag{70}$$

where C_1, C_2, C_3, and C_4 are constant for each computational cell and are functions of the scattering, absorption, and emission coefficients of all particles contained in that particular cell. Fluxes E_1, E_2, E_3, and E_4 are the four components of the specific intensity and are chosen in the direction of the two orthogonal coordinate directions; i.e., E_1 and E_2 are fluxes in the positive and negative direction of the longitudinal axis of the reactor, respectively, and E_3 and E_4 are fluxes outward and inward, respectively, along the radial coordinate. The addition of particle emission to the Chu and Churchill[12] formulation is straightforward and is included in C_4. Thus the four-flux model consists of four simultaneous, first-order, differential equations, and except for the special cases treated by Chu and Churchill, must be solved numerically.

Gosman and Lockwood[7] have developed a solution technique to solve a similar set of radiation intensity equations, limited to isotropic scattering with constant extinction coefficients. There is some question concerning whether or not they are solving for Cartesian fluxes in a cylindrical coordinate system; however, their solution algorithm has been utilized since it employs the same TDMA technique chosen for the hydrodynamic equations. The theoretical formulation is based on the earlier Chu and Churchill paper.[12]

Three different treatments, in order of increasing computational complexity, are incorporated in module QRAD. The first two consider a two-flux model (i.e., one-dimensional) while option 3 is the general four-flux model. If the coefficients in the two-flux model,

$$dE_1/dx = -C_1E_1 + C_2E_2 \tag{71}$$

$$dE_2/dx = -C_1E_2 + C_2E_1 \tag{72}$$

are constant (option 1), an analytical solution is possible[12]:

$$E_1 = [e^{-px} - G^2 e^{-p(2T-x)}]/(1 - G^2 e^{-2pT}) \tag{73}$$

$$E_2 = G[e^{-pT} - e^{-p(2T-x)}]/(1 - G^2 e^{-2pT}) \tag{74}$$

where

$$p = (C_1^2 - C_2^2)^{1/2}$$

$$G = (C_1 + C_2 - p)/(C_1 + C_2 + p)$$

and T is the thickness of the medium.

Model 2 utilizes the general varying coefficients obtained from module COAL and requires a numerical solution of Eqs. (71) and (72). Model 3 is the solution of the four-flux model by the method proposed by Gosman and Lockwood.[7]

Following Gosman and Lockwood,[7] a net radiation flux Q and total flux F can be defined as

$$Q_x = E_1 - E_2, \qquad F_x = (E_1 + E_2)/2 \tag{75}$$

$$Q_r = E_3 - E_4, \qquad F_r = (E_3 + E_4)/2 \tag{76}$$

such that the four first-order flux equations in Eqs. (67)–(70) may be combined to yield two second-order equations:

$$(d/dx)[(\Gamma/2)(dF_x/dx)] - (C_2 - C_1)F_x - (2C_3F_r + C_4) = 0 \tag{77}$$

$$(1/r)(d/dr)[(\Gamma/2)(d/dr)(rF_r)] - (C_2 - C_1)F_r - (2C_3F_x + C_4) = 0 \tag{78}$$

In the present formulation the heat-transfer expression is

$$Q = (Q_x, Q_r) = 2\Gamma \operatorname{div}(F) \tag{79}$$

where $\Gamma = -1/(C_1 + C_2)$ is independent of the coordinate (r) and is therefore equal for both components, in contrast to the Gosman and Lockwood formulation.[7]

Integrating Eqs. (77) and (78) separately over the main computational cells gives the following finite-difference equations

$$(A_E + A_W - S_P^{F_x})F_{x_P} = A_E F_{x_E} + A_W F_{x_W} + S_U^{F_x} \tag{80}$$

$$(A_N + A_S - S_P^{F_r})F_{r_P} = A_N F_{r_N} + A_S F_{r_S} + S_U^{F_r} \tag{81}$$

which may be solved by the TDMA algorithm.

Various boundary conditions can be considered for these equations, including a specified wall temperature or specified wall heat flux.

The C's in Eqs. (67)–(70) are functions of the absorption and scattering coefficients which are given by (see Chapter 5)

$$a = \varepsilon(\pi/4) \sum_i n_i d_{p_i}^2 \tag{82}$$

$$s = (1 - \varepsilon)(\pi/4) \sum_i n_i d_{p_i}^2 \tag{83}$$

where ε is the particle surface emissivity and n_i and d_{pi} are the number density and diameter of the size class, respectively.

Each computational cell will have different absorption and scattering coefficients, which are determined by the various particle trajectories intercepting the cell. Instead of storing the different number densities and particle diameters throughout the entire flow domain, $\Sigma_i\, n_i\, d_p^2$ is accumulated and stored by module COAL along a trajectory computed by PSIC.

This technique of only saving the effects of a given process on the remaining equations follows from the Lagrangian treatment, and enables an enormous reduction in computer storage requirements. The required storage to completely couple the solid- and gas-phase equations are: one array each for mass, u-momentum, v-momentum, and energy; one array for radiation, and N arrays in the general case for species concentrations of the mass transfer from the coal to the gas, where N is the number of distinct species considered as products from pyrolysis and char oxidation.

Since the radiation flux equations are solved by TDMA in a Eulerian sense and the required coefficients of these flux equations are accumulated in the Lagrangian sense, with the solution of the solid-phase equation, module QRAD has been placed in the middle of Figure 2, signifying that the radiation solution involves both portions.

1.5. Solution Strategy

The basic strategy proposed in this generalized pulverized-coal combustion and gasification model is to employ different computational schemes, each appropriate to the nature of the particular conservation equations being solved, in order to solve all the complete conservation and transport equations with maximum efficiency. Furthermore, a modular approach is utilized, where each separate group of related conservation equations being solved by the same numerical algorithm is coded into a separate module completely independent of others, and linked only by the Fortran COMMON blocks.

This approach allows new and improved submodels to be tested by merely replacing a given module or by addition of new ones. In some applications, various modules may not be needed and may be eliminated in a straightforward fashion by simply removing that block of cards and "commenting" the FORTRAN call to these modules. Indeed, such a technique was utilized in the development of this generalized model and resulted in many useful entities such as: program FC, consisting of FLOW and CHEM, which will solve a gas-phase combustion problem; program FP, FLOW + PSIC, which is a fully coupled two-phase flow treatment; program FR, FLOW + QRAD, where various radiation models and parameters may be tested; and finally a "driver deck" to test and compute quantities of interest in CREK, TPROPS, and COAL.

All dependent variables and properties of interest are logically organized into FORTRAN-labeled COMMON blocks for use by any module. The fundamental concept of the model is that these latest determined values held in COMMON are to be utilized whenever they are required; for example, a given module will take as input the various distributions of these latest values which are required and update and store those variables for which that module is responsible. This convenient treatment follows from the fact that the equations are elliptic and that a great effort has been made to code the modules and their linkage to minimize stability problems. The particle equations in PSIC are parabolic but are reintegrated every time new trajectories are wanted after changing the field values of independent variables.

Figure 3 illustrates the strategy of the solution procedure. The MAIN program (see Figure 2) controls the frequency and order of calling each module comprising the generalized model. Thus the computational framework is such that all the modules are treated as "black boxes" which must perform three general tasks: (1) read and process (initialize) any data; (2) assign an initial distribution to its variables and properties (after all modules have processes their respective input data); and (3) perform solutions of the required transport equations to update the quantities required as output from that module.

Input Data. MAIN calls each module separately with a flag (in what follows, a "flag" will mean a logical variable which must be specified), ordering that data be read and processed. Some data are assigned in BLOCK DATA, others are assigned in the modules; what is important is that input data processing is completely up to the user; and since coding is specific to the module being utilized, only data pertinent to that module are processed. FLOW needs to be called before CHEM during data processing, since the reactants and chemical state at the inlet ports depend on the pressure and mass-flow rates at these ports, the quantities of which properly belong to module FLOW.

Initialization. Again, the computational framework is such that each module is called with a flag to signify the initialization of its variables, consistent with the physics contained in the module. The initialization is as follows: FLOW assigns the Hagen–Poiscuille flow distribution for the velocities u and v, assigns a constant specified reference pressure for P, and the specified inlet turbulence intensity and its dissipation rate are assigned for for k and ε over the entire field.

CHEM computes the nonreacting, equilibrium, and stirred-reactor kinetic

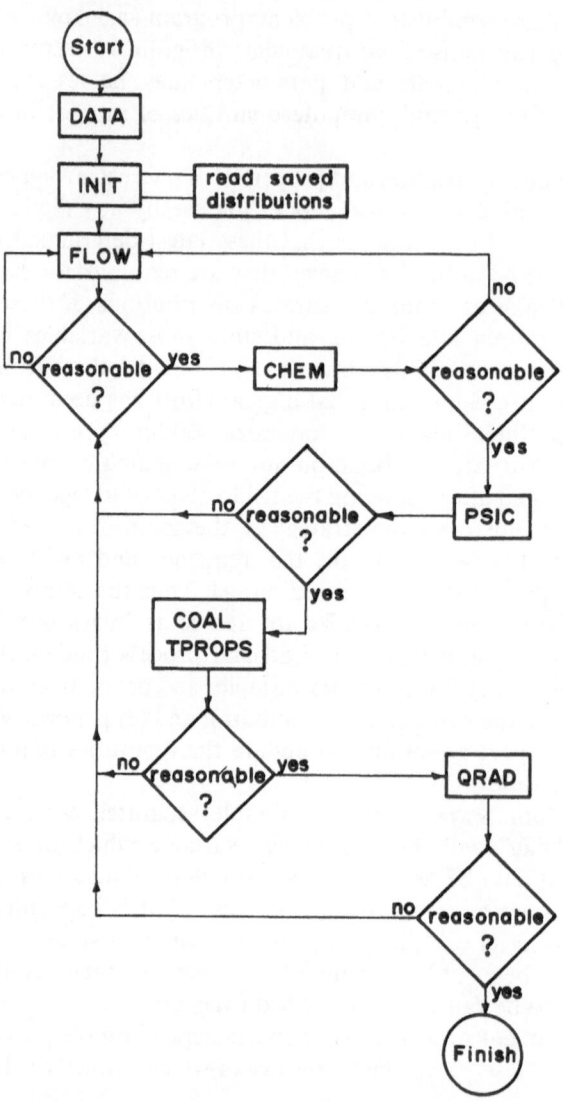

Figure 3. Flowchart illustrating the solution strategy.

solution corresponding to the total feed and total volume of the reactor, and assigns either one of these distributions of T and σ_k throughout the field depending on whether a cold, equilibrium, or kinetic solution is to be performed, respectively. If a PDF approach is considered, initial values of $\bar{f}$ and g are assigned and the curve-fits of $\int_{f_-}^{f_+} \sigma_i^*(f)\,df$ are obtained for the given feed.

QRAD assigns the analytic, two-flux solution for F_x and F_r throughout the field, based on some average scattering and absorption coefficients. PSIC, COAL, and TPROPS consist of parabolic and algebraic equations which do not require a distribution throughout the domain, but only initial conditions which are processed and assigned during the data-reading step.

The program has internal provision for saving answers from a given run to be utilized to restart another run at a future time. Modules FLOW, CHEM, and QRAD then have an option to read these saved values, followed by reassigning the inlet conditions, since they may be different.

General solution. First, the hydrodynamic solution is performed, where the velocity components, pressure, effective gas viscosity, turbulent energy, and dissipation equations are solved by module FLOW, utilizing the density (or mixture molecular weight and temperature) initially assigned and with no interaction with the solid coal particles. All the source terms describing the solid particle mass and momentum coupling have been initialized to zero by PSIC and remain at that value until PSIC is called. These hydrodynamic equations are solved one by one until a coarse convergence criterion is satisfied and a specified number (usually 30) of iterations are performed, returning control back to MAIN. These partial solutions are only tentative and will be updated repeatedly to account for the changes in the values of all the other variables. In the beginning, it is especially important not to spend much computational time in obtaining a solution, but rather to move all variables slowly in the direction of the complete solution. For this reason, the convergence criteria are relaxed and some of the modules are equipped with options allowing various levels of description and complexity to be employed in the beginning, to keep computational times down. A more subtle reason for employing approximate models in the beginning is because of stability problems; a few are discussed below.

Given a reasonably converged flow field, module CHEM may now be called to update the gas temperature (enthalpy), chemical species, mixture molecular weight, and fluid mass density. Two different techniques are possible with CHEM in the beginning, depending on whether the feed is premixed or nonpremixed. In the former case, the combustion or hot "upper-branch" solution (see Chapter 4, Section 4) is initially assigned corresponding to either a global equilibrium or kinetic solution, and the beginning iterations start directly computing the requested equilibrium or kinetic solution. Such a technique with the latter case of nonpremixed feeds leads to stability problems and slow convergence at first, which may be eliminated by the following method. An adiabatic, nonreacting solution is performed by CHEM for a specified number of iterations, which results in a more realistic distribution of fuel–air equivalence ratio in the absence of chemical reactions. An automatic switch then causes CHEM to obtain the other hot "upper-branch" (equilibrium or kinetic) solution corresponding to this "lower-branch" solution of nonpremixed combustion (see Chapter 4). From then on, CHEM at first seeks for an upper-branch combustion solution; if none exists, then the lower-branch solution is found. CHEM makes only one pass over the flow domain solving the thermochemical equations simultaneously before returning to MAIN.

The linkage of FLOW + CHEM is the most difficult portion of the generalized model. Many numerical stability problems were encountered and solved in an earlier application[9,10] of FLOW and CHEM combination. The complete gas-phase solution is obtained by superiterating between FLOW and CHEM until a stationary state is reached.

A call to PSIC determines the trajectories of the coal particles and the amount of coupling between the phases. At first, only the momentum coupling will be considered by calling (PSIC-COAL) (i.e., PSIC without calling COAL) and FLOW. This coupling is usually weak; however, if there exists a large velocity difference between the coal and gas in the inlet ports, a few iterations between FLOW and PSIC, and possibly with CHEM, may be required. At this time, FLOW + CHEM + PSIC + COAL may be attempted. Model 2 in TPROPS should be utilized, along with fast approximate techniques built into PSIC. If the solution is converging and the results seem encouraging, QRAD may be added to the iteration loop. Finally, the most accurate formulations of each module will be utilized and a final stationary state satisfying all equations and all boundary conditions is obtained.

1.6. Three-Dimensional Time-Dependent Analysis

Extensions of the generalized model to a time-dependent three-dimensional formulation is straightforward and consists of extending the various modules.

Many methods are available[3] for solving three-dimensional and/or time-dependent formulations of the hydrodynamic equations. The author has derived[9] the unsteady three-dimensional form of the Eulerian equations now utilized in FLOW and CHEM. The unsteadiness was treated as essentially a succession of steady states, allowing the same solution algorithm to be utilized as now coded. However, the reader should choose one of the many methods described by Roache[3] for transient, multidimensional problems. This choice should be based on the familiarity of the user's research team with the various methods available.

Extending PSIC to three dimensions consists of adding the third-component particle velocity and interpolating the three-dimensional, gas-phase velocities. For a given time step, FLOW + CHEM are solved, followed by PSIC computing the movement of all the coal particles during this time step along trajectories. The number of new trajectories to be computed depends on the unsteadiness of the feed and gas-phase solution.

Extending QRAD to a six-flux method is straightforward, and unsteadiness requires the scattering and absorptions coefficients, as well as boundary conditions, to be updated during every time step. COAL and TPROPS are identical for any dimensionality, steady or nonsteady.

The generalized model can, in principle, be easily extended to an unsteady three-dimensional formulation, but would require much more computation time, as well as more complexity with respect to presentation of results.

2. Notation

A	Total coefficient (convection plus diffusion) (kg s^{-1})	p	Pressure [Pa (N m^{-2})]
A_j	Area of coal particle (m^2)	$P(f)$	Probability density function
B	Blowing parameter	Pr	Prandtl number
B	Arrhenius exponent-on-ten (arbitrary)	q, Q	Heat-transfer rate (J m^{-3} s^{-1})
C_D	Particle aerodynamic drag coefficient	r	Radial coordinate (m)
		r	Mass transfer from particle (kg s^{-1})
C_P	Specific heat at constant pressure (J kmol^{-1} K^{-1})	r	Chemical species destruction rate (kmol m^{-3} s^{-1})
d	Dissipation rate of turbulence energy (m^2 s^{-3})	R	Reaction rate (kmol m^{-3} s^{-1})
		Re	Reynolds number
d	Particle diameter (m)	s	Entropy (J kmol^{-1} K^{-1})
D_{im}	Molecular diffusivity in mixture (m^2 s^{-1})	S^ϕ	Gas source term for variable ϕ (variable)
E_i	Radiation flux (J m^{-2} s^{-1})	$_pS^\phi$	Particle source term for ϕ (variable)
F	Radiation flux sum $F_x=(E_1+E_2)/2$ (J m^{-2} s^{-1})	Sc	Schmidt number
$\bar{f}$	Time-mean mass fraction of fuel	T	Temperature (K)
g	Time-mean value of square of difference of instantaneous f from time-mean f	T	Thickness (m)
		T	Activation temperature (K)
		u	Axial velocity (m s^{-1})
$\mathbf{g}$	Gravitational vector (m s^{-2})	v	Radial velocity (m s^{-1})
h	Static intensive enthalpy (J kg^{-1})	$\mathbf{v}$	Velocity vector (m s^{-1})
h^0_{fi}	Heat of formation (J kmol^{-1})	x	Axial coordinate (m)
l_p	Particle internal energy (J kg^{-1})	x	Mole fraction
JJ	Total number of reactions	X	Particle mass entering port
k	Thermal conductivity (J m^{-1} s^{-1} K^{-1})	Y	Particle mass with diameter d
		z	Numerical coefficients (variable)
m	Mass (kg)	σ_k	Mole numbers of species k (kmol kg^{-1})
M	Molecular weight (kg kmol^{-1})	σ_m	Reciprocal of mixture mean molecular weight (kmol kg^{-1})
n	Particles per trajectory		
N	Arrhenius exponent-on-temperature (variable)	σ_s	Turbulent Schmidt number
		ϕ	General variable
		ϕ_{ij}	Mixture viscosity parameter
Nu	Nusselt number	Φ	Viscous dissipation function (kg m^{-1} s^{-3})

Subscripts

d	diffusion
d	direction (N, S, E, W) from node point P in grid
g	gas
	normal to boundary
P	particle
p	node point in computational grid
r	radial
s	surface
x	axial

μ	viscosity
0	without blowing

Superscripts

*	equilibrium values
h	enthalpy
m	mass
u	axial velocity
v	radial velocity
σ_k	species k
ϕ	general variable

3. References

1. D. A. Bueters, J. G. Cogoli, and W W. Habelt, Performance prediction of tangentially fired utility furnaces by computer model, in *Fifteenth Symposium (International) on Combustion*, pp. 1245–1260, The Combustion Institute, Pittsburgh, Pa. (1974).
2. A. Lowe, T. F. Wall, and I. McC. Stewart, A zoned heat transfer model of a large tangentially fired pulverized coal boiler, in *Fifteenth Symposium (International) on Combustion*, pp. 1261–1270, The Combustion Institute, Pittsburgh, Pa. (1974).
3. P. J. Roache, *Computational Fluid Dynamics*, Hermosa Publishers, Albuquerque, New Mexico (1972).
4. S. V. Patankar and D. B. Spalding, A calculation procedure for heat, mass and momentum transfer in three-dimensional parabolic flows, *Int. J. Heat Mass Transfer* **15**, 1787 (1972).
5. A. D. Gosman, W. M. Pun, A. K. Runchal, D. B. Spalding, and M. Wolfshtein, *Heat and Mass Transfer in Recirculating Flows*, Academic Press, New York (1969).
6. A. D. Gosman and W. M. Pun, *Lecture Notes for Course Entitled "Calculation of Recirculating Flows,"* Imperial College of Science and Technology, London, Report NO. HTS/74/2 (1973).
7. A. D. Gosman and F. C. Lockwood, Incorporation of a flux model for radiation into a finite-difference procedure for furnace calculations, in *Fourteenth Symposium (International) on Combustion*, pp. 661–671, The Combustion Institute, Pittsburgh, Pa. (1972).
8. C. T. Crowe, M. P. Sharma, and D. E. Stock, *The Particle-Source-in-Cell (PSI-CELL) Model for Gas-Droplet Flows*, ASME 75-WA/HT-25 (1975).
9. J. J. Wormeck, Computer Modeling of Turbulent Combustion in a Longwell Jet-Stirred Reactor, Ph.D. Dissertation, Washington State University, Pullman, Wash. (1976).
10. J. J. Wormeck and D. T. Pratt, Computer modeling of turbulent combustion in a Longwell jet-stirred reactor, in *Sixteenth Symposium (International) on Combustion*, pp. 1583–1592, The Combustion Institute, Pittsburgh, Pa. (1976).
11. D. T. Pratt and J. J. Wormeck, *CREK: A Computer Program for Calculation of Combustion Reaction Equilibrium and Kinetics in Laminar or Turbulent Flows*, Washington State University, Pullman, Wash., Report No. WSU-ME-TEL-76-1 (1976).
12. C. M. Chu and S. W. Churchill, Numerical solution of problems in multiple scattering of electromagnetic radiation, *J. Phys. Chem.* **59**, 855–863 (1955).
13. B. E. Launder and D. B. Spalding, *Mathematical Models of Turbulence*, Academic Press, New York (1972).
14. G. L. Mellor and H. J. Herring, *A Study of Turbulent Boundary Layer Models, Part I, Mean Velocity Field Closure; and Part II, Mean Turbulent Field Closure*, Princeton University, Report No. 914 (1970).

15. J. J. Wormeck and D. T. Pratt, Computation of Turbulent Reacting Flows: Comparison of Various Mixing/Chemistry Interaction Models (1978), to be published.

16. C. T. Crowe, private communication (1977).

17. C. T. Crowe, A Computational Model for the Gas-Droplet Flow Field in the Vicinity of an Atomizer, Paper No. 74–23, Western States Section Meeting, The Combustion Institute, Seattle, Wash. (1974).

18. E. P. Bartlett, R. M. Kendall, and R. A. Rindal, *An Analysis of the Coupled Chemically React-ing Boundary Layer and the Charring Ablator, Part IV, A Unified Approximation for Mixture Transport Properties for Multicomponent Boundary Layer Applications*, NASA CR-1063, Itec Corp., Vedya Division, Palo Alto, CA (1966).

19. J. O. Hirshfelder, C. F. Curtis, and R. B. Bird, *Molecular Theory of Gases and Liquids*, John Wiley and Sons, New York (1954).

20. R. A. Svehla and B. J. McBride, *Fortran IV Computer Programs for Calculation of Thermo-dynamic and Transport Properties of Complex Chemical Systems*, Report NASA TN D-7056 (1973).

21. R. B. Bird, W. E. Stewart, and E. N. Lightfoot, *Transport Phenomena*, John Wiley and Sons, New York (1960).

22. S. Gordon and B. J. McBride, *Computer Program for Calculation of Complex Chemical Equilibrium Compositions, Rocket Performance, Incident and Reflected Shocks and Chapman–Jouquet Detonations*, Report NASA SP-273 (1971).

23. R. C. Reid and T. K. Sherwood, *The Properties of Gases and Liquids-Their Estimation and Correlation* McGraw-Hill Book Co., 2nd edition, New York (1966).

Part V
Appendixes

Conversion Factors

The following table expresses the definitions of miscellaneous units of measure as exact numerical multiples of coherent SI units, and provides multiplying factors for converting numbers and miscellaneous units to corresponding new numbers and SI units.

The digits of each numerical entry following E represent a power of 10. An asterisk preceding each number expresses an exact definition. For example, the entry "*2.54E-2" expresses the fact that 1 inch $= 2.54 \times 10^{-2}$ meter, exactly, by definition. Most of the definitions are extracted from National Bureau of Standards documents. Numbers not preceded by an asterisk are only approximate representations of definitions, or are the results of physical measurements.

This appendix was abstracted from *The International Systems of Units—Physical Constants and Conversion Factors*, E.A. Mechtly, Second Revision, NASA SP-7012, Washington, D.C. (1973). Permission to use this material was obtained from the Scientific and Technical Information Office, NASA, Washington, D.C.

Table A.1

To convert from:	to:	multiply by:
atmosphere	newton/meter^2	*1.013E5
bar	newton/meter^2	*1.00E5
British thermal unit (mean)	joule	1.05587E3
British thermal unit (thermochemical)	joule	1.054350E3
British thermal unit (39°F)	joule	1.05967E3
British thermal unit (60°F)	joule	1.05468E3
calorie (International Steam Table)	joule	4.1868
calorie (mean)	joule	4.19002
calorie (thermochemical)	joule	*4.184
calorie (15°C)	joule	4.18580
calorie (20°C)	joule	4.18190
calorie (kilogram, International Steam Table)	joule	4.1868E3

continued overleaf

Table A.1 (*Continued*)

To convert from:	to:	multiply by:
calorie (kilogram, mean)	joule	4.19002E3
calorie (kilogram, thermochemical)	joule	*4.184E3
Celsius (temperature)	kelvin	$t_K = t_C + 273.15$
centimeter of mercury (0°C)	newton/meter^2	1.33322E3
centimeter of water (4°C)	newton/meter^2	9.80638E1
electron volt	joule	1.6021917E-19
erg	joule	*1.00E-7
Fahrenheit (temperature)	kelvin	$t_K = (5/9)(t_F + 459.67)$
Fahrenheit (temperature)	Celsius	$t_C = (5/9)(t_F\text{-}32)$
fluid ounce (U.S.)	meter^3	*2.95735295625E5
foot	meter	*3.048E-1
foot of water (39.2°F)	newton/meter^2	2.98898E3
gallon (U.S. dry)	meter^3	*4.40488377086E-3
gallon (U.S. liquid)	meter^3	*3.785411784E-3
horsepower (550 foot lbf/second)	watt	7.4569987E2
inch	meter	*2.54E-2
kilocalorie (International Steam Table)	joule	4.1868E3
kilocalorie (mean)	joule	4.19002E3
kilocalorie (thermochemical)	joule	4.184E3
kilogram mass	kilogram	*1.00
kilogram force (kgf)	newton	*9.80665
lbf (pound force, avoirdupois)	newton	*4.4482216152605
lbm (pound mass, avoirdupois)	kilogram	*4.5359237E-1
liter	meter^3	*1.00E-3
micron	meter	*1.00E-6
mile (U.S. statute)	meter	*1.609344E3
pascal	newton/meter^2	*1.00
poise	newton second/meter^2	*1.00E-1
pound force (lbf avoirdupois)	newton	*4.4482216152605
pound mass (lbm avoirdupois)	kilogram	*4.5359237E-1
quart (U.S. dry)	meter^3	*1.101220942715E-3
quart (U.S. liquid)	meter^3	9.4635925E-4
Rankine (temperature)	kelvin	$t_K = (5/9)t_R$
slug	kilogram	1.45939029E1
ton (long)	kilogram	*1.0160469088E3
ton (metric)	kilogram	*1.00E3
ton (short, 2000 lb)	kilogram	*9.0718474E2
Torr (0°C)	newton/meter^2	1.33322E2

Physical Parameters for Prediction of Transport Coefficients

Contents

Table B.1. Intermolecular Force Parameters and Critical Properties[a]

Substance	Molecular weight M	Lennard-Jones parameters[b]		Critical constants[c]		
		σ (Å)	ε/K (K)	T_c (K)	p_c (atm)	$V_c \times 10^3$ (m³ kmol⁻¹)
		Light elements				
H_2	2.016	2.915	38.0	33.3	12.80	65.0
He	4.003	2.576	10.2	5.26	2.26	57.8

continued overleaf

Table B.1 (Continued)

Substance	Molecular weight M	Lennard-Jones parameters[b]		Critical constants[c]		
		σ (Å)	ε/K (K)	T_c (K)	p_c (atm)	$V_c \times 10^3$ (m^3 kmol^{-1})
Noble gases						
Ne	20.183	2.789	35.7	44.5	26.9	41.7
Ar	39.944	3.418	124.	151.	48.0	75.2
Kr	83.80	3.498	225.	209.4	54.3	92.2
Xe	131.3	4.055	229.	289.8	58.0	118.8
Simple polyatomic substances						
Air	28.97[d]	3.617	97.0	132.[d]	36.4[d]	86.6[d]
N_2	28.02	3.681	91.5	126.2	33.5	90.1
O_2	32.00	3.433	113.	154.4	49.7	74.4
O_3	48.00	—	—	268.	67.	89.4
CO	28.01	3.590	110.	133.	34.5	93.1
CO_2	44.01	3.996	190.	304.2	72.9	94.0
NO	30.01	3.470	119.	180.	64.	57.
N_2O	44.02	3.879	220.	309.7	71.7	96.3
SO_2	64.07	4.290	252.	430.7	77.8	122.
F_2	38.00	3.653	112.	—	—	—
Cl_2	70.91	4.115	357.	417.	76.1	124.
Br_2	159.83	4.268	520.	584.	102.	144.
I_2	253.82	4.982	550.	800.	—	—
Hydrocarbons						
CH_4	16.04	3.822	137.	190.7	45.8	99.3
C_2H_2	26.04	4.221	185.	309.5	61.6	113.
C_2H_4	28.05	4.232	205.	282.4	50.0	124.
C_2H_6	30.07	4.418	230.	305.4	48.2	148.
C_3H_6	42.08	—	—	365.0	45.5	181.
C_3H_8	44.09	5.061	254.	370.0	42.0	200.
$n\text{-}C_4H_{10}$	58.12	—	—	425.2	37.5	255.
$i\text{-}C_4H_{10}$	58.12	5.341	313.	408.1	36.0	263.
$n\text{-}C_5H_{12}$	72.15	5.769	345.	469.8	33.3	311.
$n\text{-}C_6H_{14}$	86.17	5.909	413.	507.9	29.9	368.
$n\text{-}C_7H_{16}$	100.20	—	—	540.2	27.0	426.
$n\text{-}C_8H_{18}$	114.22	7.451	320.	569.4	24.6	485.
$n\text{-}C_9H_{20}$	128.25	—	—	595.0	22.5	543.
Cyclohexane	84.16	6.093	324.	553.	40.0	308.
C_6H_6	78.11	5.270	440.	562.6	48.6	260.

continued

Table B.1 (Continued)

	Molecular weight M	Lennard-Jones parameters[b]		Critical constants[c]		
Substance		σ (Å)	ε/K (K)	T_c (K)	p_c (atm)	$V_c \times 10^3$ (m^3 kmol^{-1})
		Other organic compounds				
CH_4	16.04	3.822	137.	190.7	45.8	99.3
CH_3Cl	50.49	3.375	855.	416.3	65.9	143.
CH_2Cl_2	84.94	4.759	406.	510.0	60.	—
$CHCl_3$	119.39	5.430	327.	536.6	54.	240.
CCl_4	153.84	5.881	327.	556.4	45.0	276.
C_2N_2	52.04	4.38	339.	400.	59.	—
COS	60.08	4.13	335.	378.	61.	—
CS_2	76.14	4.438	488.	552.	78.	170.

[a]Table used with permission from Bird *et al.*[1]
[b]Values of σ and ε/K are from J. O. Hirschfelder, C. F. Curtiss, and R. B. Bird, *Molecular Theory of Gases and Liquids*, John Wiley and Sons, New York (1954). The above values are computed from viscosity data and are applicable for temperatures above 100 K.
[c]Values of T_c, p_c, and V_c are from K. A. Kobe and R. E. Lynn, Jr., *Chem. Rev.* **52**, 117–236 (1952); and *American Petroleum Institute Research Project*, Volume 44, (F. D. Rossini, ed.), Carnegie Institute of Technology (1952).
[d]For air, the molecular weight M and the pseudocritical properties T_c, p_c, and V_c have been calculated from the average composition of dry air, as given in *International Critical Tables*, Volume I, p. 393 (1926).

Table B.2. Lennard-Jones Potentials as Determined from Viscosity Data[a]

Molecule	Compound	$b_0 \times 10^3$ (m^3 kmol^{-1})	σ(Å)	ε/k (K)
A	Argon	46.08	3.542	93.3
He	Helium	20.95	2.551[c]	10.22
Kr	Krypton	61.62	3.655	178.9
Ne	Neon	28.30	2.820	32.8
Xe	Xenon	83.66	4.047	231.0
Air	Air	64.50	3.711	78.6
AsH_3	Arsine	89.88	4.145	259.8
BCl_3	Boron chloride	170.1	5.127	337.7
BF_3	Boron fluoride	93.35	4.198	186.3
$B(OCH_3)_3$	Methyl borate	210.3	5.503	396.7
Br_2	Bromine	100.1	4.296	507.9
CCl_4	Carbon tetrachloride	265.5	5.947	322.7
CF_4	Carbon tetrafluoride	127.9	4.662	134.0
$CHCl_3$	Chloroform	197.5	5.389	340.2
CH_2Cl_2	Methylene chloride	148.3	4.898	356.3
CH_3Br	Methyl bromide	88.14	4.118	449.2
CH_3Cl	Methyl chloride	92.31	4.182	350
CH_3OH	Methanol	60.17	3.626	481.8
CH_4	Methane	66.98	3.758	148.6
CO	Carbon monoxide	63.41	3.690	91.7

continued overleaf

Table B.2 (Continued)

Molecule	Compound	$b_0 \times 10^3$ (m^3 kmol^{-1})	σ(Å)	ε/k (K)
COS	Carbonyl sulfide	88.91	4.130	336.0
CO$_2$	Carbon dioxide	77.25	3.941	195.2
CS$_2$	Carbon disulfide	113.7	4.483	467
C$_2$H$_2$	Acetylene	82.79	4.033	231.8
C$_2$H$_4$	Ethylene	91.06	4.163	224.7
C$_2$H$_6$	Ethane	110.7	4.443	215.7
C$_2$H$_5$Cl	Ethyl chloride	148.3	4.898	300
C$_2$H$_5$OH	Ethanol	117.3	4.530	362.6
C$_2$N$_2$	Cyanogen	104.7	4.361	348.6
CH$_3$OCH$_3$	Methyl ether	100.9	4.307	395.0
CH$_2$CHCH$_3$	Propylene	129.2	4.678	298.9
CH$_3$CCH	Methyl acetylene	136.2	4.761	251.8
C$_3$H$_6$	Cyclopropane	140.2	4.807	248.9
C$_3$H$_8$	Propane	169.2	5.118	237.1
n-C$_3$H$_7$OH	n-Propyl alcohol	118.8	4.549	576.7
CH$_3$COCH$_3$	Acetone	122.8	4.600	560.2
CH$_3$COOCH$_3$	Methyl acetate	151.8	4.936	469.8
n-C$_4$H$_{10}$	n-Butane	130.0	4.687	531.4
iso-C$_4$H$_{10}$	Isobutane	185.6	5.278	330.1
C$_2$H$_5$OC$_2$H$_5$	Ethyl ether	231.0	5.678	313.8
CH$_3$COOC$_2$H$_5$	Ethyl acetate	178.0	5.205	521.3
n-C$_5$H$_{12}$	n-Pentane	244.2	5.784	341.1
C(CH$_3$)$_4$	2,2-Dimethyl propane	340.9	6.464	193.4
C$_6$H$_6$	Benzene	193.2	5.349	412.3
C$_6$H$_{12}$	Cyclohexane	298.2	6.182	297.1
n-C$_6$H$_{14}$	n-Hexane	265.7	5.949	399.3
Cl$_2$	Chlorine	94.65	4.217	316.0
F$_2$	Fluorine	47.75	3.357	112.6
HBr	Hydrogen bromide	47.58	3.353	449
HCN	Hydrogen cyanide	60.37	3.630	569.1
HCl	Hydrogen chloride	46.98	3.339	344.7
HF	Hydrogen fluoride	39.37	3.148	330
HI	Hydrogen iodide	94.24	4.211	288.7
H$_2$	Hydrogen	28.51	2.827	59.7
H$_2$O	Water	23.25	2.641	809.1
H$_2$O$_2$	Hydrogen peroxide	93.24	4.196	289.3
H$_2$S	Hydrogen sulfide	60.02	3.623	301.1
Hg	Mercury	33.03	2.969	750
HgBr$_2$	Mercuric bromide	165.5	5.080	686.2
HgCl$_2$	Mercuric chloride	118.9	4.550	750
HgI$_2$	Mercuric iodide	224.6	5.625	695.6
I$_2$	Iodine	173.4	5.160	474.2
NH$_3$	Ammonia	30.78	2.900	558.3
NO	Nitric oxide	53.74	3.492	116.7
NOCl	Nitrosyl chloride	87.75	4.112	395.3

continued

Table B.2 (Continued)

Molecule	Compound	$b_0 \times 10^3$ (m³ kmol⁻¹)	σ(Å)	ε/k (K)
N_2	Nitrogen	69.14	3.798	71.4
N_2O	Nitrous oxide	70.80	3.828	232.4
O_2	Oxygen	52.60	3.467	106.7
PH_3	Phosphine	79.63	3.981	251.5
SF_6	Sulfur hexafluoride	170.2	5.128	222.1
SO_2	Sulfur dioxide	87.75	4.112	335.4
SiF_4	Silicon tetrafluoride	146.7	4.880	171.9
SiH_4	Silicon hydride	85.97	4.084	207.6
$SnBr_4$	Stannic bromide	329.0	6.388	563.7
UF_6	Uranium hexafluoride	268.1	5.967	236.8

[a]R. A. Svehla, NASA Technical Report R-132, Lewis Research Center, Cleveland, Ohio (1962); table used with permission from Reid and Sherwood.[2]
[b]$b_0 = \frac{2}{3}\pi N_0 \sigma^3$, where N_0 is Avogadro's number.
[c]The potential σ was determined by quantum mechanical formulas.

Table B.3.   Stockmayer-Potential Parameters[a]

Molecule	Dipole moment μ (debyes)	σ (Å)	ε/k (K)	$b_0 \times 10^3$ (m³ kmol⁻¹)	δ_{max}
H_2O	1.85	2.52	775	20.2	1.0
NH_3	1.47	3.15	358	39.5	0.7
HCl	1.08	3.36	328	47.8	0.34
HBr	0.80	3.41	417	50.0	0.14
HI	0.42	4.13	313	88.9	0.029
SO_2	1.63	4.04	347	83.2	0.42
H_2S	0.92	3.49	343	53.6	0.21
NOCl	1.83	3.53	690	55.5	0.4
$CHCl_3$	1.013	5.31	355	189	0.07
CH_2Cl_2	1.57	4.52	483	117	0.2
CH_3Cl	1.87	3.94	414	77.2	0.5
CH_3Br	1.80	4.25	382	96.9	0.4
C_2H_5Cl	2.03	4.45	423	111	0.4
CH_3OH	1.70	3.69	417	63.4	0.5
C_2H_5OH	1.69	4.31	431	101	0.3
n-C_3H_7OH	1.69	4.71	495	132	0.2
i-C_3H_7OH	1.69	4.64	518	126	0.2
$(CH_3)_2O$	1.30	4.21	432	94.2	0.19
$(C_2H_5)_2O$	1.15	5.49	362	209	0.08
$(CH_3)_2CO$[b]	1.20	3.82	428	70.1	1.3

continued overleaf

Table B.3 (Continued)

Molecule	Dipole movement μ (debyes)	σ (Å)	ε/k (K)	$b_0 \times 10^3$ (m³ kmol⁻¹)	δ^{max}
CH_3COOCH_3	1.72	5.04	418	162	0.2
$CH_3COOC_2H_5$	1.78	5.24	499	182	0.16
$CH_3NO_2{}^b$	2.15	4.16	290	90.8	2.3

[a]L. Monchick and E. A. Mason. *J. Chem. Phys.* **35**. 1676 (1961); table used with permission from Reid and Sherwood.[2]
[b]From G. A. Bottomley and T. H. Spurling. *Austral. J. Chem.* **16**, 1 (1963). Monchick and Mason show that $\sigma = 4.50$ A; $\varepsilon/k = 549$ K.

Table B.4. *Values of the Collision Integral Ω_v for Viscosity Based on the Lennard-Jones Potential[a]*

$T^* = kT/\varepsilon^b$	$\Omega_v{}^b$	kT/ε	Ω_v	kT/ε	Ω_v
0.30	2.785	1.65	1.264	4.0	0.9700
0.35	2.628	1.70	1.248	4.1	0.9649
0.40	2.492	1.75	1.234	4.2	0.9600
0.45	2.368	1.80	1.221	4.3	0.9553
0.50	2.257	1.85	1.209	4.4	0.9507
0.55	2.156	1.90	1.197	4.5	0.9464
0.60	2.065	1.95	1.186	4.6	0.9422
0.65	1.982	2.00	1.175	4.7	0.9382
0.70	1.908	2.1	1.156	4.8	0.9343
0.75	1.841	2.2	1.138	4.9	0.9305
0.80	1.780	2.3	1.122	5.0	0.9269
0.85	1.725	2.4	1.107	6.0	0.8963
0.90	1.675	2.5	1.093	7.0	0.8727
0.95	1.629	2.6	1.081	8.0	0.8538
1.00	1.587	2.7	1.069	9.0	0.8379
1.05	1.549	2.8	1.058	10	0.8242
1.10	1.514	2.9	1.048	20	0.7432
1.15	1.482	3.0	1.039	30	0.7005
1.20	1.452	3.1	1.030	40	0.6718
1.25	1.424	3.2	1.022	50	0.6504
1.30	1.399	3.3	1.014	60	0.6335
1.35	1.375	3.4	1.007	70	0.6194
1.40	1.353	3.5	0.9999	80	0.6076
1.45	1.333	3.6	0.9932	90	0.5973

continued

Table B.4 (Continued)

$T^* = kT/\varepsilon^b$	$\Omega_v{}^b$	kT/ε	Ω_v	kT/ε	Ω_v
1.50	1.314	3.7	0.9870	100	0.5882
1.55	1.296	3.8	0.9811	200	0.5320
1.60	1.279	3.9	0.9755	400	0.4811

[a]Table used with permission from Reid and Sherwood.[2]
[b]Hirschfelder, Curtiss, and Bird[3] use the symbol T^* for kT/ε and $\Omega^{(2,2)*}$ for Ω_v. Bromley and Wilke[4] use $(kT/\varepsilon)^{1/2}V/W^2(2)$ for $f_1(kT/\varepsilon)$. More complete tables of these functions are given in the two references cited.

Table B.5. Values of the Collision Integral Ω_D Based on the Lennard-Jones Potential[a]

kT/ε^b	$\Omega_D{}^b$	kT/ε	Ω_D	kT/ε	Ω_D
0.30	2.662	1.65	1.153	4.0	0.8836
0.35	2.476	1.70	1.140	4.1	0.8788
0.40	2.318	1.75	1.128	4.2	0.8740
0.45	2.184	1.80	1.116	4.3	0.8694
0.50	2.066	1.85	1.105	4.4	0.8652
0.55	1.966	1.90	1.094	4.5	0.8610
0.60	1.877	1.95	1.084	4.6	0.8568
0.65	1.798	2.00	1.075	4.7	0.8530
0.70	1.729	2.1	1.057	4.8	0.8492
0.75	1.667	2.2	1.041	4.9	0.8456
0.80	1.612	2.3	1.026	5.0	0.8422
0.85	1.562	2.4	1.012	6	0.8124
0.90	1.517	2.5	0.9996	7	0.7896
0.95	1.476	2.6	0.9878	8	0.7712
1.00	1.439	2.7	0.9770	9	0.7556
1.05	1.406	2.8	0.9672	10	0.7424
1.10	1.375	2.9	0.9576	20	0.6640
1.15	1.346	3.0	0.9490	30	0.6232
1.20	1.320	3.1	0.9406	40	0.5960
1.25	1.296	3.2	0.9328	50	0.5756
1.30	1.273	3.3	0.9256	60	0.5596
1.35	1.253	3.4	0.9186	70	0.5464
1.40	1.233	3.5	0.9120	80	0.5352
1.45	1.215	3.6	0.9058	90	0.5256
1.50	1.198	3.7	0.8998	100	0.5130
1.55	1.182	3.8	0.8942	200	0.4644
1.60	1.167	3.9	0.8888	400	0.4170

[a]From J. O. Hirschfelder, C. F. Curtiss, and R. B. Bird, "*Molecular Theory of Gases and Liquids*," John Wiley and Sons, Inc. New York (1954); table used with permission of Reid and Sherwood.[2]
[b]Hirschfelder used the symbols T^* for kT/ε and $\Omega^{(1,1)*}$ in place of Ω_D.

Table B.6.   Collision Integrals Ω_v for Viscosity as Calculated by Use of the Stockmayer Potential[a,b]

T^* \ δ	0^c	0.25	0.50	0.75	1.0	1.5	2.0	2.5
0.1	4.1005	4.266	4.833	5.742	6.729	8.624	10.34	11.89
0.2	3.2626	3.305	3.516	3.914	4.433	5.570	6.637	7.618
0.3	2.8399	2.836	2.936	3.168	3.511	4.329	5.126	5.874
0.4	2.5310	2.522	2.586	2.749	3.004	3.640	4.282	4.895
0.5	2.2837	2.277	2.329	2.460	2.665	3.187	3.727	4.249
0.6	2.0838	2.081	2.130	2.243	2.417	2.862	3.329	3.786
0.7	1.9220	1.924	1.970	2.072	2.225	2.614	3.028	3.435
0.8	1.7902	1.795	1.840	1.934	2.070	2.417	2.788	3.156
0.9	1.6823	1.689	1.733	1.820	1.944	2.258	2.596	2.933
1.0	1.5929	1.601	1.644	1.725	1.838	2.124	2.435	2.746
1.2	1.4551	1.465	1.504	1.574	1.670	1.913	2.181	2.451
1.4	1.3551	1.365	1.400	1.461	1.544	1.754	1.989	2.228
1.6	1.2800	1.289	1.321	1.374	1.447	1.630	1.838	2.053
1.8	1.2219	1.231	1.259	1.306	1.370	1.532	1.718	1.912
2.0	1.1757	1.184	1.209	1.251	1.307	1.451	1.618	1.795
2.5	1.0933	1.100	1.119	1.150	1.193	1.304	1.435	1.578
3.0	1.0388	1.044	1.059	1.083	1.117	1.204	1.310	1.428
3.5	0.99963	1.004	1.016	1.035	1.062	1.133	1.220	1.319
4.0	0.96988	0.9732	0.9830	0.9991	1.021	1.079	1.153	1.236
5.0	0.92676	0.9291	0.9360	0.9473	0.9628	1.005	1.058	1.121
6.0	0.98616	0.8979	0.9030	0.9114	0.9230	0.9545	0.9955	1.044
7.0	0.87272	0.8741	0.8780	0.8845	0.8935	0.9181	0.9505	0.9893
8.0	0.85379	0.8549	0.8580	0.8632	0.8703	0.8901	0.9164	0.9482
9.0	0.83795	0.8388	0.8414	0.8456	0.8515	0.8678	0.8895	0.9160
10.0	0.82435	0.8251	0.8273	0.8308	0.8356	0.8493	0.8676	0.8901
12.0	0.80184	0.8024	0.8039	0.8065	0.8101	0.8201	0.8337	0.8504
14.0	0.78363	0.7840	0.7852	0.7872	0.7899	0.7976	0.8081	0.8212
16.0	0.76834	0.7687	0.7696	0.7712	0.7733	0.7794	0.7878	0.7983
18.0	0.75518	0.7554	0.7562	0.7575	0.7592	0.7642	0.7711	0.7797
20.0	0.74364	0.7438	0.7445	0.7455	0.7470	0.7512	0.7569	0.7642
25.0	0.71982	0.7200	0.7204	0.7211	0.7221	0.7250	0.7289	0.7339
30.0	0.70097	0.7011	0.7014	0.7019	0.7026	0.7047	0.7076	0.7112
35.0	0.68545	0.6855	0.6858	0.6861	0.6867	0.6883	0.6905	0.6932
40.0	0.67232	0.6724	0.6726	0.6728	0.6733	0.6745	0.6762	0.6784
50.0	0.65009	0.6510	0.6512	0.6513	0.6516	0.6524	0.6534	0.6546
75.0	0.61397	0.6141	0.6143	0.6145	0.6147	0.6148	0.6148	0.6147
100.0	0.58870	0.5889	0.5894	0.5900	0.5903	0.5901	0.5895	0.5885

[a]L. Monchick and E. A. Mason, *J. Chem. Phys.* **35**, 1676 (1961); table used with permission from Reid and Sherwood.[(2)]
[b]$T^* = kT/\varepsilon$. $\delta = $(dipole moment)$^2/2\varepsilon_0\sigma^3$.
[c]The values of Ω_v in this column differ slightly from values in Table B.1 at low values of T^*.

Notation for the Tables

T	Temperature (K)	Ω	Collision integral
T^*	kT/ε		
V	Molar volume (m^3 kmol^{-1})		

T Temperature (K)

T^* kT/ε

V Molar volume (m^3 kmol^{-1})

δ Dipole moment parameter

ε/k Potential parameter (K)

σ Collision diameter (Å)

Ω Collision integral

Subscripts

c critical

D diffusion

μ viscosity

References for the Tables

1. R. B. Bird, W. E. Stewart, E. N. Lightfoot, *Transport Phenomena*, John Wiley and Sons, New York (1960).
2. R. C. Reid and T. K. Sherwood, *The Properties of Gases and Liquids*, 2nd edition, McGraw–Hill Book Co., New York (1966).
3. J. O. Hirschfelder, C. F. Curtiss, and R. B. Bird, *Molecular Theory of Gases and Liquids*, John Wiley and Sons, New York (1954).
4. L. A. Bromley and C. R. Wilke, *Ind. Eng. Chem.* **43**, 1641 (1951).

Derivation of Proposed Four-Flux Radiation Model*

In the following analysis, it was assumed, for the sake of simplicity, that the gas medium is totally transparent, so that the absorption coefficient and scattering coefficient are solvely functions of the number of particles in a unit volume of the gas–particle cloud. The contribution of absorption by gaseous components of the medium, though small compared to the contribution by particles, can be added later as a correction.

In order to formulate a model that represents the actual conditions that exist in a pulverized-coal flame, namely, the presence of "large" particles that scatter radiation, an effort is made to treat scattering as anisotropic. This is achieved by introducing the forward-, backward-, and sidewise-scattering components, which represent the fraction of scattered radiation in each of the corresponding directions.

The forward-scattering component is defined as

$$f = \pi \int_{-\pi/2}^{+\pi/2} P(\theta) \sin \theta \, d\theta \cos^2 \theta$$

$$= 2\pi \int_{0}^{\pi/2} P(\theta) \sin \theta \cos^2 \theta \, d\theta \tag{1}$$

where the notation is that of Chapter 5. The backward-scattering component is defined as

$$b = \pi \int_{\pi/2}^{-\pi/2} P(\theta) \sin \theta \, d\theta \cos^2 \theta$$

*Sneh Anjali Varma, Graduate Research Assistant, Department of Mechanical and Industrial Engineering, University of Utah, Salt Lake City, Utah

$$= 2\pi \int_{\pi/2}^{\pi} P(\theta) \cos^2 \theta \sin \theta \, d\theta \tag{2}$$

and the sidewise-scattering component is

$$s = (1 - f - b)/4 \tag{3}$$

For the isotropic case, the above fractions degenerate to $f = b = s = \frac{1}{6}$. For detailed discussion, see reference 6 of Chapter 5.

The phase function, $P(\theta)$, is a function of particle size, index of refraction, and wavelength of radiation. For spherical particles, the angular distribution is symmetrical about the direction of the incident radiation.

To begin with, a six-flux model is based on drawing energy balances for six discrete components of intensity of radiation in six orthogonal directions, and then, by invoking the condition of axial symmetry, reducing the equations to a four-flux model. In order to accommodate the geometry of the combustor of interest, a cylindrical-polar coordinate system is employed.

Consider a small volume element dV located at a point $P(\mathbf{R}, \mathbf{\Omega})$, as shown in Figure 1. The intensity of $I(\mathbf{R}, \mathbf{\Omega})$ is represented by six discrete components in the direction of, and opposite to the direction of, the three major directions, which are axial, radial, and angular.

Now, consider the energy balance for the intensity vector I_z^+, directed in the positive z-direction. Change in intensity during passage through the small volume is given as

$$(I_z^+ + dI_z^+/dz) - I_z^+ = dI_z^+/dz \tag{4}$$

Loss in intensity due to absorption by the matter in the small volume is

$$-K_a I_z^+ \tag{5}$$

and loss in intensity due to scattering by the matter in the small volume is

$$-K_s I_z^+ \tag{6}$$

The fraction of scattered radiation in the direction of the forward-directed intensity vector is $f K_s I_z^+$ and the rest is scattered in other directions, and thus considered lost. The net lost scattered radiation is then

$$-(1 - f) K_s I_z^+ \tag{7}$$

Similarly, there will be an addition to the forward-directed intensity by the backward-scattering component of intensity vector I_z^-, moving in the opposite direction. This increase is

$$+ b K_s I_z^- \tag{8}$$

The intensities traveling perpendicular to the z-direction will contribute a fraction of their out-scattered radiation to the positive z-direction, given

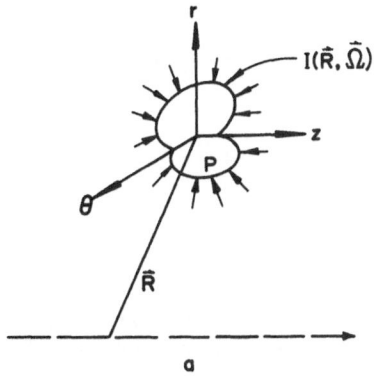

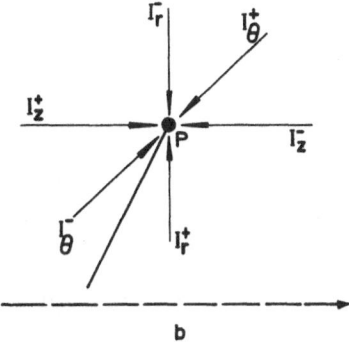

Figure 1. Volume element for radiative analysis. (a) The angular distribution of intensity. (b) The distribution of intensity in six discrete components.

in terms of s, as

$$+ K_s s(I_r^+ + I_r^- + I_\theta^+ + I_\theta^-) \tag{9}$$

There will also be a contribution to intensity in the positive z-direction by the radiation emitted by matter in the small volume. Considering this emitted radiation to be uniformly distributed in all directions, it may be represented by six equal, discrete components in the six orthogonal directions. Thus, the contribution of emitted radiation in the positive z-direction is

$$+ (K_a/6)I_b(T) \tag{10}$$

Summing up all the above terms, an energy balance for radiative transport in the positive z-direction is given as

$$dI_z^+/dz = -K_a I_z^+ -(1-f)K_s I_z^+ + bK_s I_z^- + SK_s(I_r^+ + I_r^- + I_\theta^+ + I_\theta^-)$$
$$+ (K_a/6)I_b(T) \tag{11}$$

or, combining terms:

$$dI_z^+/dz = [K_a + (1-f)K_s]I_z^+ + bK_s I_z^- + SK_s(I_r^+ + I_r^- + I_\theta^+ + I_\theta^-) + K_a/6[I_b(T)] \quad (12)$$

Replacing $(K_a + K_s)$ by K_t and K_s/K_t by W_0, the above equation can be rewritten as

$$(1/K_t)(dI_z^+/dz) = -(1 - W_0 f)I_z^+ + W_0 b I_z^- + W_0 S(I_r^+ + I_r^- + I_\theta^+ + I_\theta^-)$$
$$+ (1/6)(1 - W_0)I_b(T) \quad (13)$$

By writing similar energy balance equations in the other directions, the following equations can be obtained for each direction:

$$-(1/K_t)(dI_z^-/dz) = -(1 - W_0 f)I_z^- + W_0 b I_z^+ + W_0 S(I_r^+ + I_r^- + I_\theta^+ + I_\theta^-)$$
$$+ (1/6)(1 - W_0)I_b(T) \quad (14)$$

$$(1/K_t)[d(I_r^+ \cdot r)/dr] = r[-1(1 - W_0 f)I_r^+ + W_0 b I_r^- + W_0 S(I_z^+ + I_z^- + I_\theta^+ + I_\theta^-)$$
$$+ (1/6)(1 - W_0)I_b(T)] \quad (15)$$

$$-(1/K_t)[d(I_r^- \cdot r)/dr] = r[-(1 - W_0 f)I_r^- + W_0 b I_r^+ + W_0 S(I_z^+ + I_z^- + I_\theta^+ + I_\theta^-)$$
$$+ (1/6)(1 - W_0)I_b(T)] \quad (16)$$

$$(1/rK_t)(dI_\theta^+/d\theta) = -(1 - W_0 f)I_\theta^+ + W_0 b I_\theta^- + W_0 S(I_z^+ + I_z^- + I_r^+ + I_r^-)$$
$$+ (1/6)(1 - W_0)I_b(T) \quad (17)$$

$$-(1/rK_t)(dI_\theta^-/d\theta) = -(1 - W_0 f)I_\theta^- + W_0 b I_\theta^+ + W_0 S(I_z^+ + I_z^- + I_r^+ + I_r^-)$$
$$+ (1/6)(1 - W_0)I_b(T) \quad (18)$$

The assumption of axial symmetry results in the following conditions:

$$I_\theta^+ = I_\theta^- \quad (19)$$

and

$$dI_\theta^+/d\theta = dI_\theta^-/d\theta = 0 \quad (20)$$

By applying these two conditions in the last two equations of the six-flux model, there results

$$I_\theta^+ = I_\theta^- = [W_0 S/(1 - W_0 f - W_0 b)](I_z^+ + I_z^- + I_r^+ + I_r^-)$$
$$+ (1/6)[(1 - W_0)/(1 - W_0 f - W_0 b)]I_b(T) \quad (21)$$

Substituting these expressions for I_r^+ and I_r^- in one of the remaining flux equations gives

$$(1/K_t)(dI_z^+/dz) = I_z^+\{-(1 - W_0 f) + [2W_0^2 S^2/(1 - W_0 f - W_0 b)]\}$$
$$+ I_z^-\{W_0 b + [2W_0^2 S^2/(1 - W_0 f - W_0 b)]\}$$
$$+ (I_r^+ + I_r^-)\{W_0 S + [2W_0^2 S^2/(1 - W_0 f - W_0 b)]\}$$

$$+I_b[(1/6)(1-W_0)]\{1+[2W_0S/(1-W_0f-W_0b)]\} \tag{22}$$

$$(1/K_t)(dI_z^+/dz)=C_1I_z^+ +C_2I_z^- +C_3(I_r^+ +I_r^-)+C_4I_b(T) \tag{23}$$

where

$$C_1=-(1-W_0f)+[2W_0^2S^2/(1-W_0f-W_0b)] \tag{24}$$

$$C_2=W_0b+[2W_0^2S^2/(1-W_0f-W_0b)] \tag{25}$$

$$C_3=W_0S+[2W_0^2S^2/(1-W_0f-W_0b)] \tag{26}$$

and

$$C_4=[(1-W_0)/6]\{1+[2W_0S/(1-W_0f-W_0b]\} \tag{27}$$

The complete four-flux model can now be written as

$$(1/K_t)(dI_z^+/dz)=C_1I_z^+ +C_2I_z^- +C_3(I_r^+ +I_r^-)+C_4I_b(T) \tag{28}$$

$$-(1/K_t)(dI_z^-/dz)=C_1I_z^- +C_2I_z^+ +C_3(I_r^+ +I_r^-)+C_4I_b(T) \tag{29}$$

$$(1/K_t)[d(I_r^+ \cdot r)/dr]=r[C_1I_r^+ +C_2I_r^- +C_3(I_z^+ +I_z^-)+C_4I_b(T)] \tag{30}$$

and

$$-(1/K_t)[d(I_r^- \cdot r)/dr]=r[C_1I_r^- +C_2I_r^+ +C_3(I_z^+ +I_z^-)+C_4I_b(T)] \tag{31}$$

Derivation of Eulerian Finite-Difference Equations*

In this appendix, the general Eulerian finite-difference equation is derived in two-dimensional, steady, cylindrical coordinates. The flow field of interest is subdivided into computational cells by some grid pattern. Figure 1 displays a typical internal main cell where the ϕ-equation [Eq. (2), Chapter 14] can be integrated over the volume obtained by rotating the area represented by the dashed lines about the symmetry axis to give

$$\int_{x_w}^{x_e} \int_{r_s}^{r_n} \int_0^1 \{(\partial/\partial x)(\rho u r \phi) + (\partial/\partial r)(\rho v r \phi) - (\partial/\partial x)[r\Gamma(\partial\phi/\partial x)]$$

$$- (\partial/\partial r)[r\Gamma(\partial\phi/\partial r)] - rS^\phi\} \, dx \, dr \, d\xi_3 = 0 \qquad (1)$$

where the third coordinate, ξ_3, has the integration limits of 0 to 1 radians for convenience instead of 0 to 2π radians, because of assumed axial symmetry.

Considering the first convection term in Eq. (1)

$$\int_{x_w}^{x_e} \int_{r_s}^{r_n} \int_0^1 (\partial/\partial x)(\rho u r \phi) \, dx \, dr \, d\xi_3$$

and noting that all properties are uniform in the third direction, performing two formal integrations gives

$$\int_{r_s}^{r_n} [\rho u r \phi]_w^e \, dr$$

where e and w represent the expression to be evaluated at the east and west faces, respectively. As with any finite-difference development, the derivation

---

*John J. Wormeck, Senior Engineer, Department of Mechanical and Industrial Engineering, University of Utah, Salt Lake City, Utah

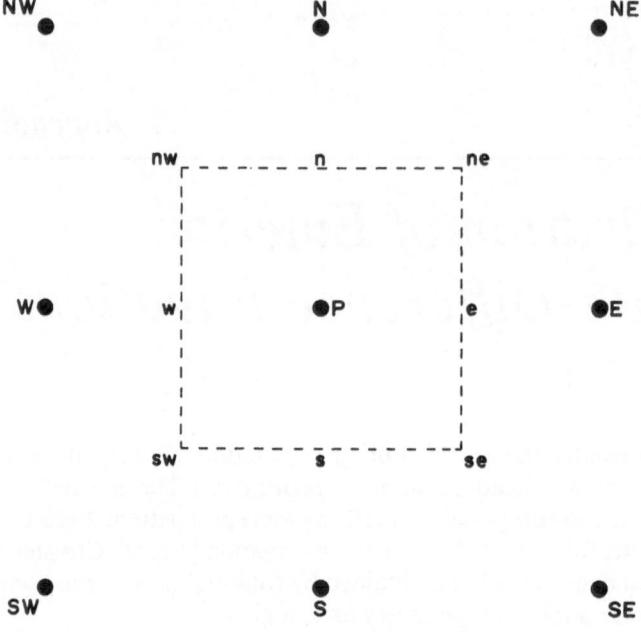

Figure 1. Illustration of the grid symbols for a computational cell.

is somewhat arbitrary; the following method yields the most accurate results. From the mean-value theorem:

$$\int_{r}^{r+\Delta r} f(r)\, dr \sim f(\bar{r})(\Delta r) \tag{2}$$

where $r < \bar{r} < r + \Delta r$, and as $r \to 0$ convergence is assumed. Applying this theorem to the first convection term yields

$$(\rho u r \phi)_e (r_n - r_s) - (\rho u r \phi)_w (r_n - r_s)$$

where each subscript represents evaluation at that particular face. Grouping the geometric terms gives

$$(\rho u \phi)_e A_e - (\rho u \phi)_w A_w \tag{3}$$

with the following definitions

$$A_e = r_e (r_n - r_s)$$
$$A_w = r_w (r_n - r_s)$$

which are the areas of the east and west faces of the cell, respectively, as shown by considering

$$A_e = \int_{r_n}^{r_n} \int_0^1 r \, dr \, d\zeta_3 = (r_n^2 - r_s^2)/2$$

or

$$A_e = (r_n - r_s)(r_n + r_s)/2 = r_p(r_n - r_s)$$

Furthermore, since $r_e = r_w$, these areas are equal and only one symbol will be used:

$$A_{ew} = r_p(r_n - r_s) \tag{4}$$

The numerical procedure TEACH[1] employs a staggered grid system,[2] where the velocities are stored midway between the grid lines; that is, at the exact locations which are required. The first convection term [Eq. (3)] becomes

$$\rho_e \phi_e u_E A_{ew} - \rho_w \phi_w u_p A_{ew}$$

Convection coefficients are defined as

$$C_E = \rho_e u_E A_{ew}$$
$$C_w = \rho_w u_p A_{ew}$$

which gives the mass flux through the face corresponding with the subscript.

Both ρ and ϕ are defined at the main grid nodes and some sort of interpolation is needed to determine their values at the faces midway between node points. The practice with TEACH is to linearly interpolate dependent variables and use simple averaging for fluid properties; thus

$$\rho_e = (\rho_P + \rho_E)/2, \qquad \rho_w = (\rho_w + \rho_P)/2 \tag{5}$$

and

$$
\begin{array}{lll}
P < \phi < E: & \phi = (1 - f_E)\phi_P + f_E\phi_E, & f_E = (x - x_P)/(x_E - x_P) \\
W < \phi < P: & \phi = (1 - f_w)\phi_w + f_w\phi_P, & f_w = (x - x_w)/(x_P - x_w) \\
P < \phi < N: & \phi = (1 - f_N)\phi_P + f_N\phi_N, & f_N = (r - r_P)/(r_N - r_P) \\
S < \phi < P: & \phi = (1 - f_S)\phi_S + f_S\phi_P, & f_S = (r - r_S)/(r_P - r_S)
\end{array}
\tag{6}
$$

Using these relationships, the convection coefficients are

$$C_E = (\rho_P + \rho_E)u_E A_{ew}/2$$
$$C_w = (\rho_P + \rho_w)u_P A_{ew}/2 \tag{7}$$

and the first convection term becomes, upon substitution:

$$C_E f_E \phi_E - C_w(1 - f_w)\phi_w + [C_E(1 - f_E) - C_w f_w]\phi_P \tag{8}$$

Similarly, the second convection term in Eq. (1) is

$$\int_{x_w}^{x_e} \int_{r_s}^{r_n} \int_0^1 (\partial/\partial r)(\rho v r \phi) \, dx \, dr \, d\xi_3$$

Again, two integrations can be performed formally to give

$$\int_{x_w}^{x_e} [\rho v r \phi]_s^n \, dx$$

and from the mean value theorem

$$(\rho v \phi)_n r_n (x_e - x_w) - (\rho v \phi)_s r_s (x_e - x_w)$$

where the geometric terms are

$$\begin{aligned} A_n &= r_n (x_e - x_w) \\ A_s &= r_s (x_e - x_w) \end{aligned} \tag{9}$$

and in this case are different, and hence the convection coefficients are defined as

$$\begin{aligned} C_N &= \rho_n v_n A_n = (\rho_N + \rho_P) v_N A_n / 2 \\ C_S &= \rho_s v_s A_s = (\rho_S + \rho_P) v_P A_s / 2 \end{aligned} \tag{10}$$

Therefore the final form of a second convection term of the ϕ-equation becomes

$$C_N f_N \phi_N - C_S (1 - f_S) \phi_S + [C_N (1 - f_N) - C_S f_S] \phi_P \tag{11}$$

Considering the diffusion terms in Eq. (1) separately

$$\int_{x_w}^{x_e} \int_{r_s}^{r_n} \int_0^1 \{ -(\partial/\partial x)[r\Gamma(\partial\phi/\partial x)] - (\partial/\partial r)[r\Gamma(\partial\phi/\partial r)] \} \, dx \, dr \, d\xi_3 \tag{12}$$

and integrating twice gives

$$\int_{r_s}^{r_n} -[r\Gamma(\partial\phi/\partial x)]_w^e \, dr - \int_{x_w}^{x_e} [r\Gamma(\partial\phi/\partial r)]_s^n \, dx$$

Using the same technique as presented for the convection terms, these last integrals can be evaluated as

$$-\Gamma_e(\partial\phi/\partial x)_e r_e(r_n - r_s) + \Gamma_w(\partial\phi/\partial x)_w r_w(r_n - r_s)$$

$$-\Gamma_n(\partial\phi/\partial r)_n r_n(x_e - x_w) + \Gamma_s(\partial\phi/\partial r)_s r_s(x_e - x_w)$$

As expected, the same geometric quantities appear as the convection terms; substituting Eqs. (4) and (9) yields

$$-\Gamma_e(\partial\phi/\partial x)_e A_{ew} + \Gamma_w(\partial\phi/\partial x)_w A_{ew} - \Gamma_n(\partial\phi/\partial r)_n A_n + \Gamma_s(\partial\phi/\partial r)_s A_s$$

The derivatives at the four faces must be expressed in terms of variables at main node points. Employing central differences (which are second-order accurate)[3] gives

$$-\Gamma_e[(\phi_E - \phi_P)/(\delta x_{PE})]A_{ew} + \Gamma_w[(\phi_P - \phi_W)/(\delta x_{PW})]A_{ew}$$

$$-\Gamma_n[(\phi_N - \phi_P)/(\delta y_{NP})]A_n + \Gamma_s[(\phi_P - \phi_S)/(\delta y_{PS})]A_s$$

where the δx and δy stand for coordinate distance between the node points indicated by their corresponding subscripts.

Diffusion coefficients can be defined as

$$D_E \doteq \Gamma_e(A_{ew})/(\delta x_{PE}) = (\Gamma_P + \Gamma_E)A_{ew}/(2\,\delta x_{PE})$$
$$D_W = \Gamma_w(A_{ew})/(\delta x_{PW}) = (\Gamma_P + \Gamma_W)A_{ew}/(2\,\delta x_{PW})$$
$$D_N = \Gamma_n(A_n)/(\delta y_{NP}) = (\Gamma_P + \Gamma_N)A_n/(2\,\delta y_{NP})$$
$$D_S = \Gamma_s(A_s)/(\delta y_{PS}) = (\Gamma_P + \Gamma_S)A_s/(2\,\delta y_{PS})$$

Thus the diffusion terms can be expressed as

$$-D_E(\phi_E - \phi_P) + D_W(\phi_P - \phi_W) - D_N(\phi_N - \phi_P) + D_S(\phi_P - \phi_S) \tag{13}$$

where the similarity with the control volume formulation is noted. The exchange coefficients and geometric quantities are contained in the diffusion coefficients D, while the difference in ϕ which drives the diffusion is explicitly shown with the correct sign, such that ϕ enters the cell when the ϕ-difference is negative.

Finally, considering the source terms in Eq. (1):

$$\int_{x_w}^{x_e} \int_{r_s}^{r_n} \int_0^1 rS^\phi \, dx \, dr \, d\xi_3$$

One of the major techniques responsible for success of the TEACH formulation is to express this source term as linear in the dependent variable. Thus

$$\int_{x_w}^{x_e} \int_{r_s}^{r_n} \int_0^1 rS^\phi \, dx \, dr \, d\xi_3 = S_U^\phi + S_P^\phi \phi_P \tag{14}$$

which defines two source term coefficients, S_U^ϕ and S_P^ϕ. If the source term happens to be nonlinear in terms of the dependent variable, ϕ_P, the technique calls for just ϕ_P to be factored from the expression (if possible) and to appear with the S_P^ϕ coefficient in Eq. (14); i.e., ϕ_P appears implicitly, while the remaining factored expression involving ϕ_P will be considered as known (explicit—based on old values) and lumped together in the S_U^ϕ coefficient of Eq. (14). Furthermore, as shown in the stability analysis,[2] S_P^ϕ must be negative to guarantee stable convergence.

In general, S^ϕ will be functions of all the dependent variables and other fluid properties as well as various types of derivatives involving these quantities. When integrating this source term, the value prevailing at the cell center will be used for all quantities and any derivatives will be evaluated by central differencing. Therefore, the source term is considered constant, giving

$$\int_{x_w}^{x_e} \int_{r_s}^{r_n} \int_0^1 rS^\phi \, dx \, dr \, d\xi_3 = S^\phi(\Delta V) = S_U^\phi + S_P^\phi \phi_P \tag{15}$$

where ΔV is the volume of the cell.

Upon substitution of these newly defined coefficients [Eqs. (8),(11), (13), and (15)] into Eq. (1) the general ϕ-equation becomes

$$[C_E f_E - D_E]\phi_E - [C_W(1 - f_W) + D_W]\phi_W + [C_N f_N - D_N]\phi_N - [C_S(1 - f_S) + D_S]\phi_S$$
$$+ [C_E(1 - f_E) + D_E - C_W f_W + D_W + C_N(1 - f_N) + D_N - C_S f_S + D_S]\phi_P$$
$$= S_U + S_P \phi_P$$

Adding and subtracting $C_E - C_W + C_N - C_S$ from the expression in brackets preceding ϕ_P and rearranging to obtain common expressions gives

$$[(C_E - C_W + C_N - C_S) + D_W + (1 - f_W)C_W + D_E - f_E C_E$$
$$+ D_S + (1 - f_S)C_S + D_N - f_N C_N]\phi_P$$
$$= [D_W + (1 - f_W)C_W]\phi_W + [D_E - f_E C_E]\phi_E + [D_S + (1 - f_S)C_S]\phi_S$$
$$+ [D_N - f_N C_N]\phi_N + S_U + S_P \phi_P$$

It is convenient to define new total coefficients to replace these common expressions:

$$A_E = D_E - f_E C_E$$
$$A_W = D_W + (1 - f_W)C_W$$
$$A_N = D_N - f_N C_N \tag{16}$$
$$A_S = D_S - (1 - f_S)C_S$$

In terms of these total coefficients, the finite-difference form of the ϕ-equation becomes

$$[C_E - C_W + C_N - C_S + A_E + A_W + A_N + A_S]\phi_P = A_E \phi_E + A_W \phi_W + A_N \phi_N$$
$$+ A_S \phi_S + S_U + S_P \phi_P \tag{17}$$

Numerical stability considerations lead to two further modifications of the above equation, which finally reduces to

$$[A_E + A_W + A_N + A_S]\phi_P = A_E \phi_E + A_W \phi_W + A_N \phi_N + A_S \phi_S + S_U + S_P \phi_P \tag{18}$$

The above analysis is for illustrative purposes only; the interested reader is referred to a lengthy but exhaustive treatment[2] of the procedure including the formulation of both velocity and pressure equations, boundary conditions, solution of finite-difference equations, underrelaxation, grid definition, and extensions to three-dimensional time-dependent, general orthogonal coordinates systems.

References

1. A.D. Gosman and W.M. Pun, Lecture notes for course entitled "Calculation of Recirculating Flows," Imperial College of Science and Technology, London, Report No. HTS/74/2 (1974).
2. J.J. Wormeck, Computer Modeling of Turbulent Combustion in a Longwell Jet-Stirred Reactor, Ph.D. Dissertation, Washington State University, Pullman, Washington (1976).
3. P.J. Roache, *Computational Fluid Dynamics*, Hermosa Publishers, Albuquerque, New Mexico (1972).

Index